HOW DETROIT BECAME THE AUTOMOTIVE CAPITAL

*Peggy,
Best of luck!
Robert Szudarek*

INTRODUCTION

Trains changed peoples' concepts of time and space but had limited application; you had to travel according to the train schedule and go where the tracks ran. Bicycles came into popularity in the 1880's starting with the high-wheeler. People had more freedom of travel with a bicycle and clubs were formed that sponsored tours to nearby communities such as Port Huron, Michigan, 57 miles away from Detroit. Going uphill and downhill on dirt roads was not a trip for the short winded. Most of the time they returned to Detroit by steamship on Lake St. Clair. Finally, the automobile arrived and gave people real freedom to travel.

The 100th anniversary of the automobile industry in the United States is celebrated in 1996, and countless magazine, newspaper and TV documentaries are being prepared. Across the United States, there will be events to celebrate the occasion. Michigan even has a special license plate with the 100 year theme. The subject of how Detroit became the Automobile Capital of the United States has never been fully explained. There are many articles and references on the subject, but none that actually give the reader a full account.

Why Detroit became the Automobile Capital of the United States is an amazing story. The New England States were in the forefront of the Industrial Revolution in the United States, and Detroit was not a good bet with a ranking of 16th in industrial output. For the first time, the epoch of how Detroit became the Capital of the Automobile Industry in the United States is explained in detail. To appreciate the successes, it is important to understand the failures.

How Detroit became the Automobile Capital is presented in a chronological sequence, encompassing all of the automobile manufacturers in Detroit and surrounding suburbs. It also has 49 additional topics that are woven in-between the automobile manufactures. The origin of Detroit and the key events, topology, and personnel that led up to the automobile era are presented first. The first car in Detroit is profiled, followed by the second and third. There are over 500 time period photographs of factories, both inside and outside, personnel, cars, and logos (including explanations). Company slogans are also listed.
There are many anecdotes, interesting explanations and derivations of terminology. Driving on the right-hand side of the road and why the steering wheel is on the left side is explained. Before the history of the first electric self-starter is discussed, the reader is taught the basics about hand-cranking.

The detail on traffic control, the first gasoline stations, selling, service parts, and roads, as examples, give the reader a real understanding of the whole industry. When you finish reading this book you will really appreciate the talented people that created Detroit's automobile industry.

Names that are in bold will help the reader follow common threads that weave in and out throughout the book,. This is a sign that the individual had a key role in Detroit's automotive history, and that the name may be encountered again in another event. Words are shown in *italics* to represent specific speech used in the time period of discussion. Technical detail is used sparingly but adequately to understand the automobiles. All material is in chronological order to the extent that is suitable for continuity. There are 105 automobile companies that are profiled which are inclusive of Detroit and surrounding communities that sold vehicles on a commercial basis. There are also 49 additional profiles interleaved that blend the topics together.

CONTENTS

The Founding of Detroit 1
"Large Rivers Happen To Run Close To Large Cities" 3
"Detroit, The City Beautiful" 5
Belle Isle 8
Belle Isle Bridge 9
The Columbian Exposition 10
John Lauer 11
Charles Brady King 12
Henry Ford 14
Barton L. Peck 17
Spanish American War 18
Bicycles 19
Emergence Of The "Buzzwagon" 21
The Name "Automobile" 23
Olds Motor Works 24
Packard Motor Car Company 30
Detroit Automobile Company 39
The Grosse Pointe Sweepstakes Race 41
Oliver Edward Barthel 46
Maxwell-Briscoe Motor Company 47
Buick Motor Company 50
Marr Autocar Company 53
Rules Of The Road 54
Steering 55
Northern Motor Car Company 56
Hussey Automobile And Supply Company 58
Henry Martyn Leland 59
Cadillac Motor Car Company 62
Ford Motor Company 68
The Selden Patent 71
C. E. Whiffler Automobile Company 74
Welch-Detroit Automobile Company 75
Wheeler Manufacturing Company 77
Dingfelder Motor Company 79
Hammer-Sommer Auto Carriage Company 80
Reliance Automobile 82
Huber Automobile Company 84
Commercial Motor Vehicle Company 85
Imperial Automobile Company, Ltd. 86
Gilmore Motor Works 87
The Ford Piquette Factory 88
Wayne Automobile Company 91
C. H. Blomstrom Motor Car Company 93
Detroit Auto Vehicle Company 96
Massnick-Phipps Manufacturing Company 98
Ried Manufacturing Company 99
Pungs-Finch Auto And Gas Engine Company 100
Electric Power 102
Little Four Automobile Manufacturing Company 103
Eureka Manufacturing Company 103
Early Gasoline Engines 104
Mixers 104
The Walker Motor Car Company 105
Detroit Automobile Manufacturing Company 106
Aerocar Company 108
Motorcar Company 110
Lozier Motor Company 116
Rands Manufacturing Company 120
Marvel Motor Car Company 121
Dragon Automobile Company 123
E. R. Thomas-Detroit Company 124
Chalmers-Detroit Company 125
Pontiac Motor Company 129

Griswold Motor Company 130
C. H. Blomstrom Manufacturing Company 131
Streetcars, Busses, And Subways 133
Traffic Control 135
Fee & Bock Automobile Company 137
Regal Motor Car Company 138
Detroit Electric Car Company 140
The Hotel Ponchartrain 146
Manufactured Versus Assembled Cars 148
Mass Production 148
Albert Kahn 149
Ford Highland Park Factory 152
Everitt-Metzger-Flanders Company 157
Herreshoff Automobile Company 162
Hupp Motor Car Company 164
General Motors Company 171
Commercial Vehicles 175
GMC Trucks 176
Detroit-Dearborn Motor Car Company 177
Anhut Motor Car Company 178
Demotcar Company 180
Carhartt Automobile Corporation 181
Krit Motor Car Company 182
Abbott Motor Car Company 185
Paige-Detroit Motor Company 189
Hudson Motor Car Company 192
United States Motor Company 198
Alden Sampson Manufacturing Company 200
The Starting Crank 202
Brush Runabout Company 203
Templeton-DuBrie Car Company 207
The Puritan Machine Company, Ltd. 209
Beyster-Detroit 210
Warren Motor Car 211
Sibley Motor Car Company 213
Owen Motor Car Company 215
Metzger Motor Car Company 217
The Steely Auto Engine Company 219
Wheels 219
Van Dyke Motor Car Company 220
Hupp-Yeats Electric Car Company 221
R. C. H. Corporation 221
Crowe Motor Car Company 224
Paint 224
B.O.S.S. Company 225
DeSchaum Motor Car Company 226
Day Automobile Company 228
Miller Car Company 230
Grinnel Electric Automobile Company 323
Brigg-Detroiter Company 233
King Motor Car Company 235
Automobile Racing 238
Louis Joseph Chevrolet 241
Chevrolet Motor Company 243
Society Of Automotive Engineers 247
Century Electric Motor Car Company 248
Colonial Electric Car Company 249
Self Starter 250
Maxwell Motor Company 252
Keeton Motor Car Company 255
American Voiturette Company 255
Church-Field Motor Company 258
S. & M. Motor Company 259
Roads 261
The Lincoln Highway 264
Read Motor Car Company 266
Wahl Automobile Company 267
Monarch Motor Company 269

Cycle Cars 271
Mercury Cyclecar Company 272
Cricket Cyclecar Company 274
Saxon Motor Company 275
Detroit Cyclecar Company 278
Princess Cycle Car Company 279
Malcom-Jones Cyclecar Company 281
Downing Cyclecar Company 282
La Vigne Cyclecar Company 283
Traveler Motor Car Company 285
Rex Motor Company 286
Excel Distributing Company 287
National United Service Company 288
Automobile Cycle Car Company 289
Grant Brothers Automobile Company 290
Sharp Engineering & Manufacturing Company 292
Gadabout Motor Corporation 293
Light Motor Car Company 295
Highwheelers 295
C. J. Fischer Company 296
American Motorette Company 297
Dodge Brothers Motor Car Company 298
Columbian Electric Vehicle Company 303
Storms Electric Car Company 304
Pilgrim Motor Car Company 305
The Detroit Athletic Club 306
Eclipse Machine Company 307
Aland Motor Car Company 308
Bour Davis Motor Company 309
Doble-Detroit Steam Motors Company 311
Libert Motor Car Company 313
Columbia Motors Company 316
E. A. Nelson Motor Car Company 318
Harroun Motors Corporation 320
Apex Motor Car Company 321
The Fisher Brothers 323
Kessler Motor Company 325
Gray Motor Corporation 327
Wills Sainte Claire Motor Company 329
Detroit Steam Motor Corporation 333
Lincoln Motor Company 334
Detroit Air Cooled Car Company 337
Commerce Motor Car Company 338
Edward Vernon Rickenbacker 339
Rickenbacker Motor Company 341
Davis Steam Motors, Incorporated 344
Gasoline 344
Tires 346
Chrysler Corporation 347
Falcon Motors Corporation 353
Graham-Paige Motor Corporation 354
Kaiser-Frazer Corporation 358
Edsel Ford 362
Detroit Automobile Golden Jubilee 367
The Automotive Capital 370

DETROIT COMPANIES THAT DID NOT GO BEYOND THE PROTOTYPE STAGE

Aetna Investment Co. 1910
Alger, R.A, and F.M. 1902
American Electromobile Co. 1906-1907
American Machine Manufacturing company 1907-1908
Atlas-Detroit Motor Co. 1916
Auto Parts Manufacturing Co. 1909
Baileyman, G.D. Co. 1912
Barlow Steam Car Co. 19-1922
Detroit Bi-Car Co. 1912
Bingman P & Son 1912
Blodgett Engineering & Tool Co. 1921
Bloom, A.J. 1914 (brother-in-law of R.C. Hupp)
Bremac Motor Corp. 1932
Brooks Automobile Co. 1905-1908
Brotherton, N.T. 1909-1910
Carson Motor Co. 1920
Chief Motor Co. 1911
Collins Motor Car Co. 1921-1923
Colonial Automobile Co. 1917
Colonial Motors 1922
Corrick, J.B. 1914
Cygnet Motor Company 1912
Dealers Vehicle Company 1900
Detroit Boat Works 1910
Detroit Body Co. 1914
Detroit Motor Co. 1922-1923
Detroit Steam Motor Corp. 1922-1923
D.H.K. Motor Car Co. 1909
Dodge Motor Car Co. 1914
Dodgeson Motors 1926
Dual Motors Corp. 1955-1958
Eagle Electric Automobile Co. 1915-1916
Edison-Ford Electric 1914
Edward VerLinden 1922-1923
Electromobile, H. F. Co. 1906-1907
Engler 1913-1914
Fanvien Motor Co. 1910
Faulkner-Blanchard Motor Car Co. 1910
Fuller Electric Car Co. 1914
Gates, A.J. Co. 1928-1930
George Breed-Machnet 1914
George Hilsendegen 1899
Globe Motor Car Co. 1910
Goodspeed-Detroit Manufacturing Co. 1913
Govreau-Nelson Engineering Works 1926-1927
Hamilton-Grapes 1921-1922
Harry Kerston 1917
Hastings Motor Car Co. 1910
Hawk Motor Car Co. 1914
H.C. Motor Car Co. 1916
Hooper, A.H. Co. 1914
Horton Autoette Manufacturing Co. 1911
House 1920
Huron Motor Car Co. 1915
John Tijaarda 1931
Kelly 1895-1901
Kenneth L. Morehouse 1929
Kermath Motor Car Co. 1907-1908

Kinsey Motor Co. 1911
Kirby Motor Car Co. 1911
Kissel Motor Car Co.
Knapp, A. C. Co. 1912
Kosmath Co. 1916
Krass, J. H. 1913
Lafayette Automobile Co. 1904
Leslie Motor Car Co. 1916
L'Esperance Motor Co. 1911
Lotz Automobile Co. 1910
Mackey Automobile Co. 1902-1903
Malcolmson 1906
McHardy-Peterson Car Works 1904
Masterbuilt Six 1926
Michigan Motor & Machine Co. 1904
Nash Auto-Car Co. 1906-1911
Nelson-Breman-Peterson 1914-1915
Nielson Motor Car Co. 1906-1907
Oldfield Motors Corp. 1917
Osborne, C. O. 1914
Patrick Sullivan 1898
Philip Bingman 1903-1904
Pittmans & Dean 1914
Raymond Russel 1946
Red Shield Hustler Power Co. 1911
Republic Motor Co. 1913
Robinson Motor Car Co. 1910
Roger J. Sullivan & Co. 1904
Roscoe C. Hoffman 1931
Savage Motor Car Co. 1914
Schneider, J. P. 1902-1904
Simon E. C. Cyclecar Company of Detroit 1914
Simpson, W.G. Consulting Engineer 1906
Stout Engineering Laboratories 1935-1946
Small Motor Car Co. 1910-1911
Sports Car Development Corp. 1954
S.R.K. Motor Co. 1915
Still Motor Co. 1900
S.R.K. Motor Co. 1915
Swift 1910
Sydney D. Walon 1920
T.H.T. 1910
Toledo Autocycle Co . 1914
Trott Automobile Co. 1917
True Blue Motor Co. 1910
Tyro Manufacturing Co. 1912
Universal Car Equipment Co. 1917
Universal Service Co. 1917
University Motor Car Co. 1910
Vulcan Motor Car Co. 1911
Wachman Auto Manufacturing & Supply Co. 1908
Wagenhals Motor Co. 1910-1915
Warren Noble 1914
Watt Motor Co. 1910
Whitney Motor Co. 1910
William B. Engler 1913-1914

THE FOUNDING OF DETROIT

Antoine Laumet De La Mothe Cadillac

Michigan is an Ojibway word meaning "a place for catching fish," and has the second longest coastline in the contiguous United States. Detroit is situated on the bottom of an ancient lake named Maumee that covered the area thousands of years ago. Wa-we-a-tun-ong was a Native American name meaning "bend in the river" for the area we know as Detroit.

Detroit was founded because Louis XIII, the King of France, decided to wear a beaver skin hat at age nine in 1610 and started a craze as the story goes. (He was crowned king in 1610, and his mother served as regent during his minority until 1617.) The peltry harvests in Europe could not meet the demand. They were a necessesity in the unheated houses to fight chill and dampnes. As the supply diminished, their value began to grow. Explorations of the American Continent led to the discovery of vast quantities of beavers that could support the insatiable demand. Nowhere were there more streams with beavers than the Great Lakes region. The Indians referred to the "Detroit" area as the "country of the beaver." The French similarly called it the area the home of the beaver, Teuscha-Gronde. With the ever increasing demand for beaver skins, the French set up a military post at Mackinac, Mich. The commander of the outpost was **Antoine Laumet De La Mothe Cadillac,** who was born on March 5, 1658, at St. Nicholas de la Grave, in what later became the Department of Tarn and Garonne, France.

While the French were trading blankets, knives, hatchets, kettles, and brandy for beaver skins, the English began trading similar goods from outposts to the South. Instead of brandy, they were making rum with molasses in the colonies much faster and cheaper than the brandy imported from France. Because the Indians received more rum from the English, there was alarm among the French over the fate of their empire.

Cadillac returned to France and petitioned King Louis XIV to support a new outpost south of Mackinac where the English were encroaching on their fur trading. After returning to Mackinac, **Cadillac** was notified that Louis XIV finally acted and the order came direct from Count Ponchartrain, minister of marine in the royal blue cabinet to Govenor Frontenac, of New France. **Cadillac** was summoned to Montreal to fill out an expedition whose purpose was to found a settlement in the southern portion of Michigan called "The Strait," or in French "Le Detroit." A fleet of 25 canoes set out on June 5, 1701, on a seven week journey through perilous lakes and streams. The party landed on the west bank of a narrow Strait, later called the Detroit River, (near the foot of Shelby Street) on July 24, 1701. A stockade made of tall log pickets was erected, enclosing about an acre of land. Because Count Ponchartrain had supported Cadillac's initiative to King Louis, the settlement was named Fort Ponchartrain. There were originally four streets: Main E.W. was called Rue Saint Annes, later renamed Jefferson, Rue Saint Louis, Rue Saint Jacques, and Rue Saint Joseph.

Streets added later included **Cadillac** Square, **Cadillac** Boulevard, Lamothe Street, **Frontenac**, Ponchartrain, Champlain, **Marquette**, **La Salle**, and Ce' Loron (the only successor to Cadillac with a street name).

Cadillac was appointed governor of Louisiana in 1710, and had no further contact with Detroit. He died on October 15, 1730, in Castelsarrasin, France.

In 1751 the name of Fort Ponchartrain took the name of the river and was called Fort Detroit. After the English took over Quebec in 1763 with a peace treaty signed, the French lost all of their Great Lakes territory. Major Robert Rogers was sent from Montreal to take over Detroit soon afterwards. Rogers and 200 soldiers paddled up the St. Lawrence River to Lake Ontario, crossed that body of water, and reached Fort Niagara. They carried their boats around the Falls and continued their journey across Lake Erie until they reached the area where Cleveland was founded. They met a band of Indians who were led by Chief **Pontiac**, who after deliberating, gave his permission for the English to continue. The Indians were great friends of the French and could not understand why the English were taking their lands away from them. The French flag was lowered for the last time over Detroit on November 20, 1760.

The Indians sold land to the new settlers, but complained they were being cheated by the fur traders. Chief **Pontiac** was respected by the French, but treated with indifference by the English, and he was not pleased. He sent messages throughout the Northwest Territory to drive the English away. After **Pontiac** made several attempts to massacre the English, he was refused entry to Fort Detroit. **Pontiac** laid siege on Fort Detroit for four months until relief came from Fort Niagara. On July 31, 1763, the gates of the fort swung open and the soldiers marched out in silence to hunt down Chief **Pontiac** About a mile and a half up river, Parent's Creek flowed through a rough hollow and entered the Detroit River. The Indians were everywhere around the small French farm houses outside the fort. Near the mouth of the creek was a bridge where the soldiers were marching. As they got halfway, there was a great yell and after six hours there were 59 killed and wounded soldiers compared to 15 to 20 Indians. The Creek became known as Bloody Run, and years later was drained and used for Elmwood Cemetery. A tablet was laid in front of the Detroit Stove works on Jefferson Avenue where the greater portion of the fighting took place. **Pontiac** made peace with the English, but in 1769 was killed by an Indian from the Illinois tribe.

In the fall of 1778 the English decided to build a fort to defend the Detroit stockade. The fort was entirely outside the stockade and was known as the second terrace. It was built by Major Lernoult, and named Fort Lernoult.

In 1783, after the Revolutionary war was over and peace was finally signed, the United States came into possession of Michigan. The English, however, held possession of certain lake forts, including Mackinac and Detroit, for some time afterwards. The Union Jack was lowered over Detroit for the last time on July 11, 1796. Two days later, Colonel Hamtramck came to Detroit and took possession of the fort. Two weeks later, General Anthony **Wayne** reached Detroit and remained all summer. He was in charge of the whole western American army. In 1787 General **St. Clair** was appointed governor of the Northwest Territory, comprised of Ohio, Indiana, Illinois, Wisconsin, and Michigan. The Indians tried to drive the Americans away from Detroit, the capital of the Northwest Territory. General **St. Clair** was taken by surprise and defeated. President Washington selected General "Mad" Anthony **Wayne** to stop the disorders. Before he arrived, the people of Detroit named the county in his honor. After he arrived, the Indians were severely punished and General **Wayne** made a treaty to end the trouble for a time at least.

Detroit was incorporated as a city in 1802, consisting of 20 acres with 300 crude homes, and fort Lernoult. In 1805 sparks from a pipe belonging to John Harvey's bakery set fire to a pile of hay, and the entire city was destroyed. People were homeless and hungry. Father Gabriel Richard appealed to the French farmers for food. People began moving away. Assistance from Congress, and from as far away as Montreal, came forth to help in the rebuilding process. The brave spirit of the people was made famous by the Detroit motto: Speramus meliora, resurget cineribus. "We hope for better things; it shall arise again from the ashes." Father Gabriel Richard, a French priest brought the first printing press in the Northwest Territory to Detroit. The Reverand John Montieth, a Presbyterian, worked closely with Father Gabriel, preaching the word to all faiths.

Within a few weeks after the fire, the people of Detroit had elected their own officials. At the same time in Washington, plans had been completed to govern the region as a territory. So the people of Detroit were to rebuild under the leadership of men they had not chosen. General William Hull was the new governor, Stanley Griswold was the new secretary, and Augustus B. Woodward was one of three judges. Woodward was a friend of Thomas Jefferson and an acquaintance of Ben Franklin. Father Richard, Reverend Montieth, and Augustus Woodward organized schools, a library, gave lectures, and founded the University of Michigan.

In 1812 the English waged a war, mainly at sea, to retain the Northwest Territory. In September 1813 Commodore Perry gained naval victory over the British near Put-In-Bay, Lake Erie, and sent his famous message "we have met the enemy and they are ours." After the close of the war, the name of Fort Lernoult was changed to Fort Shelby, in honor of the Kentucky governor who marched with his troops to the relief of Detroit. The fort was torn down in 1826.

"LARGE RIVERS HAPPEN TO RUN CLOSE TO LARGE CITIES"
-Unknown Pundit

1837- corner of Griswold and Jefferson

In the early 1800s most people were confined to a 20 mile range from their house, and many never went beyond in a lifetime. Detroit's earliest industry was the tanning business, which at first converted deer hides into buckskins. As the fur industry began falling off, Harvey Williams established Detroit's iron industry in 1820 when he set up a small foundry. He began casting much needed plows and eventually founded the Detroit Iron Works Company at the foot of First Street. After the chartering of the Detroit and Pontiac Railway in 1830 and the Michigan Central Railway in 1832, the building of railways commenced. In 1836 Judge William Austin Burt produced the first typewriters. He also invented the solar compass and equinoctial sextant. Michigan was allowed into the Union in January 1837, with Detroit as its capital.

The curfew bell would ring in old Detroit at 9:00 p.m. The night watch would shout "Nine o'clock and all's well-curfew." This was the English form of two French words, "couvre feu" (cover the fire). This was the signal to seize the fire shovel and push the glowing coals into the fireplace against its backlog. The coals would be covered with ashes to retard combustion and insure keeping a fire until dawn. There were no matches in those days. The family would then go to bed, knowing the ritual of curfew meant the day was over.

Captain Eber Brock Ward was Detroit's first great industrialist. He was a president of railroads, an owner of rolling mills, mines and transportation companies, a bank director and shipbuilder, a plate glass manufacturer, and the owner of vast tracts of timber lands. He turned out the first Bessemer steel in Wyandotte, Mich., in 1864, and made the first steel rails in America. He was Detroit's first millionaire.

During the 1860s Detroit became a major producer of paints and varnishes, tobacco, drugs, chemicals, stoves, shoes, ships, soaps, and a seed producer. In addition, Gernhard Stroh immigrated to Detroit.

RAILROADS

When Michigan became a state in 1837 it was well recognized that it was wasteful to move commerce along its rivers, and it was too far to circumnavigate the Lower Peninsula to go across the state. But because it had been only a territory, there were no railroad connections beyond the state borders. In fact, the first railroad west of Albany, N. Y., the Erie Kalamazoo, just started service in 1836 between Adrian, Mich., and Toledo, Ohio. The legislature voted 15 million dollars for building railroads. The plan was to build one line from Port Huron to Grand Haven and call it the Michigan Northern. A second line would run from Detroit to St. Joseph and would be called the Michigan Central. A third line would run from Detroit to New Buffalo and would be called the Michigan Southern.

The Michigan Central was started first, but by 1846 the state was out of money after laying over 200 miles of track, and decided to sell the railroads. John Brooks put a deal together with John Forbes from Boston and an attorney from Detroit named **James Frederick Joy** (who became one of the founders of the Republican party in Jackson, Mich.). Together, they placed $500,000 down and assumed control of the railroads. The Michigan Central tracks reached Niles by 1848, and officials decided that Chicago should be the destination. By 1850 the Central reached Michigan City. In order to run to Chicago there was competition with the Michigan Southern, and mistrust from the states of Indiana and Illinois over the rights of shipping commerce. **James Joy** hired an attorney named Abraham Lincoln to help settle some of the disputes, and was successful in getting the Michigan Central into Chigago.

While working for the Michigan Central, in the 1870s, Detroit native **Elijah McCoy,** invented a lubrication system for locomotives. Before this invention trains were constantly stopping for lubrication. Other companies tried to compete with the McCoy system, but nothing came close. This led to the expression "The Real McCoy." Elijah was a prolific inventor with over 50 patents to his name, and others that he gave away.

Other developments in Detroit included the train car with on-board diners and sleepers. **George Pullman** studied riverboats with dining and sleeping provisions, and tried the concept with railroad cars on the Michigan Central. He had factories in Detroit for ten years before relocating to Chicago. Detroiter **William Davis** invented the refrigerator car using ice. He was funded by a Detroit meat packer named **George Hammond,** who had a city in Indiana named in his honor. (George built the first skyscraper in Detroit called the Hammond Building in 1890 on the corner of Fort and Griswold).

Besides railroad cars, lumber became an important industry. Michigan ranked sixth in U.S. production in 1850, and then third in 1860, and first from 1870 to the end of the century, thereafter losing rank as the timber depleted. Foundries, machine and boiler works, sprang up to support mining industries, ship building, and sawmills. In 1870 there were over 1,500 sawmills in the state. The logging industry also required railroads, and there were over 500 different railroads in Michigan at various times, established to transport the lumber industry. Between lumber, mining and railroads, many people were able to reap huge dividends, giving Detroit a healthy economic base, and investment capital for the emerging automobile industry.

Union Depot-Detroit, Michigan

"DETROIT, THE CITY BEAUTIFUL"
"DETROIT, WHERE LIFE IS WORTH LIVING"
-Twin Slogans

Detroit was a bustling city in the 1890s, with a population of over 200,000 people. There was an assortment of amusements, bicycles built for two, and ferry boats going from the foot of Woodward to Belle Isle where you could watch a band concert. At night the city's first electric sign, using a myriad of light bulbs, read "QUEEN ANNE SOAP." There were newsboys on the street corners calling the papers, carrying Bixby boxes over their shoulders, ready to give you a shoe shine.

Few cities anywhere else could boast such a high percentage of homeowners. Detroit was a city filled with trees. Vestiges of village life survived in Detroit. Outside many homes, hung a tin cup hanging by the well, and passersby were welcome to slake their thirst. Cattle were seen grazing in vacant lots in the best parts of town. Most homeowners had chickens, gardens, and a small orchard. Gardeners and farmers everywhere knew of Detroit's Dexter M. Ferry's Seed Company, as the largest of its kind in the world.

Many businessmen went home at noon to eat dinner with their families, where they enjoyed the family circle to the full. The well-heeled businessmen were members of the original Detroit Athletic Club, formed the year Detroit won its first baseball series in 1887. In 1892 **Harry Jewett** won the world championship for running the 220 yard dash in 21.8 seconds. When the Spanish-American War broke out, half of the membership of the DAC joined up and served on the Yosemite. They lost interest in athletics when they returned.

In 1895 the Hester sisters opened a boarding house at 350 Cass Avenue, where home cooking was in evidence. University of Michigan graduates began boarding there, and soon the word spread and it became a regular spot to get started in Detroit. Close by the boarding house, around the corner on Henry Street, was a rooming house operated by a bicycle rider named **Barney Oldfield**. Henry Andre had a feed store on Gratiot, and Anthony Lingeman had a butcher shop. One of his customers was Anthony Muir, who manufactured a five-cent cigar he called "Tony's Ponies," in deference to his racing team of black mares. Anthony's son, Joe, established a seafood restaurant on Gratiot Avenue.

One night near closing time, a newly married couple came to Fred Sanders' confectionery store for a drink of cream soda. Fred discovered his cream was sour and it was too late to purchase any. The couple lived around the block from his store and he wanted to give them a good impression. In desperation, he substituted ice cream and the couple enjoyed it immediately. Word quickly spread as far away as New York. His shop became known as "that little soda place back there in Detroit."

In 1891 a merchant named **Joseph Lowthian Hudson** moved to a new building at Farmer and Gratiot. By 1946 it filled the entire block bounded by Farmer, Gratiot, Grand River, and Woodward. The 25 story building was the tallest department store in the world with almost 2,000,000 square feet of floor space.

A Detroiter named Hiram Walker located a distillery across the Detroit River in Canada. Detroit was known for its tobacco products and was sometimes called the "Tampa of the North." Detroit had the largest stove manufacturer in the world that produced the Garland. Parke-Davis became one of the largest pharmaceutical companies in the world. Detroit was a major producer of railroad cars. The Detroit Drydock Company, along with several other shipyards in the area, launched more ships annually than the ship yards of any other area in the United States. Many marine engine manufacturers, machine shops, foundries, brass and bronze product producers, and paint and varnish companies were formed to support the shipbuilders.

*1890-
Ferry landing at
the foot of Woodward*

courtesy MSHC

J. L. Hudson

Sanders shown with striped awnings. Note wood block walking pavement.

courtesy MSHC

Frederick Sanders

BELLE ISLE

The Indians called it wah-na-be-see (Swan Island), and the French called it Ile Aux Cochons (island for pigs). Early settlers claimed that Cadillac granted the island to Detroit in order that cattle could be kept there and prevented from straying off into the forest. Hogs ran wild on the island, multiplied and ate up the rattle snakes. After Michigan became a territory and was not under French rule, it was called Hog Island. At one time, the entire island was owned by Captain George McDougall, who bought it in 1789 from the Ottawa and Chippewa Indians. He paid eight barrels of rum, three rolls of tobacco, six pounds of vermillion paint and one belt of wampum. McDougall then built a house on the island. In 1794, his sons sold the island for 1694 pounds 10 shillings. Like the rest of the territory around Detroit, Hog Island had its share of bloodshed. One of the first acts of hostility in the **Pontiac War** was the massacre of six men on the island.

In the nineteenth century, even though it was privately owned, people used the island for such things as picnics and even gun duels. During a picnic in 1845, a group decided to give the island a better name and "Belle Isle" was chosen in honor of their friend Isabella Cass, daughter of the former Governor of Michigan Territory (1813-31), John Cass.

Detroit purchased the island from Barnubus Campau in 1879 for $200,000 against the protest of many of its prominent citizens who thought it was a waste of money. **Fredrick Law Olmstead**, who designed New York City's Central Park and the **Columbian Exposition**, was commissioned to make Belle Isle into a park. The 700 acre island was increased to 800 acres by adding long points at each end when dredges were used to make the channel deeper, as ships were continually getting larger. Five miles of lagoons intermingled with 20 miles of winding road, with bridges everywhere for road crossings. Through the years, pavilions, boat clubs, museums, an aquarium, zoological gardens and the Scott Fountain were added. Every conceivable pastime was available such as: boating, canoeing, and bathing on its gleaming beaches. There was golf, tennis, baseball, picnicking and athletic fields for almost any sport. For peace and solitude there was the woodland dell and forest trail. The drive around the island was more than five miles long. Belle Isle became the most popular park in the United States and some said it was more well known than Detroit. *(picture below circa 1948- courtesy MSHC)*

BELLE ISLE BRIDGE

The first Belle Isle bridge was built in 1889. It was a half mile long and cost $246,269. The bridge was a swinging drawbridge constructed of steel with a creosote block pavement. On April 27, 1915, a "steam roller" crossed the Belle Isle Bridge towing a steel cart full of hot coals. As cars crossed the bridge, oil drippings accumulated into the blocks making them flammable. As the steam roller returned from the island in the afternoon, the watchman saw blazing fires at the island side, and turned in an alarm. The last car to cross was driven by J. H. Burns.

Fireboats, fire companies, and rescue volunteers were all summoned. The girders twisted and fell into the river. At the peak of the fire, the daughter of Col. Herman T. Kallman, who was the supervisor during the construction of the bridge, attempted to run onto the burning bridge. After she was stopped by a policeman, she exclaimed, "I loved the bridge and I wanted to be the last one to set foot on it."

courtesy MSHC

THE COLUMBIAN EXPOSITION

NAMED FOR THE 400TH ANNIVERSARY OF THE DISCOVERY OF AMERICA

Truly one of the most significant events in the 19th century was the Columbian Exposition, better known as the Chicago World's Fair, which took place in 1893. It was four times bigger than any previous fair with over 27 million people attending between May and October 1893. New York City politicians made an unsuccessful attempt to host the exposition, but were beaten, and gave Chicago the nickname "Windy City," in deference to the Chicago politicians.

Down the midway of the exposition, was a 250 foot high revolving wheel designed by George Ferris specifically for the exposition. It was intended to be the focus of attention, just like the Eiffel Tower was at the Paris Exposition of 1889. In machinery hall, there was a 14,000 horse power engine, and there was a 30 foot high replica of a stove exhibited by Detroit's Michigan Stove Company.

There were bicycles, horse carriages and boats; in the electrical building were electric vehicles. There was only one steam car and one gasoline car, exhibited by the Daimler Motor Company in the Transportation Building, and driven at the fair by **Ransom Olds**. Karl Benz and Gottlieb Daimler surely must have met each other at the fair. They were both from the vicinity of Stuttgart, Germany and developed the world's first and second automobiles. Years later, their companies merged into Daimler-Benz. **W. C. Durant**, Thomas B. Jeffrey, George N. Pierce, and Alexander Winton also attended the fair.

Henry Ford saw the New York Central's locomotive "Red Arrow" and "999" which had been clocked at 112 miles per hour. (Locomotives were becoming so numerous that numbers started being used instead of names). At the Daimler exhibit, **Henry Ford** studied a two wheeled horse cart with an otto cycle engine mounted to the side used for pumping water.

Charles King was working at the fair displaying a steel brake beam for railroad cars and a pneumatic hammer, the first of its kind, which won the fair's highest award. **King** and **Ford** saw a two cycle gasoline engine for marine use made by **Clark Sintz** of Grand Rapids, Mich. Elwood Haynes of Kokomo, Ind., used this engine for his first car in 1894 which became the second commercial car in the United States, following the Duryea wagon. The **Sintz** Company later became the foundation of the Pungs-Finch Automobile Company in Detroit.
(Clark Sintz at left)

After the Columbian exposition, the Garland stove sat in front of the Michigan Stove Company. Directly underneath this for years stood the stump of a tree that marked the place where chief Pontiac waged one of the bloodiest battles ever fought in this section of the country.

JOHN LAUER

John Lauer was a self-made man, born in Germany in 1842. He immigrated to the United States with his parents in 1850, and settled in Milwaukee. He served his apprenticeship as a machinist, and later moved to Chicago. He opened a shop there, and married Anna Scherbarth. Their first son, John P. Lauer, was born in 1869. Lauer lost everything in the Chicago fire of 1872. He moved to Detroit in 1874 and was working for $1.75 a day when he read an advertisement in a newspaper that Jas. A. Forester wanted a foreman. He saved his money and within a year owned the shop at 114 St. Antoine Street. His shop supplied many early automobile producers such as Charles King, Henry Ford, and Jonathan Maxwell.

Lauer tried his luck in gold mines and lost most of his fortunes. He died on July 2, 1906.

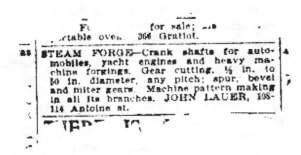

CHARLES BRADY KING AND DETROIT'S FIRST AUTOMOBILE

Charles Brady King was born in an army post, Camp Reynolds, Angel Island, San Francisco Bay, Calif., on February 2, 1868. His mother was Matilda C. (Davenport) and his father was General John Haskell King. **Charles** lived in one army post after another. One result of his nomadic experience was an early interest in transportation. As a little boy he would try to figure out a way to beat travel schedules of 10 miles per day with horses, and he saw new railroads being laid across the raw western landscape.

When he was 14 years old, his father retired and the family moved to Detroit, his mother's home town. He attended Trinity College at Port Hope, Ontario, for two years, then went to Cascadilla School in Ithaca, N. Y. From there he started at Cornell University in 1887 and stayed for two years until his dad passed away. He left school and got his first job as a draftsman with the Michigan Car Company in Detroit. In 1893 he worked for the Russel Wheel and Foundry Company, who selected him to take charge of its exhibit at the Transportation Building at the **Chicago World's Fair**. This gave him a chance to exhibit a pneumatic hammer, used for riveting and caulking, that he designed in 1890.

Another invention of King's was a brake beam for railroad cars, which was sold to the American Brake Beam Company of Chicago. This sale helped to finance the Charles B. King Company, organized in 1894. It was located at 110-112 St. Antoine Street, Detroit, for the purpose of manufacturing pneumatic hammers and marine engines. The address was that of John Lauer's machine shop, where **King** developed the first horseless carriage made in Detroit, in a second floor office.

He was developing a carriage with the first four cylinder cast *en-bloc* engine in the world. He was also making a carriage with a steering wheel on the left hand side. He had hoped to race it in Chicago in 1895, but was unable to finish it in time. In 1896 he used a delivery wagon constructed by the Emerson & Fisher Company of Cincinnati. That firm of carriage builders was trying to develop self propelled vehicles, but was unsuccessful. The Emerson & Fisher Company loaned King an incomplete, experimental, iron-tired wagon for testing. **King** removed the four cylinder engine from his incomplete carriage and installed it in the Emerson & Fisher carriage. He added Hyatt bearings, and a foot operated speed control. Then he added a water tank over the cylinders, a gasoline tank and a muffler. The carriage weighed 1300 lbs, and the engine developed three horsepower that could propel it up to eight miles per hour. It was ready for testing in March 1896. A number of trial runs made at night went unnoticed by the press. He usually didn't startle or alert his neighbors because they were so accustomed to the racket made by his pneumatic hammers.

On March 6, 1896 at 11:00 pm, the doors of the Lauer Foundry opened to a light snow. **Charlie King** and his assistant, **Oliver Barthel**, pushed the carriage onto St. Antoine Street. The engine sputtered to life and **King** headed south toward Jefferson Avenue, with **Barthel** following on a bicycle. He made a right turn and meandered down Jefferson, avoiding horse drawn wagons. He passed restaurants, stores, and a bicycle factory before making a second right turn at Smart's corner onto Woodward Avenue. As he traveled north on Woodward he passed the Merril block on his right, followed by the Sun Newspaper building, and T. B. Rayl Hardware Store. The engine began to backfire and died in front of the Russel house. With **King** underneath the carriage trying to fix the problem, a bicyclist, believed to be **Henry Ford**, was watching him closely. **King** restarted the engine and completed a journey of several miles before returning to the **Lauer** garage.

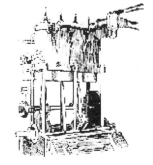

*John Lauer Machine shop
112-114 St. Antoine St., Detroit
Picture taken after the building
was renumbered and occupied by
another firm.*

*Charles B. King in his office and
drafting room on the second floor
of Lauer's shop.*

*The office was located behind the
two center windows.*

Picture taken by Alvord in November 1896 in front of Capitol Park on Griswold. Notice that King is seated on the L.H. side of the vehicle. He was the first to design this type of seating arrangement. Oliver Barthel is seated next to him.
courtesy AAMA

HENRY FORD AND DETROIT'S SECOND AUTOMOBILE
JUNE 1896

Henry Ford was born in 1863 on a farm in Dearborn, Michigan. He was the son of William Ford and Mary Litogot Ford. At 12 years of age, he was inspired to create mechanically operated transportation when he first saw a steam tractor made by Nichols and Shepard of Battle Creek, Michigan. Although his father wanted him to be a farmer, Henry left in 1879 to work in Detroit for the Michigan Car Company Works, which built streetcars using a crude form of an assembly line. Shortly after he started, he repaired some equipment that the veteran machinists couldn't handle and was castigated for being a show-off. He left after only six days on the job. His father came to the rescue and took Henry to their friends who owned the **James Flower & Brothers** machine shop at Woodbridge and Brush streets in Detroit. Henry started out working on a small milling machine, then lathes and drill presses, and in nine months he left the company and returned to the farm to help with the harvest. One Sunday, Henry was driving his sister Margaret home from church in Dearborn, as he was reflecting on the text of the sermon which had been "hitch your wagon to a star" and said that he was going to do that for the common person. He had been thinking of transportation that was easier with less work than the horse and buggy. He returned to Detroit in the fall and went to work for the Detroit Dry Dock Engine Works which had the largest collection of machinery for many miles around. Henry continued to learn more about machines, and although he was too young for many of the more dangerous operations, he was an avid watcher. To make ends meet, Henry was also working at a jewlery shop at 390 Michigan Ave. repairing watches and clocks. A friend at the Dry Dock Works named Samuel Townsend showed "*Hank*" a magazine from England called the World of Science that had an article on the invention of the internal combustion engine by the German engineer Dr. Nicolaus A. Otto. He continued to read the World of Science magazines as well as Scientific American. Henry left the Dry Docks Company in the summer of 1882 and returned to the farm until December of 1884 when he enrolled in Goldsmith's Bryant & Stratton Business University, located in Mechanics Hall on Griswold Street. He studied mechanical drawing, bookkeeping, and business practice.

Henry Ford and **Clara Jane Bryant**, the daughter of a local farmer, were married on April 11, 1888. Henry worked on a 40 acre farm that his father gave him, clearing trees and selling the wood for three years. On an Autum day in 1891, Henry came home after seeing an Otto engine in Detroit. He told Clara that he believed an Otto engine could be used to make a horseless carriage. He sketched out a rough plan of a vehicle and showed Clara who was very supportive. He called her "Clarie" in public, but in private he called her "Believer." Henry went on to explain that to make a good engine, he would need to learn about electricity. He felt they should move to Detroit so that he could work for the Edison Illuminating Company, and in September 1891 they were in a hay wagon with their furniture on their way to Detroit.

Henry was hired as an engineer for the Edison Company before they left Dearborn, and he started work at 6:00 pm September 25, 1891, the same day they arrived in Detroit. He started at a substation at Woodward and Willis, then moved to the main station at Washington Boulevard and State Street in November 1893. During the winter of 1892-93, Henry taught a metal-working class at the YMCA building on the corner of Griswold and Grand River, and one of his students was **Oliver Barthel**. More important than the money that he made was the access to the school shop and machines for making metal parts.

Edsel Bryant Ford, the only child of Henry and Clara was born on November 6, 1893. He was named after a childhood friend of Henry's, Edsel Ruddiman. Shortly afterwards, Henry was promoted to Chief engineer, and the family moved to 58 Bagley street, which at the time was one of the four thousand residences in Detroit that had direct current electricity. He immediately began building his own gasoline engine, and on Christmas eve 1893, he clamped it to the kitchen sink, connected the household electricity to the spark plug and grounded the engine to the sink. He asked Clara to interrupt her cooking and preparation for the following Christmas day guests, and help him by pouring gasoline into the intake of his engine. Henry turned the flywheel and the engine roared with flames and smoke billowing from the exhaust. It shook and vibrated the kitchen sink, and after thirty seconds, it was shut off and never started again.

Just five days later on December 29, 1893, Henry began to build his first car. Every spare dollar went into tools and parts for an engine. He had to make a crankshaft, camshaft, push-rods, valves, pistons and rings. Part of Henry's job at the Edison Company was being available to maintain or repair equipment failures, so when everything was in working order, he was free to work on his engine. He also worked on the car in his backyard at 58 Bagley street. Felix Julien, who lived in the other side of the duplex house the Ford's lived in, removed his wood and coal from his side of the woodshed to give Henry more room to work in, and a phone line was installed so Henry could be notified if there was a problem at the Edison plant.

He continued through 1894 and 1895, and eventually gathered help from **Oliver Barthel**, **"Charlie" King**, who was technically Barthel's boss, **Spider Huff**, **Jim Bishop**, and many others. On one occasion, Henry was at **John Lauer's** machine shop, and saw an article in the American Machinist magazine that described the first Kane-Pennington motor, which was light, and powerful. Although **Henry** didn't copy the engine, he learned from it because it showed how a designer solved some of the problems that many inventors were pondering. The article was somewhat critical and said *the machine work on the Pennington motors is not of high grade; there is nothing modern and nothing special in the way of tools in the shops where these motors were built; the performance of the motor is due solely to design, not workmanship.*

By the end of 1895, Henry's carriage was on stands in the shed. He named it the Quadricycle (which was one of the names submitted in the Chicago-Times contest in 1895 to replace the name "horseless carriage"). Most of the iron work came from Barr & Dates, at the corner of Park Place and State. Springs were made by the Detroit Steel and Spring Works, a seat was purchased from the C.R. Wilson Carriage Company, and a $15 line of credit was given by the Charles A. Strelinger Company for hardware and tools.

Henry was ready for his first trial run in early June 1896. **Jim Bishop** and **Henry** brought the quadricycle down off the stands and discovered it was too wide to pass through the door. Henry took an axe to the bricks and doubled the size of the opening, and they pushed the quadricycle onto the cobblestone alley. **Clara** was standing outside in a light rain with a shawl over her head and **Jim Bishop** was ready on

Painting of Henry Ford by Charles King   courtesy AAMA
Henry turned left here towards Bagley Street.

58 Bagley residence is on the left and next to an alley. Henry came through the alley, onto the street and turned to his right.

his bicycle. Henry turned on the battery current and turned the flywheel. The engine came to life and **Henry** drove up the alley onto Bagley Avenue and around the corner onto Grand River Avenue.
The quadricycle had two forward speeds but no reverse; it had to be pushed backwards by hand. It had no brakes either and was stopped by "killing" the engine. But it was fast, with four horsepower capable of propelling it 25 mph. Henry didn't get far that night. A nut came off a valve-stem and the machine came to a halt. **Bishop** got off his bicycle and helped push the quadricycle over to the Edison plant for repairs. From there, Henry drove home and went to bed.

Photoghraph taken in 1911 of the coal shed behind 58 Bagley-Note the oversize RH door. courtesy AAMA

1902- Henry Ford in the Quadricycle courtesy AAMA

1902- Henry Ford in the Quadricycle courtesy NAHC

BARTON L. PECK AND DETROIT'S THIRD AUTOMOBILE

Barton Peck was the son of George Peck who made a fortune as a clothes merchant. George was also the president of the Detroit Illuminating Company, so he was indirectly **Henry Ford's** boss. George Peck established **Barton** in a machine shop at 81 Park Place. He manufactured wholesale electrical furnace supplies there, and produced the first porcelain spark plug.

In 1897 **Barton L. Peck** built the third car made in Detroit. The car was a "Victoria" with a 4-cylinder engine of his own design. One Sunday morning James Bishop was riding with **Peck** when all of a sudden the gas tank, located under the seat, exploded, sending Bishop onto the street and burning his trousers and legs. Bishop got up and called the fire department. After they put out the fire, he and **Peck** pushed the machine back home. On another occasion Peck exceeded the speed limit of 15 mph and received one of the first automobile speeding tickets in the Untied States.

During the fall of 1898, **Peck** established the Detroit Horseless Carriage Company. By the latter part of 1899 he was unable to develop a commercially acceptable vehicle and decided to pursue other interests.

Peck was a speed buff who came up through the ranks from bicycles. He bought the first Curtiss flying-boat and moved to Florida, where he purchased property to operate his airplane. He eventually sold the property for a sizable profit.

SPANISH AMERICAN WAR

"REMEMBER THE MAINE"

On February 15,1898 a terrific explosion killed over 250 sailors on the U.S.S. Maine. The ship was moored in Havana harbor in Cuba to protect United States citizens from the Spanish atrocities that were reported by New York newspapers. To this day, no one knows what or who caused the explosion. Although President McKinnley never intended to start a war, the first United States war outside its boundries was declared against the Spanish forces in Cuba.

Before the war the Michigan Naval Reserve was established, in large part by Cyrus Lathrop and **Truman Newberry**, both lieutenants during the war. With l600 miles of coastline around Michigan, they succeeded in getting the navy to fund a reserve unit.

When President McKinley asked for volunteers, the Michigan Naval Guard was one of the first units to join up for active duty. Many of the reserves were college graduates from many different schools, with the highest concentration from the University of Michigan.

They were looked upon as pantywaist, sons of the rich. This was at a time when college graduates did not exert much respect. They served on the U.S.S. Yosemite. Among the volunteers were the chief of maintenance, **Charles King** and a business associate of King since 1894, **Henry B. Joy**, brother-in-law of Truman Newberry. **Harry M. Jewett** and **Fred Paige** also served on the Yosemite. **Frederick J. Sibley**, Henry S. Sibley, and Alexander H. Sibley served, as well as Frank G. Smith, **Frederick P. Smith** and George H. Smith.

While the Yosemite was at sea, band concerts where held on Belle Isle. One evening, **Homer Warren** sang "The Sword of Bunker Hill" in his deep baritone voice at the concert.

The Yosemite went on to fame and drove the ship "Antonio Lopez" to destruction on the beaches of Puerto Rico in June of 1898. This helped make this a short, 90-day war. The men in this unit became a tight knit group, proud of their accomplishment. But as late as 1906, Rear Admiral Coglan said naval men had more stamina than reservists. Paul F. Bagley of Detroit took exception on the front page of the Detroit News.

left- Homer Warren

FAREWELL, YOSEMITE.

*above- Henry Joy
courtesy Bentley*

BICYCLES

In the 1880s the bicycle began to rise in popularity giving people freedom of travel. The first popular bicycle was called the Highwheeler. The front wheel was as big as possible for speed, and the diameter was limited only by the length of the rider's legs. It was so odd looking that it scared horses, much the same as cars would later in the century. They were dangerous too, because when an obstruction was hit, the bike could flip over, ejecting the rider over the handlebar. This was called a "header."

In the 1890s bicycles were designed with a chain and sprocket to provide high gearing for use with two medium size wheels. This gave the same effect as the large front wheel except the rider was safer and it was called the safety bike. With this bike, the boom really began and manufacturers could barely keep up with the demand. Prices were kept high, to over $100, until the end of the century when prices came down to $30-$40.

Bicycle racing became very popular and the Detroit Wheelmen staged 25-mile "road races" on Belle Isle. **John and Horace Dodge** would often officiate these races. Bicycling was a major sport with **Tom Cooper, Eddie Bald**, and a little later, **Barney Oldfield** the leading bicycle racers in the world.

By 1900 a national club called the League of Wheelmen had over 100,000 members, and there were over 10 million bikes in the USA. In Detroit, **William Metzger** of the Huber & Metzger bicycle shop sold Stearns, Yales and the Cleveland (made by Henry Lozier) chainless brands of bikes. **W. C. Rands** sold Tribune, Pierce, and Columbia. By 1902 the bicycle business had declined and the League of Wheelmen membership was down to 10,000. Many bicycle companies and bike shop owners began to go into the automobile business.

The Detroit Athletic Club can be seen just in front of the trolley. It was located on Woodward and Garfield Avenue.

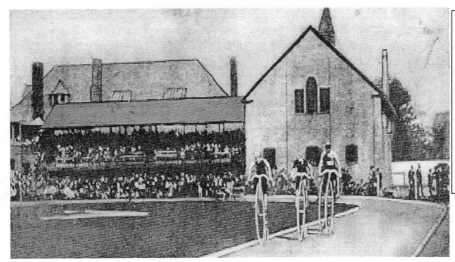

Highwheeler racing at the Detroit Athletic Club on Woodward Avenue.

*1904
Barney Oldfield*

EMERGENCE OF THE "BUZZWAGON"

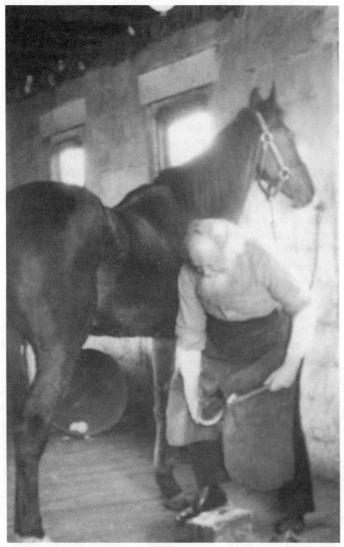

At the end of the 1800s there were few automobiles, and bicycles and horses moved at the leisurely rate of 5 mph. Settlements were distributed along waterways, or along railroads. Detroit had both. Throughout the country the population was strung out like clusters of beads on strings. Due to the limited range of the horse, it didn't pay to hold land more than 10 miles away from a railroad or a river. The horse was limited to a 12 mile circle that it could go and return in good shape.

Horses created tetanus that was carried in street dust of the horse dung. These excreta accounted for 60 percent or the filth on city streets. It irritated lungs and nasal passages, and prevented windows from being opened for proper ventilation. It became a syrupy mess when it rained. Livery stable keepers and employees contracted these diseases more than people of any other occupation.

Life in the wealthier suburbs was pleasant. Houses and estates were large, with horses kept in roomy box stalls. They were groomed by professional coachman who were proud of their calling. The carriages were magnificent specimens of the coachmaker's art. As the automobile emerged, only the wealthy could afford them. They were confident the automobile would always be the exclusive property of the rich.

There was a natural reluctance for the workmen to accept the automobile, because they couldn't afford one, and feared the unknown. There was a significant industry that revolved around the horse and carriage trade. People were afraid their knowledge base would go back to zero. Others had their fortunes tied up in various aspects of the trade. The Horse Drawn Vehicle Association and horse breeders were clearly against the automobile. The carriage makers and blacksmiths clearly profited from the automobile.

The European automobile manufactures had a head start on the United States industry, and up to 1906 most cars in the U. S. were made in France. In 1891 the Panhard et Levassor was the first automobile with an engine mounted in the front, and had the first four cylinder engine in 1896. The first car to exceed a mile-a-minute was the electric Jenatzy at 65.7 mph in 1899. In the U. S., the Baker electric went 80 mph on Staten Island in 1902, and had its wheels collapse when the brakes were jammed on. At first, electric automobiles made good sense, considering the range was 45 miles per charge, compared to a horse that was only good for a 20 mile round trip. Steam powered cars ran cleaner and quieter than *buzz wagons*, but the hissing scared people.

The first motor cars were built horse size. Horses were a certain height and to drive them the seats had to be a certain height. To accommodate the height the tread of the wagon or carriage had to be a certain width, resulting in the road standard of 56 inches. This was the horse standard. When motor cars were built the idea was to make horseless carriages, and these were first built with this idea in mind. The seats were buggy-high and buggy-wide, and the wheels were large, as in buggy practice. The vehicles were indeed horseless carriages and fit for horse speeds. As speeds were increased, it was found the weights were too high in the air and the wheels were too high. Lower constructions evolved with better weight distribution.

As the automobile emerged, design was by "rule of thumb" and "actual experiences," instead of scientific engineering. Many early automobiles were referred to as "experiments." Early science was predicated on static conditions such as bridges. Train locomotives driving on a set of smooth rails were subjected to comparatively few jolts and jounces. They required science in many parts, but empirical or practical knowledge was called upon for many parts in the shape of a factor or safety margin of varying quantity.

In the bicycle, the rule of safety was thrown aside and the factor of safety was made smaller than ever before on any machine. Many parts of the bicycle were designed by the process of elimination. Machines were built and failed, and another machine was built with strengthened parts until it worked. In this way, the lightest possible machine was arrived at. The automobile resembled a bicycle in this respect. The strains called upon it were immeasurable, incalculable, and not possible to foresee. The average automobile got into the ditch sooner or later. A wheelbarrow was not subjected to such punishment. The unexpected culvert in the country road, the rainworn gully, threw a strain upon the automobile that could be no more measured by scientific methods than the strains in a railroad collision.

For example, the wheels of early automobiles were arrived at, and not calculated. The axles were guessed at, and the wheel bearings were under a degree of scientific control, that is, the plain journals were. The bearing surfaces were derived from accepted railroad and engine construction. Springs could not be scientifically calculated, nor could the frame. One of the most common design errors was to formulate something where there was nothing to formulate. A Belgian engineer tried to design entirely from scientific formula and made a 7 hp automobile that weighed 2600 lbs. It was as useless to apply scientific formula to automobile design as it was to measure bricks with a micrometer.

In 1900 there were 10,000,000 horses in the United States, and under 5000 automobiles. By 1907 it was obvious that when the manufacturers could bring the price of the automobile down they would obviate the horse. By 1910 the car was accepted as an integral part of America. In 1913 there were 1,125,000 American made horsedrawn buggies sold. By 1919, for all practical purposes, the horsedrawn buggy was no longer used. In 1920 automobile companies were competing against each other for more sales volume instead of trying to produce more cars. Car replacement demand accounted for a larger percentage of new car sales than initial and multiple car sales combined in 1927. By the end of the 1920s the expression "look, there's a horse" was in vogue.

THE NAME "AUTOMOBILE"

In the 1890s a male driver of a self propelled vehicle was a "Chauffeur" and a female driver was a "Chauffeuse." A person who repaired automobiles was a "Mechancian," and a self propelled vehicle was called a "Horseless Carriage." People were never satisfied with that name because it described what it was not, rather than what it was. In France, the Greek word "Autos," meaning "self," and the Latin "Mobilis," meaning "moving" were contracted into the name "Automobile." The French Academy decided it was a properly constructed word and it stuck in France. Meanwhile, people in the United States were still searching for an appropriate name.

In 1895 the first automobile race in America was held in Chicago, and sponsored by the Chicago Times-Herald. The owner of the Times-Herald, Mr. H. H. Kohlsaat sponsored a contest with a $500 prize to select a proper name for the Horseless Carriage. Among thousands of suggestions, the prize was divided among three people who proposed the word "Motocycle." In October 1895 Kohlsaat started publishing America's first trade journal entitled The Motocycle Maker & Dealer. The name did not stick very long, however, and the French word "automobile" began to take hold. In January 1899 the New York Times was the first publication to use the word "automobile" in an editorial, criticizing it as a mixture of Greek and Latin origin.

In Great Britain the public never accepted the word automobile. The founder of the British Daimler Company, Mr. F. R. Simms, coined the term "motor-car." The Latin "Carrus", that describes two wheeled wagons, was shortened to "car."

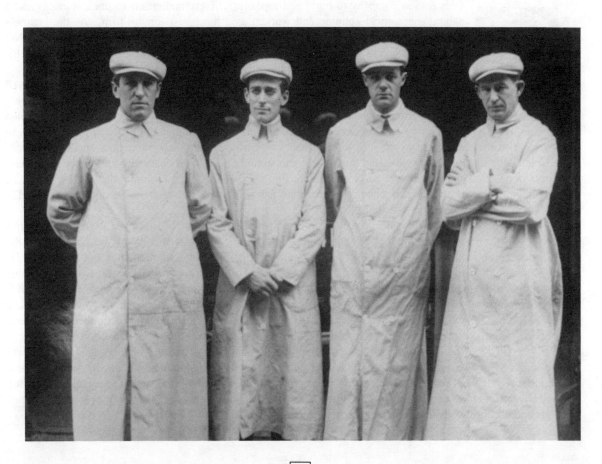

OLDS MOTOR VEHICLE COMPANY
LANSING, MICHIGAN
1897-1899

OLDS MOTOR WORKS
DETROIT, MICHIGAN
1900-1905

OLDS MOTOR WORKS
LANSING, MICHIGAN
1902-

"NOTHING TO WATCH BUT THE ROAD"

Ransom E. Olds was born in June of 1864 in Geneva, Ohio. He was the youngest of five children born to Sarah and Pliny Olds. While growing up he cared for the family horses, but disliked the chores and the smell of horses on the farm, but enjoyed helping his father repair tools in his blacksmith shop. In 1880 his family moved to Lansing, Mich., first by taking a lake steamer to Detroit, and then a steam locomotive train to Lansing. During the trip the steam powered transportation made "Ranny" dream about building a horseless carriage. After arriving in Lansing, Ransom's father and older brother Wallace formed the P. F. Olds & Son company to build and repair machinery and steam engines. Ransom had a natural mechanical aptitude and worked as a machinist for the firm. As the business grew, production capacity was about 12 engines per year with seven workers. Net income reached $7000, but financial problems persisted until **Ransom** bought his brother's share of the company, and assumed control.

In 1887 **Ransom** convinced Pliny to get a loan and expand. They moved across the street into a new two story factory with a work force of 12, and increased the manufacturing capacity to 400 engines per year. This gave Ransom the chance he dreamed of and he built a three-wheeled carriage using a one horsepower Olds steam engine. He worked long hours, and used his limited funds to buy most of the parts, such as the steel-tired wheels, whitewood body, oak frame, and steel gears, from outside manufacturers. He road tested the vehicle one summer night then tinkered with the car for several years, noting that one horsepower was insufficient. Pliny Olds once told a customer that "Ranse thinks he can put an engine in a buggy and make the contraption carry him over the roads. If he doesn't get killed at his fool undertaking, I will be satisfied."

Pliny retired in 1898 and moved to California with Sarah while **Ransom** was starting to establish himself in the business world, feeling secure enough to marry Metta Ursula Woodward in June 1889. **Ransom** built a second car in 1892 only this time he used a pair of 2 hp steam engines and did away with the transmission by connecting each one to a driving wheel. Hill climbing was awkward because the power source and the brakes were marginal. The machine weighed 1200 lbs and could obtain 15 mph on a level surface, compared to 10 mph maximum for the first vehicle.. The gasoline supply could power the vehicle for 40 miles and the water needed replenishing every 10 miles, which could be supplied from horse watering troughs scattered throughout the city. Neither of these first two machines had a reverse.

In 1897 the steam engine manufacturing business had gross sales of $42,000 and operated with two shifts. **Ransom** was driving his car around town attracting considerable attention from the business community. He was ready to start an automobile company, but didn't have the capital to expand. He started to receive offers for investment from Chicago and Detroit but he did not want to move from Lansing. When a group of Lansing businessmen made an offer, he accepted.

1891
A gas powered steam car driven by Ransom Olds
Note: fringes hanging from the top helped absorb and dissipate moisture, and swaying movement helped to shoo flies.
courtesy AAMA

Ransom Eli Olds

1896
Ransom Olds, the driver in the dark suit, manufactured this "horseless carriage," which, as he said, was "practically noiseless and impossible to explode." With him were Frank G. Clark, who built the car's body, Mrs. Metta Olds and Mrs. Hattie Clark. Mrs. clark is at the left, while Mrs. Olds is partIally obscured by her husband. Ransom is seen holding the gearshift in his right hand and the tiller in his left.
One popular magazine said in its September 1896 issue: "When it is in operation, the motion of the machinery is not in any way noticeable by the passengers. It steers as easily as a bicycle and speed as high as 18 miles an hour can be obtained."
courtesy AAMA

The first automobile company in Michigan was incorporated in 1897 with $50,000, and called the Olds Motor Vehicle Company. Edward W. Sparrow was president with 500 shares of stock. Eugene F. Cooley was VP with one share, Arthur C. Stebbens with 500 shares was secretary, and **Ransom E. Olds** with half of the 5000 shares was treasurer and general manager. Other shareholders were F. Seibly with 500, Frank Clark with 125, A. Beamer with 374, and **Samuel Latta Smith** with 500. **Smith** made his fortune in the copper industry, and had lived in Lansing before moving to Detroit in early 1897.

At the first board meeting, **Ransom** was tasked to build a perfect carriage, and when he protested, secretary Stebbens crossed out the word "perfect" in the meeting minutes and added "nearly perfect." As vice president, **Ransom** could not control the company as he did as President of the Gasoline Engine Works, and yet he was still faced with the dilemma of producing carriages in his engine factory that was working to full capacity.

In May 1899 the Olds Motor Vehicle Company combined with the Gasoline Engine Works as the Olds Motor Works, capitalized at $500,000. New officers of the company were **Samuel Latta Smith** as president, **Ransom Olds** as VP and general manager, and Samuel's son, **Fred Smith** as secretary and treasurer. It was capitalized with $50,000, with $10,000 paid in making Samuel Smith the largest stockholder, and with a provision that production would be in Detroit. Within a week of incorporating, five acres were purchased near Belle Isle in Detroit.

Samuel Latta Smith

The Olds family, which now included daughters Gladys and Bernice, moved to Detroit in time for the construction of a three story factory at 1308-1318 Jefferson and Concord, which began in June 1989 and was completed in February 1900. **This was the first factory in the world that was specifically designed for automobile manufacturing**. It had a 150 foot frontage on Jefferson and extended 350 feet toward the Detroit River, making it one of the most recognizable sights in town. The first automobiles built at the new plant were good machines selling for $2823, and although few were made, they were all sold. In the assembly hall, techniques were being tried using wooden bucks (sawhorses) on wheels so that each worker had specific parts to assemble. Parts were built in quantity, although they had to be custom machined to fit. The Olds Motor Works was the first company to apply the progressive system of assembly to manufacture an automobile.

The Olds Motor Works on Jefferson Avenue in Detroit, west of the Belle Isle Bridge   *courtesy AAMA*

Early test run   courtesy AAMA

In the fall of 1900 it is believed that **Ransom** sketched out a concept for a small automobile and said "what we want to build is a small low-down runabout that will have a shop cost around $300 and will sell for $650." After the Spanish-American War ended, Charles King sold his marine engine business to the Olds Motor Works. That provided the company with a stable income in case the automobile business floundered. King was chief engineer of the Olds Marine Engine division. He recommended **Jonathan Maxwell**, who worked for Elwood Haynes in Kokomo, Indiana, for the position of engineer at the Olds Motor Works. **Maxwell**, Horace Loomis and Milton Beck took Ransom's concept and designed the Curved Dash Oldsmobile. Among several patents for the runabout, the dash was patented for its ability to guide air around and through a copper disc radiator that was mounted horizontally, underneath the body. The Oldsmobile Regular Runabout, as the curved dash was officially called, weighed 600 lbs, had a 66 inch wheelbase, and 28 inch diameter spoked wheels. A one cylinder engine rated at 7 hp drove the rear wheels through a two-speed planetary gearbox and chain. A five gallon gas tank gave it a range of 150 miles on good roads. The mixer did not have to be shut off after use, and the air did not require manipulation. The steering lever was attached to the body, and then through a full elliptical leaf spring to the steering arm, so it absorbed shock from the road. In October 1900 the Curved Dash Olds was tested, and in November it was already being advertised.

In February 1901, with business starting to pick up, the Olds family went on a five week trip to San Diego to visit Sarah and Pliny, who had retired in 1898. Disaster struck at the Olds plant in Detroit on Saturday, March 9, 1901 at 1:35 pm when sparks from a welder touched off tanks of carbonic acid used to inflate tires, causing an explosion that turned the factory into rubble within a half hour. The alarm was sounded by **James J. Brady**, who was a timekeeper in charge of factory work and transportation. Four workers were injured with broken ankles and sprained backs as they jumped from the third-story windows, but luckily most employees had completed their customary half day shift at noon on Saturday. The building and equipment were almost a total loss, except for the foundry next door that suffered minor damage. The Olds family returned by train from California the following day, and when he heard about the fire, **Ransom** knew the Olds Motor company was in trouble because only part of the $72,000 damage was covered by insurance.

James J. Brady managed to drive a curved dash runabout out of the burning factory. It appeared that he saved the company, because it seemed all of the engineering drawings were lost to the fire, but in actuality, the prints were in a fireproof safe and were later retrieved. Based on its simplicity of construction and "sudden" popularity it was decided to make only the curved dash runabout model until the company got back on its feet. To get back into production quickly, Ransom began purchasing more parts from outside manufactures like he did with his vehicle in 1897. The **Leland Faulkoner** shop supplied motors, the **Dodge** brothers supplied transmissions, the **Briscoe** Manufacturing Company supplied radiators, and one of the **Fisher** brothers made the bodies.

James J. Brady- The timekeeper who pushed the runabout out of the blazing inferno. courtesy NEWS

An agreement was signed in August 1901 with the Lansing Businessmen's Association to return production to Lansing. A factory sight was chosen, and the Olds family returned to Lansing. The Olds Motor Works sold 425 runabouts that year, and an Olds employee, **Roy D. Chapin**, made a cross country trip from Detroit to New York City in 7.5 days to demonstrate the durability of the runabout. The Marine Engine Division was sold to the Michigan Yacht and Power Company of Detroit, and Charles King became the head of design and production.

Early in 1902 **R. E.** went to the Ormond Beach, Fla., area with a race car called the "Pirate," which was a modified runabout chassis capable of 60 mph. Olds raced **Alexander Winton,** who was driving his "bullet." **The race was a draw, but it helped establish Ormund as a leading race area**. Olds sold 2500 automobiles in 1902, and 4000 in 1903 making it the largest producer of automobiles in the world.

*Test driver **Roy Chapin** ascends a man-made hill on Bell Isle where canals were being dug.*
courtesy MSHC

In April 1903 **Fred Smith,** who was in control of the company's Detroit factories, pointed out to **Ransom** that quality was a problem, and that production of a touring car he and his brother Angus wanted to build had not begun. Later in the same month **Fred** started an experimental department in the Detroit plant. In May Ransom wrote a letter to **Fred Smith** saying, "Now if this is your policy, to do business underhanded and unbeknown to me, as you have several other things, I do not care to be associated with you." **Ransom** officially retired from the Olds Motor Works on January 5, 1904.

The Olds family took an extended vacation to visit with Pliny and Sarah in California. Upon their return they were informed that a group of **Ransom's** friends in Lansing organized the R. E. Olds Company for $500,000, with Ransom receiving controlling interest of $260,000. The Olds Motor Works "naturally" opposed the use of the **Olds** name on another company's car, so Ransom changed the name to the REO Car Company in September 1904, and immediately started construction of a new factory. The REO company built 864 cars in the 1905 fiscal year, ending in August, and Oldsmobile made 6500. By 1908 REO made 4105 cars, to Oldsmobile's 1055. Later that year Oldsmobile was purchased by Billy Durant and became part of General Motors.

OHIO AUTOMOBILE COMPANY
WARREN, OHIO
1900-1902

PACKARD MOTOR COMPANY
WARREN, OHIO
1902-1903

PACKARD MOTOR CAR COMPANY
DETROIT, MICHIGAN
1902-1954

STUDEBAKER - PACKARD CORPORATION
SOUTHBEND, INDIANA & DETROIT, MICHIGAN
1954-1956

STUDEBAKER - PACKARD CORPORATION
SOUTHBEND, INDIANA
1956-1961

"ASK THE MAN WHO OWNS ONE"

James Ward Packard graduated from Lehigh University in 1884 with a degree in mechanical engineering. In 1890 James and his brother William founded the Packard Electric Company and started making arc lamps. In 1893 the brothers were interested in making a horseless carriage, and **James** made some preliminary designs. They were interrupted by the panic of 1893, but continued to study designs, and in 1896 contacted **Charles King** regarding an engine for their carriage. In the early morning one day in 1898, **James Packard** was in Cleveland and purchased the 12th horseless carriage produced by **Alexander Winton**. He started to drive the 60 miles back to Warren, Ohio, but it quit running. He was flat on his back, covered with oil and dirt, as he tugged and pulled, toiled and cussed. He finally reached home, towed by a team of plow horses. **Packard** was a good mechanic and resolved to find out why the machine had failed. He was excited enough to return to the Winton factory and offer his ideas on how to improve the machine. He gave **Alexander Winton** his constructive criticism, and Winton retorted "Well, if you're so darn smart maybe you can build a better machine yourself." The soft spoken Packard replied, "Perhaps I could, at that."

In 1899 **Packard** hired George L. Weiss and W. A. Hatcher away from the Winton organization to help design the first Packard. The model "A" was completed on November 6, 1899, having a one cylinder engine with 7 horsepower and center chain drive to the rear wheels. The first model was so successful that friends insisted that **Packard** build duplicates. Four more were completed that year.

On September 10, 1900 the Ohio Automobile Company was formed by the Packard Brothers, Weiss, Hatcher, and James P. Gilbert of the Packard Electric Company, with an authorized capital of $500,000, with $100,000 subscribed and only $10,000 of hard capital.

The Ohio Automobile Company was a modest success, but needed substantial funding that the Packard brothers could ill afford. **James Packard** approached the Cleveland Chamber of Commerce for inducements to move his company there. They replied "nothing doing" because the Winton Company was already there, and they had just induced the largest clothes pin producer in the country to build a local plant.

In 1901 a sales agency was opened in New York to sell Packards by the firm of Adams and McMurtry. To a large extent, New York was the market place for new "horseless carriages," especially from Europe. France was the largest automobile producer in the world at the time. Because of this, **Henry B. Joy** of Detroit, Mich., went to New York to shop for an automobile. He was accompanied by his brother-in-law, Truman H. Newberry.

Henry Bourne Joy was born in 1864, the son of **James F. Joy**, who was a member of one of Detroit's oldest families. **James** built the C. B. and Q. Railroad, revived the Michigan Central Railroad, and was instrumental in raising money for the Sault Ste. Marie locks. He also built the Fort Street Union railroad station in Detroit, and had holdings in the Peninsular Car Company, mines in Utah, and real estate.

Henry Joy attended the Michigan Military Academy, the Philips Academy in Andover, Mass., the Sheffield Scientific School, and Yale University. He began his career as an office boy for the Peninsular Car Company, then became a clerk, and worked his way up to assistant paymaster. He pursued the mining business in Utah from 1887 to 1889. Then he became assistant treasurer and director of the Fort Street Union Depot Company in Detroit. In 1896 he was president and director of the Detroit Union Railroad Depot and station, and became associated with **Charles B. King** in the manufacturing of marine engines. He served in the **Spanish-American War** in 1898, then became director of Peninsular Sugar Refining in 1899. In later years he was a regent for the University of Michigan.

When **Joy** and Newberry reached New York in 1901, they were standing outside when a fire engine raced by. This was at a time when everybody went to a fire. Two Packards were parked on the street and the drivers ran to their carriages, threw on the ignition switches, and gave quick spins to the starting cranks. When **Joy** witnessed both engines come to life so quickly, he was impressed and purchased the only other Packard in the city for himself. He drove back to Detroit, and gradually showed all of his friends how reliable the Packard was. He tinkered with the car as he had done with his motor boat, and made frequent trips to the Warren factory to consult with **J. W.** on ways the engine could be improved.

By this time a bid of support to increase capitalization of the Ohio Automobile Company came from Henry Wick of Youngstown, Ohio. **Henry Joy** contacted **James Packard** and persuaded him to wait while he contacted friends in Detroit who would also be interested in investing in the Ohio Automobile Company. He enlisted Truman Newberry, whose father was in the freight-car building business, and was a customer of Russell A. Alger. Then he contacted Russel A. Alger, Jr., son of a millionaire lumberman and Republican party leader, and Phillip H. McMillan, son of millionaire Senator James McMillan. Next he contacted Dexter M. Ferry, Jr., son of a millionaire seed man and a Republican stalwart, and **Joseph Boyer** of the Boyer Machine Company. Finally he enlisted Charles A. DuCharme, son of one of the founders of the Michigan Stove Company. **Joy** succeeded in convincing these second generation railroad Republicans that the Packard was the automobile of the future. The group visited the Ohio Automobile Company, in Warren, and agreed to invest in Packard's company.

At first **Packard** wanted $125,000, but the group persuaded Packard to accept $250,000, with an additional $125,000 as an inducement to build a factory in Detroit, if needed. On October 2, 1902 the firm's name was formally changed to the Packard Motor Car Company, with **James Packard** as president. The Packard group held $125,000 of the stock and the Detroit investors held $250,000 of the stock. It soon became clear that the investors, under **Joy's** leadership, were assuming control of the company and **Packard** was becoming a figurehead.

Joy tried to sell the 1903 Packard for $7500. The company lost $200,000, but did produce the Gray Wolf, the first automobile to travel a mile in a minute. In early 1903 a decision was made to build a new factory in Detroit, and **Joy** was made general manager. A site with 40 acres on the inner beltline of the Michigan Central Railroad was purchased, and the first building fronted on a thoroughfare around the outside edge of Detroit, called the Grand Boulevard. It was a big two story wood-framed mill, with better light than the prevailing "prison workshops." The structure was designed by **Albert Kahn,** whose earliest industrial design was a small mill construction factory on Second Avenue for **Joseph Boyer** in 1900. Boyer manufactured pneumatic hammers of **Charles King's** design. **Joy** could sense the ability of **Albert Kahn** when **Boyer** introduced him in 1902.

The first automobiles produced in the Detroit factory were the Model L, which represented a radical departure from the previous Packards. It was a big *open job* with a four cylinder motor, and was made with expensive steel parts imported from France. In 1904 a yoke shaped radiator (the yoke shaped radiator and hood were called the "life lines") was adopted that continued to denote Packard throughout its history. Production of Packard trucks began in 1904, mostly for the purpose of transporting manufacturing materials for the company's own use.

After completing nine mill-type buildings for the Packard Company, **Kahn** erected building "10" in 1905. It was **the first structurally reinforced concrete factory in the world.** (Eventually the Packard Company had 74 buildings on 80 acres, plus a 500 acre test track north of Detroit). **Joy** developed precision manufacturing methods by which he was able to produce 500 automobiles in 1905 and earn back most of his losses. By 1909 the Packard was solidly established, and was a member of the so called three P's: Packard, Pierce-Arrow, and Peerless.

In 1911 **Joy** hired **Alvin Macauley** away from the Burroughs Adding Machine Company. **Macauley** was born in Wheeling, West Virginia in 1872, and attended George Washington University in Washington D. C., and Lehigh University in South Bethlehem, Pennsylvania. He studied law and practiced in Washington D. C. for five years. He was with National Cash Register from 1895 to 1901, then became general manager of American Arithmomitor Company (later the Burroughs Adding Machine Company) in Detroit starting in 1901.

Macauley then hired a Burroughs colleague, **Jesse G. Vincent,** to head up Packard engineering in 1912. The twin-six arrived in 1916 as the first series production 12-cylinder automobile in the world. **Macauley** became president in 1916, with **Joy** taking over as chairman of the board. **Joy** was promoting a merger as part of an expansion program, but the board disagreed and **Joy** left the company.

Vincent headed up the famous **Liberty engine** design for use in the First World War. It was the finest airplane engine in the world at that time, according to Orville Wright. (This was the same engine that **Henry Leland** produced after he left the Cadillac Motor Car Company.) After the war **Macauley** used $10,000,000 in war profits and preferred stock issued for an additional $7,500,000 to purchase new machine tools and new standards of precision. To make room for the huge stamping machinery, the entire truck production was scuttled in 1923. **Macauley** felt that trucks were becoming too large to display on the showroom floors.

Despite not having a giant corporation to lean on Packard survived the 1929 Depression thanks to clever marketing, and less expensive cars. In 1937 it ranked ninth in the industry. During the second world war Packard built Rolls Royce Merlin aircraft engines and other power plants for the military. It was the only "independent" to emerge from the war completely free of debt. After the war Packard continued to produce middle-price cars, which cheapened its image, although sales were high throughout the 1940s. (After the war Cadillac dropped its middle-priced car, the La Salle, and sold every high price car it could, making it the premier luxury car in the United States.)

Packard celebrated its golden anniversary in 1949, and although sales would begin to decrease in the 1950s, it continually developed high quality cars with innovations such as self adjusting suspension with torsion bars, push button drive, and a overhead valve V-8. In 1953 Packard ranked 14th in the industry with 81,000 sales, but as the overall market expanded, it did not keep pace. After the Korean War downturn the 1954 sales were down to 27,000. Although the Packard Motor Car Company was free of debt and had $8,000,000 in cash reserves, president James Nance went looking for a partner to help fund tooling, technology, and factory expansion for its new models and purchased the Studebaker Corporation. Studebaker, however, was over-extended trying to keep up with the big three, and became a detriment to Packard. A new body shell, with a hint of tail fins and three tone paint combinations, and V-8 power helped sales increase to 70,000 in 1955. Nance bailed out as sales dropped to 13,000 in the depressed 1956 market. The last "real" Packard was produced on June 2, 1956.

The Curtiss-Wright Corporation acquired Studebaker-Packard, primarily as a tax write-off, and moved its holdings to the Studebaker Southbend, Indiana facilities. The 1957 and 1958 Packards were re-badged Studebakers, and the name "Packard" disappeared from the automotive seen thereafter. The Studebaker-Packard Corporation dropped the name "Packard" in 1961, and became the Studebaker Corporation. Studebaker ended automobile production on February 26, 1966 after it moved to Canada.

Original Packard building in Warren, Ohio   courtesy AAMA

W. D. Packard in the model "A"   courtesy AAMA

The first Detroit plant was laid out in a shape of a hollow square 400 feet on all sides with windows on all eight sides. Only one entrance was provided, on the south side of building number four. All employees entered directly to the wash rooms with 600 lockers. The four buildings contained machine and hand tools, motor asembly and testing, running gears, polishing and plating, blacksmith shop, painting, varnishing, trimming, and finishing.
courtesy MSHC

1904
Henry B. Joy in a Packard Roadster   courtesy AAMA

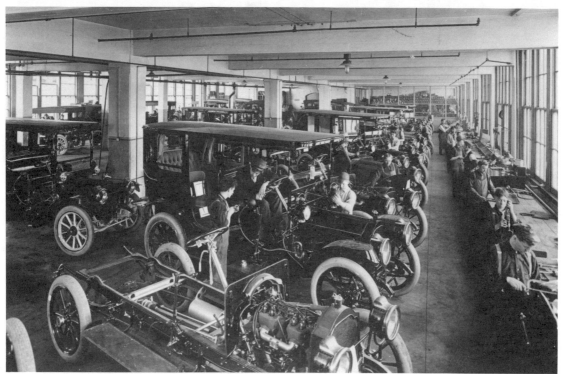

Inside view of building number 10, erected in 1905, was the first reinforced concrete factory in the world. It originally had two stories, with a third and fourth added later.

courtesy KAHN

Packards on the final test (located at Harper and Sherwood, the inside portion of the track was used as a baseball field). Each car was fitted with a weather worn body and old tires. Every chassis was tried out and tuned up. They kept the cars on the road until it was known that every adjusment was perfect and every working part was operating like a watch.

Packard   courtesy Tinder

Packard Truck   courtesy Burton

1953-
Packard Pan American (note the life line grill)

1953 Studebaker
A new Studebaker station wagon intended to celebrate the 100th anniversary of the company is shown alongside an 1852 Studebaker horse drawn wagon (the board of directors authorized the new Raymond Loewy designed car late and tooling problems persisted causing a six month delay in the new car's introduction.)   *courtesy AAMA*

1958- Packard Hawk
This was the last year for the Packard nameplate. This was a 1953 vintage Studebaker with fiberglass fins and front end, known as the "droop snoot." It was powered by a 289 cubic inch engine with a McCulloch supercharger, yielding 275 hp at 4800 rpm.
                                                                                                                                                                courtesy AAMA

Picture taken in 1995

In 1928, in memory of James Ward Packard who died that year, the Packard family crest became the official Packard emblem.

DETROIT AUTOMOBILE COMPANY
DETROIT, MICHIGAN
1899-1901

In August of 1896, **Henry Ford** was sent as a delegate from the Detroit Edison plant to the annual convention of the Association of Edison Illuminating Companies then being held in New York City. **Henry** conversed with **Thomas Edison**, and told him about his gasoline powered motor carriage. **Edison** encouraged him to continue, giving **Ford** fresh inspiration. He sold his Quadricycle to **Charles G. Annesley** for $200 to use for funding his second motor carriage.

In 1897 **William Maybury**, Detroit's most popular Democrat, became mayor. The **Maybury** and **Ford** families were good friends. William's father, Thomas, and Henry's father, William, were from the same town in Ireland and immigrated to Michigan within two years of each other. **William** was a prominent attorney before becoming mayor, and was a good friend of **Henry Ford**. (In 1893, **Maybury** wrote a letter of recommendation on Ford's behalf when he was seeking a position with the city power plant.) After **Ford** began working on his second motor carriage, **Maybury** used his business influence to help **Ford**, and as expenses mounted, he paid some of the bills.

Maybury began to get together a group of individuals in order to increase financial support for **Ford**. Everett A. Leonard and Benjamin R. Hoyt, both friends of Maybury, and Ellery I. Garfield formed together to assist **Ford**. Garfield, who worked in the Boston, Massachusetts branch of the Fort Wayne Electric Corporation was the only non-Detroiter in the group, and probably met **Maybury** in his dealings with Detroit's power companies. They signed a contract to raise capital for **Ford's** second vehicle, and to gain control over Ford and the products of his labor with the hope of becoming motor carriage producers. As time went on, Garfield grew impatient with **Ford** for taking so long to finish. It came as a relief when the second motor wagon was completed in October 1898.

In January 1899, Garfield wrote that he was seeking to interest "moneyed friends" and urged **Maybury** to use his persuasion and get **Ford** to speed up his tests. He continued writing **Maybury** to urge **Ford** to finish. Before Garfield could organize his eastern business friends, a powerful group of Detroit capitalists moved in and seized control of the Detroit Automobile Company. This was just a few weeks after another group of Detroiters had obtained the majority of stock in the Olds Motor works. **The East, which had a chance to control the Olds Company, now lost a second chance to control a key automobile producer.**

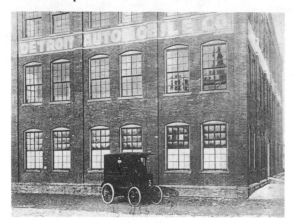

The Detroit Automobile Company was incorporated on July 31, 1899 with a capitalization of $150,000. Maybury, Leonard, Hoyt, and Garfield were among the charter investors, but the list was headed by Colonel Frank J. Hecker, whose fortune had been made in the manufacture of railroad cars. Others included William McMillan, son of senator James McMillan, Frank W. Eddy, a director of several McMillan companies, and Albert E. F. White, one of the four founders of the D. M. Ferry Seed Company, of which James McMillan was vice-president. Smaller amounts of stock were obtained by Lem Bowen, treasurer of the Ferry Seed Company, Safford S. DeLano, whose father had been associated with McMillan in the railroad car industry since the early 1860's. Others included Frank R. Alderman, Frederick S. Osborne, Clarence A. Black, Mark Hopkins, Patrick A. Ducey, and William H. Murphy. By August, **Ford** and a few workmen began to occupy a building at 1343 Cass Avenue, formerly occupied by **Maybury's Detroit Motor Company**. By the end of the year, the first machine was completed. It was a delivery wagon that was put on an experimental run by the Detroit Post Office. By April 1900, there were at least a dozen machines in various stages of completion, but no sales were made. The directors were getting desperate, and part of the building was leased to the C. R. Wilson Body Company to save expenses.

Then, with nothing to show for over a year's work and costs approaching $86,000, the directors met to plan their next move. Most wanted to throw in the towel, but Murphy and Hopkins persuaded them to reduce operations to an experimental basis again. Sledge hammers were taken to the uncompleted bodies and the remains burned in the engine room boilers. **Ford** was given a small section of the building for experimentation, but the destruction of the bodies really ended the Detroit Automobile Company. Dissolution was filed in Lansing in January 1901.

1900
Interior of the Detroit Automobile Company showing the assembly hall with 13 workers. A truck can be seen on the right. Note the overhead lineshafts used for turning lathes and grinders.   courtesy MSHC

THE GROSSE POINTE SWEEPSTAKES RACE

Henry Ford and Tom Cooper rented space in a barn from Barton Peck at 81 Park Place, between Grand River and Clifford Street, while Henry was chief engineer for the Detroit Automobile Company. After the DAC went out of business, work began at 81 Park Place on building a race car. Day after day the sounds of hammers and files against steel were heard behind the curtains that draped over the windows.

By the summer of 1901, the car was finished and named "sweepstakes." It was two cylinder, with each cylinder 7 inches by 7 inches, estimated to have 26 horsepower. The engineering was mostly the work of Oliver Barthel, along with Harold Wills and Spider Huff. Barthel's dentist, Dr. W. E. Sandborn made spark plug insulators with porcelain using Barton Peck's furnace. Up to then, layers of mica were used for spark plugs making them extremely expensive. (During World War I sillimanite was found at the top of a volcano in California by another dentist, Dr. Jeffrey. The sillimanite in a powdery form gives the porcelain its resistance to compression and heat, compared to using wet clay with varying ores.)

Henry Ford at the wheel of "sweepstakes" and his passenger Oliver Barthel. The picture was taken on East Grand Boulevard.

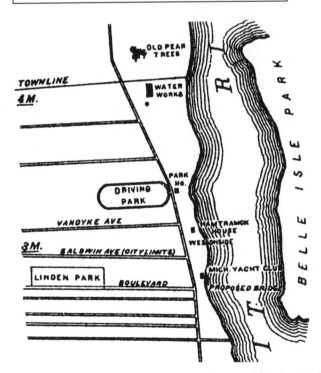

Henry Ford wanted to go 1 mile in a minute or less to break the world's record and he wanted to beat Alexander Winton on the oval track. Alexander Winton came to the United States from Scotland when he was 19 years old. He worked with marine engines in New York City. He started work on his first car in 1893, and became a race driver. He built and drove cars in the Gordon-Bennett races in France. Henry got his chance to race Winton on October 10, 1901. The race was held at the Grosse Pointe Horse Racing Track, which was paved with dirt. The race committee was headed by William E. Metzger and serving with him was the owner of the race track, Daniel J. Campau, a wealthy Detroit lawyer.

The term "Grosse Pointe" designated several localities. First, the township of Grosse Pointe started near water works and extended along the river to the northern limits of Wayne County. The driving park as shown on this map precedes the construction of the first Belle Isle Bridge.

Some of the race applicants included, Edgar Apperson, Windsor and Rollin White (who had begun producing steam cars in the White Sewing Machine factory in Cleveland).

1884- opening day at the driving park that was later used for automobile racing in Grosse Pointe

Also, **William Grant,** and **William C. Rands** from Detroit, **Roy Chapin** and **Jonathan Maxwell** from the Olds Motor Works all had steam cars. This was the first "real thing" in the line of an automobile race in the mid-west according to Motorcar magazine. There was a big parade through downtown Detroit, the day before the race at the track. Oldsmobile had curved Dash Oldsmobiles on display, along with various demonstrations, and closed for the day to allow its 200 workers to attend the race. With a boatload of people from Cleveland to see **Winton**, there were 8,000 people packing the grandstand to watch the events. The first events were dull, especially the electrics. The Baker Electric, of Cleveland, winning time was more than 30 seconds slower than the existing record of a man running a mile.

circa 1905- The front gate of the driving park

Another feature was an exhibition race between two famous bicyclists, **Tom Cooper** and **Barney Oldfield**. However, the crowd was not there to see bicycle races, but the cars were having all kinds of problems. **Bill Metzger** could see the disappointment and got **Winton** to agree to break the mile-a-minute barrier. He did a mile in one minute and twelve seconds to bring the record to Detroit. Because the first three events took longer than expected that day, the last race was reduced from 25 miles to 10 miles. Only two out of twenty-five original entrants were left for the final race. Most entrants of this "feature" race withdrew because of mechanical troubles leaving **Ford** and **Winton** to a "dual it out." Ford took a two-lap warm-up with **Tom Cooper,** to get advice on track conditions.

The employees of the Olds Motor Works were in attendance for the race. The Olds company constructed a wooden staircase with a 30 degree incline to demonstrate the power of the runabout. The see-saw pictured at the left was used to show its agility. Ransom Olds is probably driving the middle car.

Near the start of the races- note the Olds Motor Works staircase at the left.

Henry Ford and Spider Huff demonstrating racing positions

Finally they were off, and **Winton** led most of the way. Near the end, **Winton** suffered mechanical problems and **Ford,** with **Spider Huff** leaning on the curves, came "zooming" by to win in 13 minutes 23 seconds. **Ford** was hailed as one of America's front-rank "chauffeurs." **Tom Cooper** introduced **Barney Oldfield** to **Henry Ford** that day. Sweepstakes was sold to William C. Rands, who in 1902, did a half mile in 26.2 seconds.

above- Ford is in the left car with a cloth with a large number 4 on it. Winton is at the right.

Winton's car had 40 hp and Ford's had 26 hp. Winton led easily for the first 5 miles, but Ford began to catch up and finally won after Winton had engine problems.

Before the race, there was a parade of 100 cars that started up Bagg Street at 10:30 am. The tally-ho for the band was pulled with two steam carriages. Electric cars were given preference and led the march followed by gas and steam.

After the race was over, everyone seemed to leave at the same time and caused traffic congestion and scared the horses.

courtesy Free Press

Tom Cooper was a champion in his own right, who represented and advertised for various firms, which netted him prosperity. After the race at Grosse Pointe, **Cooper** and **Oldfield** went out west to February 1902 and supplied **Ford** with needed funds to develop two new racing machines. **Henry** named them **999** and **Arrow**. These were the names of New York Central Railroad locomotives that **Henry** saw at the 1893 Chicago World's Fair, which were very capable of doing 60 miles per hour and no one seemed to acknowledge that they held the world's land speed record.

At the time, the 999 was the biggest automobile built in the United States at 9 feet 9 inches long and was rated at 75 horsepower. Both **Ford** and **Cooper** drove the 999 but were afraid of it; but **Oldfield** was willing to race it. A little over a year after the **Winton** race, **Barney Oldfield** set a new American record of 5 minutes and 28 seconds for 5 miles, beating **Winton** in the race.

Ford standing next to Barney Oldfield and racer 999 at Driving Park. The car was steered with a tiller with two handles because according to Ford, you couldn't see the track through the dust and the handles let you know when you were going straight.

By the fall of 1903, the Ford Motor Company was in business and doing quite well. **Ford** still wanted to beat the land speed record. He needed a straight-away and in January 1904 went to New Baltimore, Mich., on frozen Lake St. Clair. Miles of ice were cleared and coal cinders laid on top. **Ford** used the Arrow, which was faster than 999, because of a better intake manifold which **Barney Oldfield** designed. Henry had it repainted to look like the 999 in order to amplify the fame won by **Barney Oldfield**. On January 12 **Ford** went 91 miles per hour, setting a new land speed record, or so he thought. It wasn't until January 21, after some discussion, that the judging committee of the AAA agreed that the land speed record could be set on water, solid or not.

Cooper, who still owned both 999 and Arrow, sold both vehicles and then became a general sales agent with the Matheson Company in Grand Rapids, Michigan. On November 19, 1906, Tom Cooper was killed in a collision with another car in Central Park, New York.

OLIVER EDWARD BARTHEL

Albrecht and Elizabeth Barthel lived in Augsburg, Germany, and decided to move to Detroit where there was a large German population. Detroit needed men with his industrial skills, and Albrecht got a job at the Michigan Stove Company.

They had a son, **Oliver,** who was born in Detroit and educated in the Detroit public schools. Albrecht became a representitive of the Stove Company and was sent to Europe with his family where Oliver attended schools in Germany and France.

In 1894, at 16 years old, **Oliver** got a job as a *draughtsman* at the Frontier Iron Works in Detroit, which made marine steam engines. The job did not last long, and by November he was working as a *draughtsman* and assistant for **Charles B. King.** They rented space in **John Lauer's** workshop which had several German machinists. The German language that **Oliver Barthel** learned while his father was working in Europe came in handy to help **King** explain his requirements to the machinists. **King** became the first person to build and drive a car in Detroit. While working for **King, Barthel** took home-study courses, and night classes on machine operation and repair, taught by Henry Ford.

Barthel continued with **King** until 1901, then became a consulting engineer. By this time, **Henry Ford** had been involved with the Detroit Automobile Company which was already dissolved. **Ford** wanted to design and build a race car, and hired **Barthel** to help design "Sweepstakes." After the success of his first race car, **Ford** received financial backing again and the Henry Ford Company was formed. **Barthel** was hired full-time. **Ford** continued with his preoccupation with racing and left the Henry Ford Company in March 1902. **Barthel** took over for **Ford,** but soon after, **Henry Leland** was in charge, and the Cadillac Motor Company was founded.

Barthel had made a name for himself by this time, and with the backing by a group of drugstore owners the Mowhawk Motor Company was formed in 1903. The name was soon changed to Lafayette Motor Company, since it was discovered that a Mohawk Motor Company already existed in Indianapolis. The company was unable to go into production and closed in September 1903. Next, the Davis brothers of Detroit and Alex Groesbeck, a future governor of Michigan, offered financial backing for the Barthel Motor Company. It was organized with $400,000, with James Davis as president, Charles Davis as secretary and **Barthel** as consulting engineer. By January 1904, only two cars were completed and the venture soon ended.

Next, **Oliver** worked for the Olds Motor Works developing marine engines. In 1906 he married Adele Vargason and became an assistant engineer at the Great Falls Power Company. Oliver was awarded over 30 patents on a variety of items throughout the years. He operated a consulting business in Detroit until 1955, and died in 1969.

1903- Oliver Barthel (L) and Arthur Mitchell (R) in an Olds Runabout
courtesy NEWS

MAXWELL-BRISCOE MOTOR COMPANY
TARRYTOWN, NEW YORK
1904-1910

MAXWELL-BRISCOE MOTOR COMPANY
Division of <u>United States Motor Company</u>
NEW YORK, NEW YORK
1910-1912

"Perfectly Simple-Simply Perfect"

Benjamin Briscoe

Benjamin Briscoe was born in Detroit on May 24, 1867, the son of Joseph A. and Sarah Smith Briscoe. Benjamin's grandfather, Benjamin Sr., was born in Newcastle County, Del., in 1812. He apprenticed at the Baldwin Locomotive Works in Philadelphia and became a highly skilled mechanic. He brought the first two locomotives the city of Detroit ever saw in the 1850s. He settled in Detroit and became a master mechanic for the Detroit and Milwaukee Railroad, and a superindendant of the D & M Railroad through the end of the 1860s. He was in real estate in the 70s, and an engineer in the late 70s. He served on the board of public works in the 1880s, and died in 1898.

Along with his son, Joseph, he contributed to the development of railroads in Michigan with inventions that improved rolling stock. After finishing common school, **Benjamin** went directly to work as a clerk at a Detroit wholesale firm. In 1885, with a capital of $472, he formed his own firm, Benjamin Briscoe and Company. He formed sheet metal into pails, sprinkler cans and tubs well enough to sell the business to the American Can Company. He subsequently developed a machine to form tubing and organized the Detroit Galvanizing and Sheet Metal Works. He made corrugated pipe, and parts for stoves. His brother joined the business in 1900, and he re-organized as the Briscoe Manufacturing Company. The company was on the brink of failure the very next year due to a bank collapse. He received modest financial support from George Russel, a Detroit banker, but was essentially turned down for the major amounts he needed. He decided to get help, "where the money was," from the J. P. Morgan & Company. He claimed that Morgan himself put up $100,000, and that "Mr. Morgan and myself became the main owners of the Briscoe Manufacturing Company."

In 1902 **Ransom E. Olds**, accompanied by a young engineer named **Jonathan Maxwell,** went to the Briscoe factory with a radiator for the Olds runabout that **Benjamin** said looked like "some antiquated band instrument." The Briscoe company, with over 500 men, and factories in Detroit and Trenton, N. J., became one of the largest manufacturers of radiators, gasoline tanks, water tanks, *bonnets*, fenders, and other sheet metal parts in the United States.

He supplied sheet metal to **David Buick** and loaned him money for development of an automobile. After several loans, with nothing to show, **Briscoe** sought the help of **Jonathan Maxwell**, who had started his career with the Haynes company, then Olds, and by then worked for the Northern Motor Car Company. **Maxwell** was asked to visit **Buick** and check on the progress. **Maxwell** reported back without giving an opinion, but offered that he had a few horseless carriage notions himself, and maybe the two of them "could 'hook up' in the automobile business."

After visiting Europe to study radiator designs, **Briscoe** returned to Detroit and found that **Buick's** car was completed. **Briscoe** decided to recoup his $3500 investment and sold the Buick company to the Flint Wagon Works, in Flint, Mich. On July 4, 1903 **Briscoe** entered into a contract with **Maxwell**, with $3000 backing from C. W. Althouse, to build a prototype car. **Maxwell** built the car in **John Lauer's** Machine Shop, and on December 25, 1903 **Briscoe** was able to view the first successful test of the car.

Maxwell factory in Kingsland Point, N.Y.   courtesy AAMA

Benjamin secured some backing from Detroit banks, but the feverish automotive developments taking place in the city placed a strain on the lending institutions. **Briscoe** "wore out several pairs of shoes" trying to secure capital in Detroit for a new automobile company. He finally went to New York to see J. P. Morgan, who raised $100,000, and in return, the car was built in Tarrytown, N. Y., at Kingsland Point. The factory was formerly occupied by John Brisbane Walker, who built the Walker steam automobile from 1899 until 1903, and was the founder of Cosmopolitan Magazine.

In 1905 more "Maxwells" were sold than the factory capacity permitted, and money was refunded to prospective purchasers. For 1906 a branch factory was established at Pawtucket, R. I., in connection with the celebrated machine shops of Brown and Sharpe, and an assembly plant was established in Chicago for Northwest distribution. Refinements were made to the car and some of the parts were strengthened in which slight weaknesses developed. The first sales agency was established in Detroit in 1906 by Alex I. McLeod, as the Maxwell-Briscoe-Mcleod Company, in a five story building at 243-246 Jefferson Avenue, formerly occupied by the Olds Motor Works. *The first floor was re-arranged so that the salesroom was entirely separate from the disagreeable noise, dirt, smell, etc., ordinarily* connected *with even the best arranged and conducted automobile garage.* The Maxwell was seen at the Detroit Auto Show in spaces on the main floor, occupying the entire large room known as the ladies parlor. The Model H touring car had 16 to 20 hp and sold for $1450. A polished chassis was at its side, open for inspection. Close along side, with its companion chassis, was the Model L tourabout with 10 hp, which sold for $780. The natty looking Dr. Maxwell with 16 hp sold for $1600 and surely caught the eye of physicians. Tradesmen and department store owners saw the Model O delivery wagon, which sold for $1400. The gentlemen's speedster runabout with 10 hp sold for $800, and the new Model M with a 30 hp four cylinder sold for $2000. It had a 104 inch wb and could fit seven passengers without crowding.

Briscoe didn't like the insecurity of leasing the Tarrytown factory, so he purchased it. He also spent over $125,000 to build a new plant in Newcastle, Ind. As the panic of 1907 loomed, **Briscoe** couldn't get much support from his backers and received admonition from his partner **Jonathan Maxwell** for taking too many risks. **Briscoe** was thinking big, and he called **Billy Durant**, who thought like he did, and suggested a merger. After a lengthy discussion, they met in the Ponchartrain Hotel in Detroit with **Ransom E. Olds** and **Henry Ford**, with **Durant** advocating a holding company, and **Briscoe** wanting to combine all operations into a central organization. **Ford** and **Olds** demanded cash for their companies of $3,000,000 each. Even with J. P. Morgan interested in the venture, **Briscoe** and **Durant** could not come up with $6,000,000. **Briscoe** and **Durant** continued the discussion and came up with the name International Motor Car Company. Word was leaked to the New York Times which reported the House of Morgan was among the underwriters of the project. The Morgans were very upset and pulled out. **Benjamin** wrote **Billy** "Why they should feel it as deeply as they do, I can't quite fathom myself."

Briscoe then merged with the Columbia Motor Car Company of Hartford, Conn., forming the base for the **United States Motor Company** in February 1910, with a capital investment of $30,000,000. Its national headquarters was at 505 Fifth Avenue, N. Y.

courtesy AAMA

1906- "Doctor's Model" Maxwell based on specifications by physicians making house calls. Price without top $1300

L. to R. in tonneau- Benjamin Briscoe, Elwood Haynes, J. D. Maxwell

BUICK MANUFACTURING COMPANY
DETROIT, MICHIGAN
1901-1903

BUICK MOTOR COMPANY
DETROIT, MICHIGAN
1903

BUICK MOTOR COMPANY
FLINT, MICHIGAN
1904-

"WHEN BETTER AUTOMOBILES ARE BUILT, BUICK WILL BUILD THEM"

David Dunbar Buick was born in Scotland in 1854. Two years later his parents immigrated to the U.S.A. and settled in Detroit. He later became a Free Press newspaper carrier in the morning and sold the Detroit Union newspaper on the street corners at night. He left Detroit and worked on a farm for a number of years, then returned as an apprentice at the James Flower & Brothers Machine Shop (**Henry Ford** also worked there) in the brass finishing trade for 12 years. He was a foreman from 1879 to 1881, then left the Flowers Bros. Company and started in the newly flourishing plumbing trade for himself. In 1884 **Buick** joined with William Sherwood and formed the Buick & Sherwood Mfg. Co. Sherwood was born in England in 1851, and educated in London. The two men met when Sherwood worked at Flowers in 1873 as a brass molder. They became manufacturers of plumbers woodwork and sanitary specialties. The shop was located at the corner of Champlain Street and Meldrum Ave. The building had a frontage of 162 feet on Meldrum and 152 feet on Champlain, with an additional 80 foot structure in the rear. They had 122 employees, with Buick as the senior member of the firm.

While **Buick** was in the plumbing business he developed and patented a process for affixing porcelain enamel to steel bathtubs. Enameled steel products were in production in Europe at that time, but the manufacturing proccess was a closely guarded secret. In 1900 indoor plumbing with hot water was becoming available in the U.S.A., but enameled tubs were not common until 1910, and one piece enameled tubs were not mass produced until 1920.

In 1900 **Buick** accepted an offer of $100,000 for his share of the Buick & Sherwood Co., including his porcelain enamel patents. He invested the money and formed the Buick Auto-Vim and Power Co. in a shop at 39 Beaubien Street. He produced an L-head engine for marine use and stationary farm use with some limited success, but became increasingly interested in automobiles. **David Buick** and his son **Tom** built their first automobile, using the L-head engine, in a barn behind the Buick house on Meldrum. **Buick** was not satisfied with the motor or the way his company was proceeding, and wanted to get into automobile manufacturing. **Walter Marr**, who had automobile experience, began working with **Buick** on a new engine which had poppet valves located in the head. Before the engine was perfected, **Marr** left due to a disagreement and organized his own auotomobile company.

In 1901 Eugene C. Richard joined Buick and continued work on the valve-in-head motor. Richard had learned about engines in Philadelphia, and worked for the Olds Motor Works in Detroit before it *burnt* down.

1902-
David Buick was anxious to produce cars and organized the Buick Manufacturing Company, located on 416-418 Howard Street in Detroit.

THERE IS NO GAS OR GASOLINE ENGINE
On earth that can compare with the Buick in simplicity, and satisfaction guaranteed; sizes from 1 h. p. up; price, 1 h. p., $100; responsible agents wanted in all parts of United States. Phone West 505. The Buick Mfg. Co., 416-18 Howard st.

In the fall of 1902 the valve-in-head engine was perfected. Unlike the L-head engines in use at that time, the valves were directly over the pistons, resulting in a more compact combustion chamber, and did away with the pockets over the intake and exhaust valves which were required for the L-head engines. As a consequence, the valve-in-head scavenged exhaust gases much faster because there was less remaining in the cylinders after each explosion to mix with the incoming new gas. This resulted in a gain of 20 percent or more power from the same sized cylinders.

With the new engine, a second car was under construction, but by this time **Buick** was in serious financial difficulty and turned to **Benjamin** and **Frank Briscoe**, who had already contributed money and materials to **Buick**. The **Briscoe** brothers owned a sheet metal company and had recently started to supply parts to the Olds Motor Works. They advanced **Buick** $650 in cash, and in return wanted the next car **Buick** built. The car was completed and tested at daybreak by **Buick** and Richards around the Buick premises in Detroit in early 1903. The **Briscoes** let him use it as a demonstrator for money raising purposes in order to put it into production. By the spring of 1903 **Buick** still had not secured financial backing.

Buick was out of money again and returned to the **Briscoes** for further help. They gave **Buick** $1500, but insisted on a re-organization. In May 1903 The Buick Motor Co. was incorporated with a capital stock of $100,000, divided up with $300 for Buick and $99,700 for the **Briscoes**. Buick had the option of purchasing the Briscoes' stock by repaying $3500 he owed them by September 1903, or else they would take over the business management of the company. Within a few weeks, the **Briscoes** were having second thoughts about the Buick Motor Co., and were becoming interested in a car designed by **Jonathan Maxwell**.

During the summer of 1903 the **Briscoes** visited a relative, Dwight T. Stone, in Flint. Stone was a realtor who introduced the Briscoes to James Whiting and other directors of the Flint Wagon Works. Early in September 1903 five directors of the Wagon Works borrowed $10,000 and bought the Buick Motor Co. from the Briscoes. The Flint Journal gave front page coverage that a splendid new manufacturing industry would be established in the city. In December 1903 the first Buick plant in Flint, next to the main plant of the Flint Wagon Works, was operating with 25 workers. At the same time the Detroit Buick plant remained in operation with **David Buick** as manager, and **Walter Marr** was back, as chief engineer. Total production for 1903 was only six cars.

The Buick Motor Company was incorporated in January 1904 with $75,000 capital. Out of a total of 6388 shares of stock, **David** and **Tom Buick** controlled 1500, but the company held the stock and took David Buick's note until his debts were paid off. **Buick** was having difficulty paying off his debt and periodically renewed the note through a bank. One one occasion, without reading the slip of paper, he signed what he thought was a renewal, but some time later, he learned he had signed a bill of sale for his stock in lieu of paying off his debt.

In 1908 **David Buick** left the Buick Motor Company and went to California where he organized an oil company. Litigation over ownership of the lands ensued and he lost everything. He returned to Michigan with his son and started a small carburetor company which failed. He secured a controlling interest in the Lorraine Motor Corp. in Grand Rapids, Mich., in 1920, and it failed. In 1922 **Buick** organized the David Dunbar Buick Corp. to manufacture a car called the Dunbar. Headquarters were in New York City, and the factory was 70 miles away in Walden, N.Y. The Dunbar was to be available in four models, costing $1000 for the *open jobs* and $1400 for closed models. The corporation was capitalized at $5,000,000, but it also failed. Buick then became a partner in a real estate firm during the Florida boom, and it failed. In 1927, at 72 years of age, **Buick** returned to Detroit and worked as an instructor and a clerk for the Detroit School of Trades. An engineer from Saskatchewan named Gordon E. Mann heard of **Buick's** circumstances and offered him his home and his care for the merit he felt was due to **Buick** which went unrecognized. **David Buick** died on March 5, 1929, in Detroit's Harper hospital.

1891- William Sherwood

1891-David Buick

Above- Buick's chief engineer, Walter L. Marr (L) and Thomas D. Buick (R), son of David Buick, in the first Flint Buick as it ended its successful Flint-Detroit round trip in July 1904. Marr said that with it's two cylinder engine "we went so fast we could not see the village six-mile-an-hour sign."   courtesy AAMA

below- 1904 Buick model B

*above-
The tri-shield marque was based on the coat of arms of David Buick. The design consists of a stag head at the top, a cross at the bottom and a diagonal bar of diamonds through the middle. The marque was first used in 1937 as a single shield. It was expanded to the tri-shield in 1959 to represent LeSabre, Invicta and Electra models.*

MARR AUTOCAR COMPANY
DETROIT, MICHIGAN
1902-1904

Walter L. Marr was born in Lexington, Mich., in 1865. He served his apprenticeship with John Walker & Sons, engineers in East Tawas, Mich., in 1882. In 1887 **Marr** began with the Wick Brothers, sawmill and steamboat engineers, in Saginaw, Mich., and stayed there until 1896. In 1888 **Marr** began his engine work by building a motor using all water cooled steel castings and hot tube ignition. It was designed by the superintendent of the Wick Bros. shops.

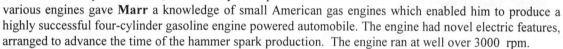

Marr left Wick Bros. in 1896 and went into business for himself in Saginaw as a bicycle manufacturer. The machinery in his shop was driven by various gas engines. First a Hercules of 2.5 hp, then a Philadelphia Otto of 7 hp, being replaced by a 7.5 hp Sintz two cycle. He moved to Detroit in 1898, and used a 6.5 hp Olds motor, followed by a 7 hp Cofield. The use of these various engines gave **Marr** a knowledge of small American gas engines which enabled him to produce a highly successful four-cylinder gasoline engine powered automobile. The engine had novel electric features, arranged to advance the time of the hammer spark production. The engine ran at well over 3000 rpm.

In 1899 **Marr** sold the bicycle manufacturing business and went to work for the Detroit Shipbuilding Company for a short time. He then began a more important engagement with the Buick Auto-Vim and Power Company. In 1900 **Marr** built his second motor wagon. It was driven by a single cylinder, water cooled, and had jump-spark ignition. **Marr** was the first of the group of Detroit experimenters to use the jump-spark. He also designed various forms of cylinder-fired experimental motors, both two and four cycle. By the time he left the Buick Auto-Vim and Power Company, he was general manager and superintendent.

He next worked for the **Olds Motor Works** in Detroit where he assembled the first two Oldsmobiles. A short time later he worked for Detroit Brass, Iron and Novelty Company to help perfect a bicycle motor. **Marr** now believed the time had come to build motor wagons. In 1901 he went to work for the American Motor Carriage Company in Cleveland, Ohio, as chief designer and superintendent. The first of these new models was placed on the road in December 1901. The wagon ran well and sold for $1000, but conditions were not satisfactory and he returned to Detroit in February 1902.

He entered into a working agreement to begin construction of a new wagon with J. P. Schneider, an automobile dealer in Detroit. The motor had a single cylinder with a 4.75 inch bore and a 6 inch stroke, with a normal speed of 700 rpm. The vehicle made its first run on the road the night of December 25, 1902. The vehicle was shown at the Chicago and Detroit Automobile Shows in February 1903.

It was received with such favor that an automobile company was formed with $100,000 capital. The Marr Autocar Company was headquarterd in Detroit, in the Chamber Building. A contract was made with the Fauber Manufacturing Company in Elgin, Ill., to construct the first lot of 100 wagons, with **Marr** as the supervising engineer. The car had adjustable steering, Timken roller bearings for the steering wheel and rear axle, and jump-spark from a battery for ignition.

In August 1904 the Elgin factory burned to the ground with 14 Marr vehicles inside. The factory was not rebuilt and **Marr** returned to work for **David Buick**. They perfected the valve-in-head engine and formed the Buick Motor Company.

RULES OF THE ROAD

In ancient times, people walking by themselves had a natural tendency to stay to the left of oncoming individuals in order to keep their right hand free in the event they were attacked. This tendency prevailed up to the time of the Roman Empire, when traffic regulations were established for wheeled carts to keep to the left when going across bridges. This early regulation was extended throughout the Roman Empire in Europe even up to the mid-1700s. After horse drawn carriages came in use by the nobility in France, pedestrians walked on the right side of the road, against the flow of traffic, for safety. As the French Revolution began to take hold in 1789, the carriages joined the pedestrians on the right-hand side of the road to reduce class distinction between nobility and peasants. As a gesture, Napoleon directed his army to march on the right-side like the peasants. As his army conquered the surrounding countries, it also instituted the right-hand rule.

England was not one of Napoleon's conquests, and it stayed with the left-hand rule. As the English Empire expanded to every continent, they inaugurated the left-hand rule everywhere, including America.

By 1750 people began moving west in the United States. Lancaster, Penn., the largest city in the United States that was not situated on a waterway, was the main point of departure. The "Conostoga" wagon with large 4 inch wide wheels was developed there. These "prairie schooners" could haul 5,000 lbs and were pulled by four or six horses. The teamsters would often ride the lead horse, which was on the left-side, or near side, for right-handed people to mount. The teamster could also ride in the wagon, walk along side, or stand on a pull-out lazy board. With four to six horses, it was difficult to see around them, so they favored riding on the right side of the road.

1840- The Lancaster Pike   courtesy Bureau of Public Roads

General Lafayette of France had served in the United States war for independence, then served with Napoleon in France, and afterwards became a statesman to the United States. He influenced the U.S. government to adopt the right-hand rule like France. In 1792 the Pennsylvania turnpike became the first road with formal legislation for driving on the right-side of the road. New York soon followed in 1804 and New Jersey in 1813.

During the First World War, it became apparent that the logistics of traveling between countries in Europe, where there was a mixture of left-hand and right-hand rules, was very inefficient. Soon after the war, most left-hand rule countries converted to the right-hand rule, including Canada. The countries that retained the left-hand rule were mainly island countries such as England, Australia, and Japan.

STEERING

The first automobiles were steered with a tiller and crossbar, with a steering ratio of 1:1, making it tiring to hold. A large arc had to be made when turning, and the tiller could be whipped out of a driver's hand on rough surfaces. The driver sat on the right side, similar to the horse and buggy practice. Transmission and secondary brake levers were usually located on the right hand side of the driver. The driver and passengers would typically enter from the unobstructed left side by walking into the street, usually in the mud with the imminent risk of being run down.

Charles King designed the Northern automobile and it was the first car with the steering wheel on the left side. This was a safer design for driver vision, as well as passengers because they now could board on the curb side. The Ford Model T had left hand drive, and the Brush attracted a lot of attention with left-hand drive with and small turning radius. Gradually, most cars switched to left-hand drive. Exceptions were usually made for expensive cars that were driven by a hired chauffeur.

Left-hand drive permitted a cleaner view of the road ahead in overtaking a slower vehicle. It permitted a backward view of the road when making a left-hand turn, or a complete turn around. Right-hand turns did not need to be guarded with the traffic rules in place.

In France, roads were narrow, especially mountain roads, so traffic could not go in both directions. When passing on-coming traffic, it was safer for the car passing to be on the inside, and the car stopping to be on the outside, near the edge. The driver had better control with the steering wheel on the right hand side to see the edge of the mountain road. The European automobiles, such as in France and Germany that were driven with the right-hand rule of the road did not convert until the 1920s.

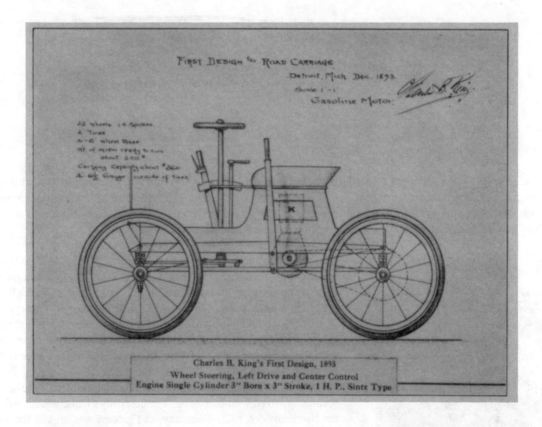

Charles King designed a motor wagon with a steering wheel on the left side in 1893. Due to lack of capital, he was not able to build it, and instead purchased a wagon steered with a tiller which became the first horseless carriage in Detroit. He got the idea from seeing a hook and ladder wagon that had a steering wheel in the rear.

NORTHERN MANUFACTURING COMPANY
DETROIT, MICHIGAN
1902-1906

NORTHERN MOTOR CAR COMPANY
DETROIT, MICHIGAN
1906-1908

"SILENT AS THE STARS"

William T. Barbour was born in Detroit in 1877, attending Detroit public schools and Phillips Academy in Andover, Mass. In 1895, he started his career as a purchasing agent for the Detroit Stove Works, becoming a vice president, and then by 1897, becoming the president of the Stove Works, with over 5000 employees. The first factory was located on Clinton Avenue and Champlain Street, Detroit, Michigan.

The Olds Motor Works factory, where **Charles King** and **Jonathan Maxwell** worked, was erected next door to the Stove Works on Jefferson Avenue in 1900. **King** left the Olds Works after the fire in 1901

Detroit Stove Works--The largest stove plant in the world

and worked for the Michigan Power and Yacht Company. Maxwell resigned in 1902 and went next door to interest **William Barbour** in organizing a new automobile company. The Northern Manufacturing Company was the result, with **William Barbour** as president, G. W. Gunderson as VP, and Victor Gunderson as secretary and treasurer. **Maxwell** was joined by **King** in the engineering department.

The first "Northern" was similar to the Olds runabout and went on sale in the spring of 1902. With **Maxwell** heading up the design, the similarity between the two automobiles came as no surprise. It had a wb of 129 inches and 28 inch clincher tires on 14-inch spoke wood wheels. The engine had one cylinder, with a 4.75 x 6 inch bore and stroke, with a normal speed of 600 rpm.

(In 1903, **Maxwell** left the Northern Company and teamed up with **Benjamin Briscoe**. They began production of the "Maxwell" in Kingston, N.J., in 1904.)

In 1906, **Bill Metzger** left the Cadillac Motor Company and became involved with Northern. The company's name was changed to the Northern Motor Car Company, and ground was broken for a new factory in Port Huron, Mich. Both of the Northern factories produced automobiles, as well as many of the components

Messrs Barbour seen demonstrating the Northern

As chief engineer for Northern, **Charles King** continued to make improvements, culminating in a number of patented features for the 1906 model year. The model K had a 119 inch wb, weighed 3200 lbs, and cost $3500. A four cylinder quadruple engine, with 50 hp, used a compressed air starter. The air pump also activated the clutch and brakes, and provided air for repairing tires. The two long muffler shells were continued in use, along with a large, 24-inch diameter flywheel with fan-vane arms. **Charles King** left Northern in 1907, and went to Europe to study automobile practices. The compressed-air features were dropped for 1908, because they were too unusual compared to other cars at the time. The Northern Company merged with the Wayne Automobile Company and, by December 1908, was discontiued. The Wayne and Northern factories became the foundation for the newly formed **E.M.F.** Company.

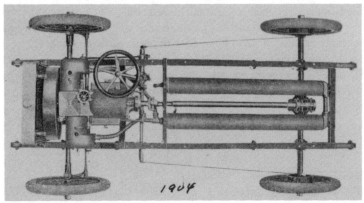

left-1904 Northern chassis with two 55.5 inch mufflers which helped make the engine silent as the stars. Note the 24 inch flywheel with 12 fan blades mounted in the front of the engine to blow dust away while driving.

left-1907 Northern on Belle Isle   courtesy AAMA

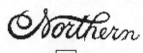

HUSSEY AUTOMOBILE AND SUPPLY COMPANY
DETROIT, MICHIGAN
1902-1903

The Hussey Automobile and Supply Company was located at the S.E. corner of Beaubien and Trombley Avenue, with Patrick L. Hussey as president. The firm developed a steering wheel that could be swung back around a pivot on the steering post column to permit free ingress and egress to the motor carriage. The wheel pivoted on a bracket extending forward from the column and at right angles. It was held in position centrally at the end of the steering post by means of a thumb screw and hinged bolt.

A runabout automobile was produced in December of 1902 to display and promote the tilt wheel. It could be ordered with or without a seven hp engine. A small production run was made until mid-1903.

After **Oliver Barthel** left the Detroit **Automobile Company**, Patrick Hussey was named manager. He was having a hard time assembling the first Cadillac using the new Leland & Faulkner engine. **Alanson Brush** was asked to take over and quickly assembled it in the Leland shop. Later, Cadillacs with right-hand drive featured the swing away steering wheel.

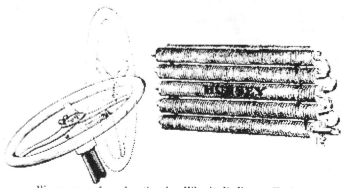

We are manufacturing Steering Wheels, Radiators, Tanks, Circulating Pumps, and other supplies. Our goods are of the highest grade, and fully covered by a guarantee. Prices are consistent with quality. Before purchasing elsewhere, be sure to get our quotations.

Hussey Automobile & Supply Company,   Detroit, Mich.

"Hussey" runabout   courtesy NAHC

HENRY MARTYN LELAND

"CRAFTSMANSHIP A CREED, ACCURACY A LAW"
-sign in the Leland Faulkoner factory

Henry Leland was born in 1843 and spent his youth on a farm in remote Vermont. As the family grew, the farm barely sustained the Leland family so they moved to Worcester, Mass., seeking higher paying work. His father, Leander Leland, got a job in a grist mill, and his mother, sister, and two older brothers found assorted work. Henry's first contact with industry began in 1857 when he and his father were both working at the **George Crompton Loom Works**. He was cutting wood to the proper length for wheel spokes all day long, but he didn't like the repetitious work. The Lelands were a very religious family that attended church services regularly. It was natural for Leander to go to the local minister, the Reverand Gerald, seeking advice about Henry's disappointment. By chance, the minister knew the superintendent of the Crompton Works, and he recommended that Henry be considered for the position of mechanic. Henry was offered an apprenticeship, but wasn't sure that he wanted to become a machinist. He decided to try it, and began working six days per week, ten hours per day.

Henry was in training when the Civil War broke out in 1861. **Abraham Lincoln** was his idol, and although he was only 17 years old and not eligible to join the military, he went to the recruiter and tried to enlist anyway. His mother's instincts led her to follow him to the recruiter where she had his name removed from the volunteer's sign-up sheet. Knowing that **Henry** wanted to help the war effort, his supervisor gave him the task of making Blanchard lathes for the Federal armories to make gun stocks with. But he wanted to work where he could do the most good for the war effort, and learned that the **U. S. Armory** in Springfield needed expert mechanics. He applied for a job and was accepted. He worked in all phases of manufacturing firearms until the war ended in 1865. The particular lesson he learned from that experience was the value of order and neatness in the shop. Everything was clean and systematic, unlike the usual cluttered factories and machine shops of the time.

With the war ended, the Springfield plant cut back production substantially, and **Henry** was discharged. With a background in textile machinery and all phases of firearms, he was hired the very next day at the **Colt Revolver Factory** in Hartford, Conn. He corresponded with Ellen Hull, whom he met while he lived in Worcester, until she finished regular-school. They became engaged, and **Henry** moved back to Worcester and found a job with **August B. Prouty**, a manufacturer of card-setting machinery for weaving. After they married, Henry found a job at the **Brown and Sharpe Company** in Providence, R.I., where he made dramatic productivity improvements in the manufacturing of sewing machines. He changed the direction of the company from making machinery for their own sewing machines to making machinery for others.

Because of **Henry Leland**, Joseph R. Brown, the inventive genius of the Brown and Sharpe Company, turned his interest from gauges and measuring devices to milling machines, grinders, and automatic gear cutters. At **Leland's** suggestion, Brown began designing the "Universal Grinder," which was completed the same year as his death. Although it was very succesful, **Leland** felt the grinder was too light. Ten years later **Leland** advised **Charles H. Norton** to re-design the grinder with heavier parts, with more coolant, and more speed. This was called the No. 1 surface grinder. In 1878 **Henry Leland** became head of the sewing machine department at Brown and Sharpe.

Throughout his working career **Henry** wanted to be in business for himself. In 1890 **Henry**, Ellen and their son **Wilfred** moved to Detroit, Mich. It was an open-shop town, with gracious tree-lined streets and water panoramas. It offered a happy location for a home, and he found a partner to help establish a business. **Charles Strelinger** sold tools and hardware in Detroit, and knew **Leland** enough to introduce him to **Robert C. Faulkoner** of Alpena, Mich. **Faulkoner** had made his money in lumbering, and was looking for an opportunity for investment.

At that time, Detroit had a population of over 200,000. A Grand Boulevard encircling the city was being considered, but it was being hotly debated because many people felt that no one would ever live that far away from the center of the city. Visible evidence that the city was thriving came with construction of the first tall building downtown, and the new Belle Isle Bridge. There were eight railroad lines, and the city was producing ships, boilers and engines. The Michigan Car Company was the largest business in Detroit. The bicycle industry was beginning to emerge, along with a plethora of other industries in exsitence and others yet to come. It was evident Detroit would need more machine shops as there were only nine in town.

On September 19, 1890 the firm of **Leland, Faulkoner,** and **Norton** was formally organized with

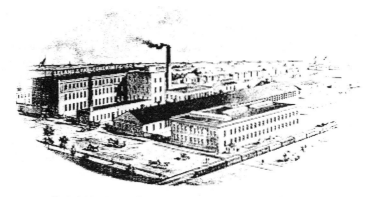

$50,000. **Faulkoner** supplied $40,000 and Leland contributed $3600. **Norton**, who was brought from **Brown and Sharpe**, was given a small amount of stock along with **Charles Strelinger**. **Faulkoner** was made president, **Leland** VP, and **Strelinger** was secretary. **Leland** was also general manager, and his first task was to write a form letter announcing their new company to all prospects in Michigan, Northern Ohio, and Northern Indiana. The fourth floor of the **Strelinger building** at Bates and Congress streets was rented, some machinery was brought in, and the shop opened for business.

Within three months the labor force increased from 12 to 60 hands, and **Henry** wanted to have his son **Wilfred** work with him. **Wilfred** was in his second year at medical school at Brown University. He had worked at Brown and Sharpe during summer vacation periods, and he agreed to spend a year with his father.

He arrived in Detroit on September 7, 1890, on his twenty-first birthday. After the year's end he decided to stay in the machinery business, and returned to Brown and Sharpe to get special training. After **Wilfred** rejoined his father, they were referred to as H. M. and W. C. They were highly regarded by their employees, including a machinist named **Horace Dodge**.

The company thrived with a variety of assembly work such as a typewriter, a motorcycle, and an automatic chicken feeder. There were a thousand other articles, and in three years the company outgrew its downtown quarters. In 1893 a factory was specifically designed for the firm, and constructed at Trombley Avenue near Dequindre. In 1894 **Norton** returned to the east coast to rejoin his family and the company's name was shortened to **Leland** and **Faulkoner**.

In the 1890s, when the demand for bicycles was reaching its highest point, some manufacturers began making chainless bicycles. They soon learned that the gears were wearing out too fast due to dust and weather conditions. They tried heat treating and quenching in water or oil, but the gears would warp. **Leland** and **Frank Ferris,** an engineer from the Pope Bicycle Company, found a solution by making the gears oversize. After heat treatments, the gears were ground to the correct size. This worked so well that **Colonel Pope** bought ten machines for his factory in Hartford, and **George N. Pierce** in Buffalo, N.Y., also bought several. The **Leland** and **Faulkoner** Company became a manufacturer when they began supplying gears to several other bicycle manufacturers.

The **Leland** and **Faulkoner** Company began to establish a reputation for building engines, starting with a steam engine for the Cadillac Hotel in Detroit. They provided hundreds of steam engines used for streetcars, and even the clock in the city hall. An order was received for internal combustion engines in 1896 to be used for "naptha launches." At first they were supplied to Frank Dimmer, and later to **Charles Strelinger**. **Leland** gained valuable experience with the marine engines through endless testing to get the most out of the engine for fixed quantity of fuel.

After the Olds Works began producing automobiles in Detroit, there were noise problems with the transmissions. After trying to solve the problems for several months without success, **Ransom Olds** asked **Henry Leland** for help, and soon L & F was producing transmissions for the Olds Motor Works. Not only were the transmissions quiet, but they were interchangeable, which was more than **Ransom** expected.

When the transmissions were built by Olds, the parts were ground, scraped, sanded and fitted together individually.

The engine department of the Olds factory was destroyed during the fire at the Olds Motor Works in March 1901. The **Dodge Brothers** had been supplying motors, but could not replace the destroyed engines fast enough to keep up with automobile production demand. In desperation, **Ransom Olds** turned to L & F again, this time to make engines, and placed an order for 2000 units. The **Dodge Brothers** and L & F were then making motors from the same design, but the results were different because each company had a different approach to manufacturing. During the first Detroit Automobile Show, held in 1901, an example of each motor was being demonstrated on a platform with a large dial indicating that both were running at the same speed, and with the same three horsepower. **Henry** and **Wilfred** were watching the motors on the second day when **Henry Ford** suggested to go up on the platform and look behind the motors. He pointed out that a brake load had been applied to the flywheel of the L & F engine to slow it down because it had eight percent more horsepower.

Although careful workmanship had increased the horsepower of the **Olds** designed motor, the **Lelands** felt the design could also be improved. They decided to design a motor with the same bore and stroke, and same cubic inch displacement, but with even higher horsepower. With the knowledge and experience gained from making marine engines, the Lelands made valves that were larger and held open longer to increase timing efficiency. The resultant engine had 10.25 horsepower, and was the same size and cost as the original motor. It was offered to the Olds Motor Works, but rejected by **Fred Smith** because he didn't want to spend money for re-tooling.

circa 1900- Robert Faulkoner is on the left and Henry Leland is on the right.   *courtesy MSHC*

HENRY FORD COMPANY
DETROIT, MICHIGAN
1901-1902

DETROIT AUTOMOBILE COMPANY
DETROIT, MICHIGAN
1902

CADILLAC AUTOMOBILE COMPANY
DETROIT, MICHIGAN
1902-1904

CADILLAC MOTOR CAR COMPANY
DETROIT, MICHIGAN
1904-

"STANDARD OF THE WORLD"

After he won the Grosse Pointe sweepstakes race, the name **"Henry Ford"** took on a new meaning and most of his backers from the former Detroit Automobile Company were willing to try again. They even named it the HENRY FORD COMPANY when they filed incorporation papers on November 30, 1901. A factory was procured at 1363 Cass Avenue, and **Henry Ford** was named chief engineer. He was given one-sixth share along with Black, Bowen, Hopkins, Murphy, and White who also had one-sixth share. The five put up $35,000 in cash on a $60,000 capitalization. Clarence Black was president and Albert White was VP as they had been for the Detroit Automobile Company. William Murphy continued as treasurer, and Lem Bowen was named secretary. **Ford** hired **Oliver Barthel** after the Grosse Pointe race as an assistant. **Barthel** soon realized Ford was consumed with trying to break speed records, and was designing a new race car. Nevertheless, **Barthel** designed a two cylinder motor intended for a production automobile. Murphy had been willing to subsidize **Ford** in his earlier racing venture but was becoming impatient and wanted to start producing automobiles in volume. **Henry** felt the pressure from Murphy, but at the same time, he was preventing him from developing new ideas. After less than a half year, **Ford** left the company on March 10, 1902, determined that he would never again *be put under orders*. He took the drawings for his new race car and a $900 cash settlement, while Murphy agreed to drop the name Henry Ford, even though it still had good market value. **Barthel** took over for **Ford**, but he was unable to pull everything together for Murphy. By August 1902, the company officers decided to liquidate.

William Murphy and Lemuel Bowen paid a visit to **Henry Leland** and asked for an appraisal of their automobile plant and equipment. **Leland** promised to go to the plant at 1363 Cass Avenue and look over the machinery. **Leland** realized that the one cylinder motor he offered to **R. E. Olds** would be suitable for the Detroit Automobile Company. After he had the appraisal figures, he took his motor and went to meet with the directors. He placed the appraisal sheet on the table and his motor beside it. He reviewed the machinery appraisal first, including some observances on the factory layouts and methods that he considered inefficient. Then he showed them the motor and explained that the parts were interchangeable: it had three times more horsepower and it could be made for less than the Olds motor. The directors were sold on the idea to stay in business, provided that **Leland** would join them in the reorganization of their firm.

At a special directors meeting on August 27, 1902, Black, Bowen, and White agreed with Murphy to reorganize the company and chose the name CADILLAC, honoring the founder of Detroit. The 200th anniversary of Detroit was celebrated in 1901 and was still fresh in everyone's mind. The name conveyed dignity, was historic, and attractive. It was perfect for both locality and timing.

Capitalization was increased to $300,000 and **Leland** was given a small block of stock, made director and awarded contracts for making motors, transmissions and steering mechanisms for the Cadillac. **Barthel** left the company and was replaced by **Allison Brush**, who had worked on the **Leland's** one cylinder motor. **Leland** began immediately to build the first Cadillac at the L & F factory. The first Cadillac had a horizontal motor with the cylinder head facing the rear, and a planetary transmission. An L-section frame was used with semi-elliptic springs, which were attached with bolted-on goosenecks. It was completed on October 20, 1902, and the first test drive was made by **Allison Brush** and **Wilfred Leland**. Two more were constructed at the Leland & Faulkner factory. Two of the automobiles were turned over to **Bill Metzger**, the sales manager for Cadillac, to display at the New York Automobile show in January 1903. The cars were an immediate success and **Metzger** took orders and down payments for over 2000 automobiles at the show.

Two models were offered, a runabout, designated Model A priced at $750, and a larger four-passenger Model B with a tonneau priced at $900. Both models were right-hand drive and steered with a wheel mounted on a post. Brass carriage lamps and a horn were extra equipment that was not included in the quoted cost.

With an output of 1895 in 1903, the Cadillac Company exceeded every automobile producer except the Olds Motor Works. In its second year, new buildings were constructed and new equipment was installed to increase production to 50 automobiles per day. On April 13, 1904, the Cadillac plant was gutted by fire, which put the company temporarily out of business. The fire was started by a bursting cap in the rivet room, igniting varnish and gasoline and causing several explosions. It burned through portions of the main factory and partially through the new south wing that was nearing completion. Along with stock, the big assembly hall was lost which was most needed during periods of rush orders. Fortunately, 500 completed automobiles and 2000 engines were in a warehouse across from the main building. The management quickly rented new quarters and the 600 workers were back to work.

The Cadillac Company could sell every car it could produce, but there were production problems that had to be remedied. Many parts were purchased from suppliers, and then put together in the order of assembly. The bodies and chassis were seldom ready for engines and transmissions, and stopped the whole assembly process. The company directors reasoned that if the Lelands could do such a good job shipping their parts with high quality, and on time, they could do the same for other parts. After things continued to get worse, the directors told the **Lelands** that "either you fellows come and run the factory for us or we will go out of business." **Henry**, who was 60 years old, and **Wilfred** agreed to manage the Cadillac factory, starting on December 26, 1904. In 1905, representatives of the Cadillac stockholders visited all of the L & F stockholders and bought their stock. **Faulkoner**, almost an invalid by then, was glad to liquidate his investment. The name of the new combination was changed to the Cadillac Motor Car Company, and capitalized at $1,500,000, with the assets of Cadillac valued at $1,000,000 and L & F at $500,000.

The first four-cylinder Cadillac was placed on the market in 1905, selling for $2000. After designing the first four cylinder Cadillac motor, Brush quit in 1905 due to disagreements with **Leland**. He was paid royalties for a number of his patents and was not to enter into any business in direct competition with the Cadillac Company for two years. The single cylinder model was still available and a total of 20,000 "one Lungers" were produced up to 1908. This helped the Cadillac Company during the 1907-1908 depression. In the fall of 1908, a new era in the company's history began when the Cadillac "thirty," with a four cylinder engine, sold for $1400. This was an unheard of price for car with its size and power.

Henry Leland imported the first set of "Jo blocks" into the United States, improving the accuracy of his company's work even further. News of the Jo blocks brought many automobile people to the Cadillac factory to see for themselves the new standard of accuracy. In February 1908, the Cadillac agent in England, Fred Bennett, and the Royal Automobile Club sponsored a standardization test for motorcars that was open to all automobile companies. Bennett supplied three Cadillacs with one cylinder engines to the Brooklands racetrack near London, England to demonstrate complete interchangeability of parts. The Cadillac Company was the only company that entered the competition. The three Cadillacs were completely disassembled, and the parts were mixed up. Then the observers removed 89 parts that were replaced by Bennett from his inventory. The automobiles were reassembled using basic tools, then driven for 500 miles around the race track showing no problems. In February 1909, the Cadillac Company was awarded the Dewar trophy, considered as the highest award given in the automobile industry. While many luxury automobiles used racing to help make their reputation, Cadillac never did, and depended solely on their reputation of excellence.

General Motors Corporation bought the Cadillac Motor Car Company in 1909, retaining **Henry and Wilfred Leland** to manage production.

1902-Early September at noon, the first Cadillac was assembled at the Leland Faulkoner factory. It was driven on its first trial run that afternoon. The first three Cadillacs were built at the Leland Faulkoner shop. *courtesy NAHC*

1902-Bill Metzger standing with one of the first three Cadillacs and prepares to go to the New York Automobile Show. The car is pictured at the Driving Park in Grosse Pointe.

1903-
A Cadillac with a detachable tonneau and driven by Alison P. Brush is shown going up Shelby Street Hill pulling 16 men. This was the first Cadillac built in 1903. (Note the Pingree Shoes advertisement on the building to the right of the Cadillac. Hazen Pingree had been mayor of Detroit and also govenor of Michigan.)   *courtesy NAHC*

1903
Model A Cadillac with a one cylinder engine with 10 hp and a 70 inch wb. Alison P. Brush shown climbing the steps of the Wayne County Building.   *courtesy NAHC*

1904- The burning of the Cadillac factory

1905- Rebuilding the Cadillac factory   *courtesy Burton*

circa 1905-
Henry Leland in center with white beard and his managers   *courtesy NAHC*

1908
Cadillac driven in an endurance run from NY to Boston courtesy AAMA

1919
Cadillac Sedan

Cadillac's coat of arms is based on the family crest of Antoine de la Mothe Cadillac. The celebrated many-quartered shield represents the Cadillac family's heroic deeds. The seven-piked coronet, garland with a laurel wreath signifies Cadillac's descent from the royal count of Toulouse. To the automaker, the crest symbolized pioneering and leadership in the automotive industry.

FORD MOTOR COMPANY
DETROIT, MICHIGAN
1903-1908

FORD MANUFACTURING COMPANY
DETROIT, MICHIGAN
1905

FORD MOTOR COMPANY
HIGHLAND PARK, MICHIGAN
1909-1919

FORD MOTOR COMPANY
DEARBORN, MICHIGAN
1919 -

"THE UNIVERSAL CAR"

After **Henry Ford** left the HENRY FORD COMPANY, he realized that competition for investment capital for new automobile companies was becoming tight. Olds, Northern, Cadillac, Packard, Reliance, Blomstrom, Buick, and Barthel had usurped the first big volley of cash. Fortunately for **Ford**, the largest coal dealer in Detroit, **Alexander Malcomson**, was an automobile enthusiast. **Malcomson** knew **Ford** from the Edison Illuminating Company when he sold them coal before they bought directly from the mines. After **Ford** left the Edison Company, he continued to purchase coal from **Malcomson** for home heating. The two men began discussing a partnership agreement to develop an automobile that could be used to attract investors. Despite the extensive holdings, **Malcomson** had very little cash. He put up $500 and promised more when he got it. Malcomson also contributed one of his employees, **James Couzens**, to manage the business end of the partnership.

A carpenter named **Albert Strelow** had a one story carriage factory that was vacant at 685 Mack Avenue, located near one of **Malcomson's** coal yards. He was asked to remodel it for automobile production, and a lease was taken out for three years. **Ford** liked the location on the east side of Detroit because of the concentration of German machinists in the area. **Ford** and an associate, **Harold Wills**, set up shop with two lathes, two drill presses, a planer, a saw, a grinding wheel, and a forge.

The first automobile was finished in December 1902. It had a 72 inch wb, a planetary transmission with two forward and one reverse speeds, chain drive, and a two cylinder motor. The partners were out of money, but it didn't faze **Malcomson**. He contracted the **Dodge Brothers** to build complete chassis with engines and transmissions, the C. R. Wilson Company to make wooden bodies, the Hartford Rubber company to supply tires, and wheels. The Ford operation would assemble the cars.

Malcomson could not make the payments to the suppliers, and he was forced to go to the bank, where his uncle, **John Gray** was president. **Gray** agreed to guarantee payments on condition that a company be formed of which he would be president and hold 10.5 percent of the stock. **Malcomson** and **Couzens** went looking for additional investors. The Ford Motor Company was incorporated on June 16, 1903. A thousand shares at $100 each were issued, with **Ford** and **Malcomson** receiving 255 shares each. **Gray** received 105 shares and was made president. Others included Charles Bennett of the Daisy Air Rifle Company, Horace H. Rackham, John Anderson, **The Dodge Brothers**, **Vernon C. Fry**, **Albert Strelow**, Charles Woodall, **James Couzens** and his sister. The company started with $28,000 in cash.

The first production automobile was the Model A, which was advertised as the Fordmobile. Because he was tired of hitching and grooming horses, **Dr. Pfennig** of Chicago purchased the first "Ford" for $850 cash. The Ford Model A and the Cadillac Model A looked similar to each other, but there were major differences. The Ford had a parallel twin engine and the Cadillac had a single cylinder. The Ford frame was a channel section and the Cadillac had an L section. Ford had full elliptic springs versus Cadillac's semi-elliptic, and Ford's spring attachment was to the frame directly over the axle, while Cadillac used bolted-on gooseneck spring hangers.

Just around the corner from the Ford factory was Eastern High School, newly constructed in 1901. The road in front of the school was macadamized, and perfect for automobile testing and to do a little speeding. Thus, the Eastern High kids were among the first to see, and hear, the pioneering Fords, and among the first to recommend "get a horse."

1903
In Malcomson's coal office on the left is John S. Gray. Alexander Malcomson is on the stool in the rear. James Couzens is in the center and an unknown individual to the far right.   courtesy MSHC

1903- The original Mack Avenue assembly plant (it was later enlarged to two stories)

1903 Model A Fords can be seen in this scaled replica of the Mack Avenue factory.
                    picture taken 1994

1904- The Mack Avenue assembly plant after the second story was added. (Note Edsel Ford and a son of one of the Ford employees sitting by the fence to the left.)

THE SELDEN PATENT

There were two themes at the great Centennial Exhibition held in Philadelphia in 1876: the celebration of 100 years of American independence, and the age of energy. A 1400 hp two cylinder steam engine, each cylinder with a bore of 3.33 feet and a stroke of 10 feet, was the prime mover, driving almost a mile of line shafting that powered many of the mechanical exhibits at the fair. The engine was developed by George H. Corliss of Providence, R. I. The opening ceremony reached its peak when President Grant and Pedro II, Emperor of Brazil, opened the throttle valves and started the Corliss engine.

One of the exhibitors was the Pennsylvania Ready Motor Company, manufacturer of the Brayton Hydro-Carbon Engine. It was the world's first practical internal-combustion engine to utilize liquid hydrocarbons as fuel. Compared to steam engines, it was quick starting, hence the Ready Motor Company appellation. The engine was designed by **George Brayton**, who had worked for the Corliss Engine Works. Nearly all attempts before **Brayton** had been to develop internal combustion engines by means of explosions of flammable mixtures, adding heat at constant volume with corresponding rapid rises in temperature. **Brayton's** knowledge of steam engines led him to design an engine employing the addition of heat at constant pressure, so that the volume gradually increased as the heat was added. The two cycle engine was noted for its simplicity and quietness. It was demonstrated in Europe at the Vienna Exposition of 1873, when negotiations began that led to its manufacture in Germany, Scotland, and **England**.

Invention implies research, and with so much at stake, it was not surprising that inventors and entrepreneurs sought credit for the development of important automobile parts and processes. Patent law helped determine the winners and losers in the automobile industry. A patent lawyer named **George B. Selden** filed a patent in his own name, based on the Brayton engine, for the automobile in 1879. Although he claimed the date of invention in 1877, at the time it was not customary to specify the date the application was filed.

The Selden patent owes much of its fame to George Day, who in 1900 was the president of the Electric Vehicle Company of Hartford. He was convinced that he could control the new motorcar industry by teaming up with **George Selden**. They soon led a lawsuit against the Winton Motor Car Company of Cleveland. After convincing **Alexander Winton** that the patent was valid, they soon enlisted other leading manufacturer's to recognize the patent and pay royalties. Day then organized the ASSOCIATION OF LICENSED AUTOMOBILE MANUFACTURERS to help make this group of companies more efficient by standardizing parts and developing specifications. **Henry Ford** saw merit in joining the A.L.A.M.

The constituents of the A.L.A.M. were by their definition, manufacturers who made all of their own automobile parts. The Ford Motor Company was considered an assembler that purchased components, and consequently was not allowed into the A.L.A.M., although it was expected to pay Selden Patent royalties. A lawsuit was filed in 1903 against Ford for selling automobiles without paying royalties. **Henry Ford** took issue immediately and fought the Selden patent.

The total number of licensed cars built from January 1, 1903, to January 1, 1906, was 41,696. They were valued at $63,141,437 with collected royalties amounting to $814,183. The majority of licensed manufacturers were producing high-priced cars while the non-licensed manufacturers were producing low-priced cars. There were 1545 automobile dealers in the U.S. in 1906. Of the total, 1250 belonged to the A.L.A.M. There were 1057 covered by the Selden patent and 34 electrics and steamers. The largest payment to A.L.A.M. was $93,855 for 8697 cars made in the second quarter of 1910 by Buick.

The Brayton engine was a two cycle, slow combustion, "constant pressure" using liquid fuel. It was, in fact, the predecessor to the Diesel engine. The Otto engine with a four cycle, explosion combustion or "constant volume" engine was used in **Ford's** cars. During the court proceedings it was discovered that **Selden** had made an entry in his diary detesting the four cycle engine.

When **Selden** filed his patent, the Brayton engine was the leading engine at the time, and his attention was drawn to its supposed advantages. It was clear that if he had chosen the Otto engine, his patent would have held. On January 9, 1911, the United States Court of Appeals handed down its decision that the famous patent did not cover the explosion or "constant volume" engine.

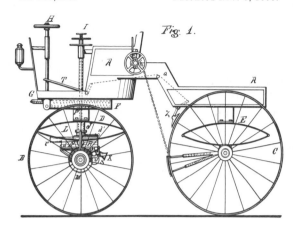

1908
The first car actually built by Selden. E. R. Patridge is at the left and George Selden is at the wheel.
courtesy AAMA

NOTICE

Users, Agents, Importers, Dealers and Manufacturers of
GASOLENE AUTOMOBILES

United States Letters Patent No. 549,160, granted to George B. Selden, Nov. 5, 1895, controls broadly all gasolene automobiles which are accepted as commercially practical. Licenses under this patent have been secured from the owners by the following-named

MANUFACTURERS

Electric Vehicle Co.
Winton Motor Carriage Co.
Packard Motor Car Co.
Olds Motor Works.
Knox Automobile Co.
The Haynes-Apperson Co.
The Autocar Co.
The George N. Pierce Co.
Apperson Bros. Automobile Co.
Searchmont Automobile Co.
Locomobile Co. of America.
The Peerless Motor Car Co.
U. S. Long Distance Automobile Co.
Waltham Manufacturing Co.

Pope Motor Car Co.
J. Stevens Arms & Tool Co.
H. H. Franklin Mfg. Co.
Charron, Girardot & Voigt Co. of America (Smith & Mabley).
The Commercial Motor Co.
Berg Automobile Co.
Cadillac Automobile Co.
Northern Mfg. Co.
Pope-Robinson Co.
The Kirk Mfg. Co.
Elmore Mfg. Co.
E. R. Thomas Motor Co.
Buffalo Gasolene Motor Co.

The F. B. Stearns Company.

IMPORTERS

having license for the importation of all makes of foreign cars:

Charron, Girardot & Voigt Co. of America (Smith & Mabley)
Central Automobile Co.

Standard Automobile Co.
E. B. Gallaher

These manufacturers are pioneers in this industry and have commercialized the gasolene vehicle by many years of development and at great cost. They are the owners of upwards of four hundred United States Patents, covering many of the most important improvements and details of manufacture. Both the basic Selden patent and all other patents owned as aforesaid will be enforced against all infringers.

No other manufacturers or importers than the above are authorized to make or sell gasolene automobiles, and any person making, selling or using such machines made or sold by any unlicensed manufacturers or importers will be liable to prosecution for infringement.

A suit was commenced on October 22d against a Dealer and against a Manufacturer infringing United States Letters Patent No. 549,160.
A suit was commenced November 5th against a Purchaser and User of an Automobile infringing the same patent.

ASSOCIATION OF LICENSED AUTOMOBILE MANUFACTURERS
7 East 42nd Street, New York.

C.E. WHIFFLER AUTOMOBILE COMPANY
DETROIT, MICHIGAN
1903-1904

Charles Whiffler formed an automobile company in June 1903 with a shop located at 12-16 Sherman Street, at the corner of Hastings. He was a mechanic who built a touring car of his own design that sold for $1000. It was powered by a two cylinder horizontal opposed engine, with virtually every part made by Whiffler. The company lasted until late 1904.

C. E. WIFFLER AUTOMOBILE CO.

MANUFACTURERS OF

Automobiles and Parts,

12-16 SHERMAN STREET,   DETROIT, MICH.

CHELSEA MANUFACTURING COMPANY
CHELSEA, MICHIGAN
1903

WELCH MOTOR CAR COMPANY
PONTIAC, MICHIGAN
1904-1911

WELCH-DETROIT AUTOMOBILE COMPANY
DETROIT, MICHIGAN
1909-1911

When he was a boy, **Allie R. Welch** built a steam engine which drove a lathe also of his own construction. He worked for a gasoline stove company in Chelsea, Mich., where he rose to the position of superintendent. In 1895 he went to Connecticut and took charge of a metal-working factory. He later established a metal novelty shop in New York. In 1898 he moved his company to Chelsea and was joined by his brother, who had a local bicycle shop. In February 1900 the brothers formed the Chelsea Manufacturing Company, maker of flashlights, pencil sharpeners, and other novelties invented by **A. R. Welch**. At the same time, they began working on a two-cylinder gasoline automobile that was completed by April 1901. The project was such a financial drain that by the fall of 1902 **A. R. Welch** went to Battle Creek, Mich., seeking further capital support. Welch was introduced to A. C. Wisner, who was interested in the details of the automobile, including a telescoping steering wheel, invented by **Welch**, that dropped out of the way for front passengers. Wisner went to Chelsea to look over the factory and the prototype automobile.

With help from Wisner, the Chelsea Manufacturing Company was re-organized with $200,000 in January 1903. Material for 100 touring cars was ordered, and the first "Welch" was completed in time for the Chicago Automobile Show in February. It was a touring car with a 78 inch wb and a 20 hp motor. There were numerous inquiries for the $2000 Welch, but only two more were completed by July. With ten more cars ready to be assembled, production came to a halt when Wisner was unable to continue financing. The federal district court declared the Chelsea Manufacturing Company bankrupt in January 1904. The Welch was an assembled automobile, with parts supplied by the **Dodge Brothers**, Brown and Sharpe, Brown-Lipe, and C. R. Wilson. Three appraisers, including **John Dodge** of Detroit, were appointed to evaluate the physical assets of the Chelsea plant.

1905 Welch showing seating arrangement

The **Welch Brothers** moved to Pontiac, Mich., and joined Arthur Pack who provided a factory and $25,000 for the formation of a new automobile company. Work began immediately and two cars were completed in time for the Detroit Automobile Show in February 1904. On March 1, 1904 the Welch Motor Car Company was finalized with Pack serving as president and treasurer, **Allie Welch** was VP, and **Fred Welch** was secretary. Production of a touring car, with a four cylinder engine rated at 36 hp, began in April 1904.

On December 23, 1906 a Welch was the first six cylinder automobile seen in Detroit. It had been shown at the Grand Central Palace Show in New York earlier that same month. The impression at first glance was that it was a very powerful machine, due to the exceptionally long *bonnet* and long body with a 138 inch wb. The six cylinders were cast in pairs with a 4 3/8 inch bore and a 5 inch stroke, producing 70 hp. The engine had hemispherical combustion chambers, overhead valves, and an overhead cam.

The transmission had one lever for three forward speeds and a second lever for two reverse speeds. The gearset operated on the individual clutch principle having its gears constantly in mesh. It had three shafts in the same plane, with the main shaft in the center and a secondary shaft at each side. The clutch consisted of multiple discs made of bronze against steel. The final drive was by a cardan shaft enclosed throughout. By 1909 the Welch touring coach was available for $9000 made with Krupp steel imported from Germany. It had accommodations for cooking, dining, and sleeping, with an ice box, and lavatory.

In June 1909 the Welch Motor Company was purchased by **Billy Durant**. A. B. C. Hardy was installed as plant manager, and the **Welch Brothers** stayed actively involved. **Durant** next purchased the former Detroit plant of the Oldsmobile Works from **Samuel L. Smith**, and set up a separate company named the Welch-Detroit Auto Company, to produce a less expensive model to broaden the market for Welch. By the fall of 1909 it was clear that the "Welch-Detroit" weakened the status of Welch. After GM purchased Cadillac in the summer of 1909, it was evident that the Welch was no longer suited for the GM product line.

The last Welch and Welch-Detroit cars were produced in 1911. Both companies became part of the newly formed Marquette Motor Company in Saginaw, Mich., but were never produced there. Along with the Welch and Welch-Detroit companies, the Rainier Motor Car Company was merged into the Marquette Company. The Rainier Company started in 1905 with a factory in Flushing N.Y. The chassis was made by Garford (who also made chassis for early Studebakers) until 1907. The company re-organized and moved to Saginaw, Mich., in 1908, and was purchased by GM in 1909. After producing the "Marquette" for a year, the company was re-organized as the Peninsular Motor Company in April 1912, and produced the "Peninsular." Production was stopped in 1913.

During the fall of 1913 **A. R. Welch** was nearing completion of his next automotive project. He designed a cyclecar with a 110 inch wb a 36 inch tread, and a body as roomy as a touring car. A reporter said, "**Mr. Welch** will make in all probability the highest priced cyclecar on the market." On November 8, 1913 **Allie** went on a duck hunting trip in a canoe. A severe storm ensued and he was never seen again.

A. R. Welch

WHEELER MANUFACTURING COMPANY
DETROIT, MICHIGAN
1903-1904

The Wheeler Manufacturing Company was formed in 1899 to manufacture bicycle and carriage equipment. The company was handled by people who had first succeeded in the retail bicycle line, and simply invested their surplus earnings. With a capitalization of $25,000, they expanded to production of automobiles in 1903. Stephen E. Hartnell was president, **W. C. Rands**, was the secretary and treasurer, and Edwin S. Anderson was the manager. There were 68 hands working at the factory, which was located at 10-16 Baltimore Avenue, in Detroit.

The Wheeler Manufacturing Company was the largest manufacturer of bicycle handlebar grips in the United States, with an output of 800,000 in 1903. They made small rattan seats, that were popular for the back of Oldsmobiles, and hampers for the rear of runabouts. They also made dos-a-dos seats for automobiles and saddles for horses.

The "Detroit" was first announced to the public at the Detroit Automobile Show in February, 1903. It was a detachable tonneau type, with a hooded front and a reachless platform frame. A single cylinder motor was suspended from an angle iron brace at the rear and drove a center chain from a planetary transmission that was attached to the differential on the rear axle. Forced circulation cooling was used, with a disc radiator made up of six pipes laying transversely, in a plane parallel to the floor, under the hood.

In mid-year, the Wheeler Company announced an improved "Detroit" automobile. It was available with one chassis, with either a tonneau or a delivery body. The body was large, with a divided front seat; there was room for five people. The gasoline and water tanks were under the hood, which distributed weight and allowed filling without difficulty. The water tank capacity was sufficient for 150 miles and the gasoline tank capacity was sufficient for 100 miles. The motor was a two-cylinder, opposed type, said to be perfectly balanced, almost entirely eliminating unpleasant vibration. The cylinders were cast in one piece, with long water jackets, without any gaskets. One oiling of the engine would last from four to six weeks. Fifteen horsepower was obtained at 1000 revolutions under brake test. The transmission had two forward speeds and one reverse. The muffler was located in the extreme back of the car, so that heat would not be felt under the tonneau floor. The muffler could be cut, by means of a pedal, and the motor could be controlled, from six to thirty-five mph, without releasing the high speed clutch. The Detroit could climb a 35 percent grade, by using the climbing gear. The Wheeler Company discontinued automobile manufacturing in late 1904.

Seventeen coats were applied in painting the Detroit. It was finished in two colors, red and green, with yellow trimmed gear. (It is seen here at the Detroit Automobile Show.)

The Wheeler Manufacturing Company occupied a two-story brick factory with 13,000 square feet of floor space. The company was capitalized with $25,000 and employed 68 hands.

Note the canopy top on the Olds runabout that is pictured on the bottom of this advertisement.

DINGFELDER MOTOR COMPANY
DETROIT, MICHIGAN
1903

Max Dingfelder had a small factory at 958 Jefferson Street in Detroit where he made small gasoline engines on the first floor, and lived on the second floor. He also had an auto garage at 41-43 Washington Avenue. In 1903 he produced a number of two passenger runabouts. The "Dingfelder" weighed 500 lbs and had a 3.5 hp one cylinder motor.

One magazine reported that "Mr. Dingfelder is held in high esteem for his honorable business methods and sterling integrity," and "the machine has been very well received by the local trade and a nice business is anticipated." His car did not survive into 1904. He did better with boats, and won a big race that was sponsored by the Detroit Yacht Club in 1906 with his auto boat "999."

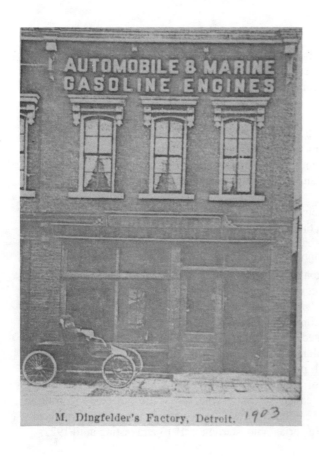

M. Dingfelder's Factory, Detroit. 1903

HAMMER-SOMMER AUTO CARRIAGE COMPANY, LTD.
DETROIT, MICHIGAN
1903-1904

HAMMER MOTOR COMPANY
DETROIT, MICHIGAN
1904-1905

SOMMER MOTOR COMPANY
DETROIT, MICHIGAN
1904-1905

After building a number of experimental cars, the Hammer-Sommer Auto Carriage Company, Ltd., located at on 573 Gratiot, was in production in August 1903. The Model A was named "Ideal" and was a two passenger runabout that sold for $700. The company was owned by Henry F. Hammer, Herman A. Sommer and William J. Sommer.

For the 1904 season the Model D, called the "Hammer-Sommer," had considerably changed from the Model A. With a detachable tonneau style of body it could seat five. The spark coils, batteries and water tank were placed under the metal hood. The engine was an opposed two cylinder 12 hp gasoline engine with a 4.75 x 5 inch bore and stroke. The motor was hung on an angle iron steel frame in the center of the chassis. A planetary transmission was used, and a gear pump was used to circulate water through the cooler.

Henry Hammer was interested in producing a larger and faster car than the Hammer-Sommer. Apparently the Sommer brothers did not agree, and the partnership was dissolved in 1904.

On December 1, 1904, it was announced that the Hammer Motor Company had been formed with $10,000 capital, $1000 paid in cash and $2000 in patents. The stockholders were Henry Hammer, Leon J. Paszski, Foster F. Allen and Harry F. Nickoalds. The main office was in the Majestic Building and the factory was located at 313-315 Riopelle. The company built auto boat engines and automobiles called the "Hammer," which used a four cylinder engine. One model called the Hammer Suburban Limousine had 32 hp and cost $4500. Automobile production was discontinued after 1905.

In 1904 the Sommer Brothers incorporated the Sommer Motor Company with $40,000 capital stock. They moved into a factory at 298-300 Columbia Street East, with Herman Sommer as president. They built a 15 hp five passenger touring car, with production capacity of 100 cars per year. Production was terminated late in 1905. The superintendent of production, Harry Tarkington, moved to Marysville, Ohio to try his hand in the manufacture of a car called the Marysville, but it did not come to fruition. Another Sommer brother, L. A. Sommer, who had been involved with the partners, became general factory and chief engineer for the Allen Motor Car Co. in Fostoria, Ohio, starting in 1913 and lasting until 1922.

Hammer-Sommer (body finish was dark green striped in gold using 18 coats of the finest coach paint)

Hammer

Sommer

RELIANCE AUTOMOBILE COMPANY
DETROIT, MICHIGAN
1903-1904

RELIANCE MOTOR CAR COMPANY
DETROIT, MICHIGAN
1904-1907

The Reliance Automobile Company was organized in late 1903 with a capital stock of $150,000, all of which was subscribed. The directors included **F. O. Paige**, president of the Detroit Automobile Club, D. O. Paige; George Wetherbee, manufacturer and wholesale woodenware; James T. Lynn, largely *interested* in gas and electric lighting; J. M. Mulkey, widely known as the president of the Detroit Salt Company; Hugh O'Conner, president Michigan Wire Cloth Company, and E. O. Abbott. The first car was designed by E. O. Abbott and W. K. Ackerman, both of whom were previously with the Cadillac Company, *the former as the factory superintendent and the latter as chief draughtsman.* Ackerman had built a number of cars of his own design starting in 1897, with air cooled and water cooled engines. A 152,000 square foot plant of the former Detroit Safe Works at 87 and 89 Fort Street East was secured for manufacturing.

The car was *driven* by a double opposed alternating four cylinder motor 4.75 x 5.5 rated at 15 hp at 1000 rpm, placed horizontally under the body. The transmission was a planetary type with chain drive connected to a spur gear differential. A feature of the motor was a centrifugal spark governor that automatically advanced the spark as the engine increased speed and retarded the spark when the engine was shut off so that there could be no kick back when cranking the motor. Throughout the construction of the motor and frame, a special lock-nut device was used. To improve reliability the motor was protected from dust and dirt by a heavy leather covering attached from the fenders to the framework. This was one of the first cars with side doors for the rear compartment instead of the single rear door, permitting the use of a rear seat with an unbroken back.

In December of 1904 the company was re-organized into the Reliance Motor Car Company with new capital of $400,000. **Paige** was replaced by J. M. Mulky as president and the workforce was increased to 300 in order to turn out 1000 cars during 1905. The company still faced problems and in late March 1905, and De Witt C. Loomis, once connected with the Carnegie Steel Company, purchased a large interest in the Reliance company and was elected president, succeeding Mulkey who resigned.

For 1906 a two cycle motor with 28 hp was used to improve reliability. The larger motor eliminated a substantial number of parts used in the four cycle engine including springs, camshaft, gears, push rods, and valves, which usually required constant grinding and seating. It had a sliding gear transmission with three speeds forward and one reverse using high grade nickel steel gears. The webs of the gears were forged integral with the shafts, eliminating the use of keys and gear loosening on the shafts. Besides the touring car, a business wagon was offered. They shared the same motor, but the car used a drive shaft to a bevel gear differential with live rear axles, whereas the wagon used double chains and a solid dead rear axle, which was much stronger for heavy loads than a live axle.

In 1907 the Reliance Motor Car Company decided the future of the automobile was in commercial vehicles. Many companies in Detroit were starting to use trucks and delivery wagons and eliminated their horse deliveries entirely. It was demonstrated that one truck would do the work of three to six horses, and as many persons, more quickly and economically. Even though trucks were geared for low speeds of 15 mph, they were still four times faster than a horse team.

The Crescent Motor Company purchased the touring car business of the Reliance Motor Car Company, and the Reliance truck business was sold to General Motors in 1909, which evolved into the G.M.C. Truck.

1905

1906- Reliance model E Touring car, speed 3 to 50 mph   2 cycle motor without valves

IDE-SPRUNG-HUBER AUTOMOBILE COMPANY
DETROIT, MICHIGAN
1902

HUBER AUTOMOBILE COMPANY
DETROIT, MICHIGAN
1903-1909

Dr. Henry G. Ide and Edmund A. Sprung built a one-off motor car in 1894 and formed the Automatous Machine Company. Sprung opened a bicycle shop on 29 Lafayette Avenue in Detroit, and he also helped Henry Ford with his Quadricycle. After the partners relocated to Detroit, the Ide-Sprung-Huber Automobile Company was organized.

Its first success was replacing an electric motor drive with a four cylinder gasoline engine in a delivery wagon owned by the Detroit News. When the delivery wagon was displayed at the Detroit Auto Show, it created quite a stir and could carry a ton at 15 mph. The wagon with electric drive weighed 5700 lbs. The weight was reduced to 3800 lbs with the gasoline engine. The engine could develop 21 hp with 3.75 x 4.5 bore and stroke. A 40 lb flywheel was used as a friction clutch. When the machine was standing still, all gear wheels were standing still. In starting up, the low speed gear was thrown in first. While running on this speed, all gears were running in mesh, although performing no work. When the next higher speed was thrown in, the slow speed gears continued to run, but ceased doing any work. The gears were made of bronze and cast steel to minimize noise.

In 1903 the Huber Automobile Company was organized with a capital stock of $100,000. Edmund Sprung was the president, joined by **Emil Huber**, Henry Ide, and Mersden Burch. A factory was secured at 248 Jefferson Avenue for production of automobiles. Soon after production started, more emphasis was given to large sight-seeing busses.

By 1906 **Emil Huber** held property on Belle Isle to carry passengers across the bridge using Huber busses. Eventually, instead of being sold, all Huber vehicles were leased for hire.

In 1907 Ide and Burch noted that the company bookkeeping was out of order, so they attempted to secure a chattel mortgage on all company property. In defense, Sprung countered that the books had been tampered with. The whole fracas led to the end of the Huber Automobile Company.

Huber left for his home state and joined the Meteor Motor Car Co. in Bettendorf, Iowa. The company was relocated to his home city of Davenport, Iowa, and just before it folded in 1910, **Huber** quit. He subsequently built a runabout with a four cylinder engine, with a 100 inch wb and an underslung frame. He drove in the Glidden Tour in the summer of 1910, hoping to find investors for a new automobile company. His plans to build the car in Davenport fell through, and he returned to Detroit to work for Hudson.

Sprung, the inventor, went into the marine motor industry. He invented the reverse gear on marine motors. He worked for Gar Wood in developing marine engines, and after eight years as an engineer on the yacht owned by **Lawrence P. Fisher**, one of the General Motors' Fisher brothers, he retired in 1933. He died in April 1950.

COMMERCIAL MOTOR VEHICLE COMPANY
DETROIT, MICHIGAN
1903-1905

The Commercial Company manufactured a 500 lb electric runabout called a businessman's carriage. It was powered by a .75 hp motor capable of 40 miles on one charge of its 10 cell battery and attained speeds up to 14 mph. The factory was located at 259-267 Franklin Street in Detroit where electric powered trucks were also produced. **Dr. J. B. Book** was president and Frederick S. Evans was secretary and manager.

A commercial vehicle, called the "Quadray," was made for the Michigan Stove Company. It had a total length of 23 feet. The vehicle weighed seven tons without the top and could carry 12 tons. It was powered by an underslung battery pack with eighty cells of 375 amp hours capacity, driving four 3.5 hp motors. It consumed 100 watts per ton mile.

Another model truck had a 16 hp gasoline motor connected directly to a 12 K.V. generator, driving four 2.5 hp electric motors. Five storage cells were used to start the generator as a motor, thus doing away with the hand cranking, and they also furnished spark for the engine. The battery was continually charged from the generator. The vehicle could travel at 12 mph with a load of three tons and consumed a gallon of gasoline at full load and full speed in an hour.

In October 1905 forclosure proceedings were instituted by Frank N. Crosley, alleging that a default was made in the mortgage payment.

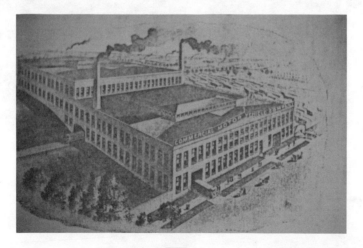

IMPERIAL AUTOMOBILE COMPANY, LTD.
DETROIT, MICHIGAN
1903-1904

The Imperial Automobile Co., Ltd. was organized in November, 1902, with $250,000 capital to manufacture light electric vehicles with Imperial motors, such as: runabouts, stanhopes, surrey and delivery wagons. Members of Detroit's prominent families with interest in the Imperial Company included **Dr. J. B. Book,** who was chairman of the board of directors, and **Daniel J. Campau,** who was a lawyer born in Detroit in 1852 and educated in Forham N.Y. **Campau** was the controlling owner of a magazine devoted to horse breeding and turf news, called the Chicago Horsemen. **Joseph Ledwinka** of Pittsburg, who later worked as an experimental engineer for the Edward G. Budd Manufacturing Company, and Rudolphus Fuller of Detroit were consulting engineers. Temporary offices were established in the Majestic Building.

The Imperial had two 1 hp motors and a 40 cell battery pack. It was priced at $950 without a top. The top cost $25 - $50 extra.

The Imperial Company was unable to market economicaly and stopped production in late 1904.

1903- Imperial

Majestic Building

GILMORE MOTOR WORKS
DETROIT, MICHIGAN
1904

George A. Gilmore formed the Gilmore Motor Works in 1904 using his cycle shop on 1174 Fort Street West. Besides automobiles and bicycles, he made tools, dies, models, and did general repairing. Generally, automobiles were built to order and the quantity was very low, but may have lasted several years. He formed the Gilmore Motor Manufacturing Company in 1913 with a capital investment of $35,000 for engine production only.

G. A. GILMORE,

SUCCESSOR TO

The Gilmore Cycle Co.

MANUFACTURER OF

AUTOMOBILES

AND

BICYCLES

MODELS, TOOLS, DIES AND GENERAL REPAIRING.

1174 Fort Street West,

Phone, W. 266-L.   DETROIT, MICH.

THE FORD PIQUETTE FACTORY

In May 1904 foundations were laid for a new plant on the NW corner of Piquette and Beaubien streets in Detroit. **Couzens** had an office on the first floor, while **Ford, Wills**, and the rest of the design staff were on the second floor. The assembly of automobiles was a primitive operation. The engines, transmissions and running gear, frames and bodies were assembled separately in one big hall and brought together by carrying all the parts to a designated spot on wooden horses, then finished by a group of 10 to 15 men.

Ford continued filling his stable of talented and brilliant personnel, and never gave anyone a title for as long as he lived. Everyone was just referred to by his name. Harry Love and C. J. Smith were added to help Harold Wills in experiment and design. **Gus Degener** and **Charles E. Sorensen** and a brilliant engineer named **Joe Galamb** were hired. **Walter Flanders** saw automobiles individually ordered to be finished to the specification of each buyer. This caused the plant workload to vary, and few orders were received in the winter, causing the plant and machinery to be underutilized for several months out of the year. He drafted a plan to pace production over 50 weeks out of the year. This he borrowed from working for the Singer Sewing Machine Company.

There were constant improvements such as Models B and C. Models D and E did not go into production, Model F did, and so on. From the beginning, relations between **Ford** and **Malcomson** were fragile, but things came to a head over the Model N. **Malcomson** wanted to follow the lead set by Olds and market a large luxury car to make more unit profit. **Ford** wanted to produce the Model N, based on **Flanders'** cost saving concepts, and sell it for $500. The board of directors agreed with **Ford,** and **Malcomson** was furious. To get around **Malcomson's** defiance, an idea was conceived to form a separate company to build the Model N, and it was named the Ford Manufacturing Company. This would provide **Malcomson** with stock dividends, but no profit in the manufacture of the car. **Malcomson** , in turn, formed the Aerocar Company to build his own car. At the same time, to appease **Malcomson,** the Ford Motor Company built the upscale Model K, which was a financial disaster.

With the Ford Motor Company split in two, sales suffered and profits declined to one-third compared to the 1904-1905 season. The two factions agreed the time had come to part, and Malcomson sold his stock to Ford for $175,000. Woodall, Fry, Bennett, and Strelow sold their Ford stock and went with **Malcomson**. Strelow invested his money in a gold mine and lost it all. Several years later he was seen in a line of men applying for work at Ford's gates. So great were the sales of the Model N, there were 6000 orders on hand and no cars to fill them. **Couzens** had to stop taking orders. The Model N was the training ground for the most famous alphabet car, the Model T. Meanwhile, a Wall Street and banking panic stemming from over-capitalization and schemes for monopolies took place in May 1907. Two thousand wealthy New Yorkers sold their cars and thousands of others canceled their orders for new ones. In Detroit, the Detroit Auto Vehicle Company, Huber, Marvel, and the Aerocar Company went out of business. **Alexander Y. Malcomson**, the founder of the Ford Motor Company and the Aerocar Company, returned full time to his coal business.

In the summer of 1906 former mayor **Pingree's** financial adviser, **Robert Oakman** , was lunching in a restaurant on Woodward Avenue. At a nearby table **Henry Ford** was discussing the company's need for a site to construct a new factory. **Oakman** was a real estate agent and suggested to Ford that he take a look at a piece of land five and a half miles from downtown Detroit that was being used for farming and horse racing.. This would become the home of the Model T. Early in 1907 **Ford** told **Joe Galamb** that he had an idea for a new car, and for **Joe** to set up a 12 by 15 foot room on the third floor in the back of the building with his drafting board and a blackboard. **Ford** brought his mother's rocking chair in for good luck. They started with the transmission, and made a half-scale brass model when the design was finished. As they continued designing the Model T, **Ford** sat in the rocking chair studying the designs on the blackboard.

Ford knew what he wanted - a car simple enough for everyone to drive. It had to be rugged, responsive, and cheap. The chassis was made with Vanadium alloy steel, that had two and a half times the tensile strength of the steels used at the time. This made the car light and strong enough to survive the country "roads." It used a planetary gear transmission operated with foot pedals, which enabled drivers to go directly from forward to reverse to rock the car out of any mudhole. The motor was a four cylinder, cast *en bloc* with a detachable head, making servicing much simpler. It also had a unique magneto system, designed by Ed Huff. The magneto was built into the flywheel, thus making the electrical ignition system part of the engine, and eliminating the need for costly storage batteries. It had large wheels to help negotiate rough roads. For simplicity, the car had no gas gauge and no fuel pump. The fuel was gravity fed (the Model T had to be driven backwards up steep hills). It used thermo-syphoning cooling, obviating the need for a water pump. If it did overheat, the sides of the hood could be opened.

In late 1907 the board of directors authorized the purchase of the former 57 acre horse racing course in Highland Park for $81,225, and in July 1908, $250,000 was appropriated to construct a plant. The first Model T rolled out of the Piquette plant in the late fall of 1908, painted brewster green with a red stripe. It weighed 1200 lbs, had 20 hp, and cost $850. The public received the car with enthusiasm, and more importantly, farmers realized for the first time that the car could replace the horse.

The Piquette Plant with Clara Ford to the right in the model N, driven by Miss Clarkson, the company office manager.

A close up view of Clara, to the right, with Miss Clarkson driving.

Clara is seen driving

Inside the plant saw bucks were used to hold the chassis. Parts from outside suppliers were brought in and delivered to the workstations.

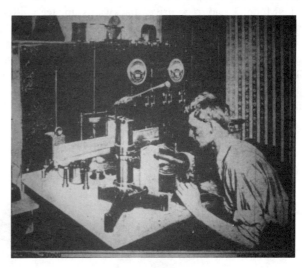

The model K- a six cylinder produced by the Ford Motor Company. (The company split and the Ford Manufacturing Company was formed to build the less expensive model N.) Henry is driving with E. Leroy Pelletier, his first advertizing manager at his side. Miss Clarkson is in the back far side; Edsel Ford, and Clara Ford are on the near side.

1906- Edward (Spider) Huff, 29 years old, seen in his laboratory. He was a consulting engineer for the Ford Motor Company and the Hawley Manufacturing Company. Nicola Tesla gave Huff the original high frequency coil at the Columbian Exposition in 1893. Huff designed the Ford magneto system and the Hawley company produced it.   courtesy NEWS

*The Ford Piquette factory was designed by Field, Hinchman and Smith. It was a three story New England Mill 56 feet x 402 feet. The foundations were laid in May 1904 at the N.W. corner of Piquette and Beaubien.
                picture taken 1994*

The original Ford logo was produced by James Couzens. He had an ink stamp kit which he had used for making business cards.

WAYNE AUTOMOBILE COMPANY
DETROIT, MICHIGAN
1904-1908

The Wayne Automobile Company was formed in 1901 by Charles E. Palms (whose grandfather was the largest landowner in Michigan); Dr. **J. B. Book** (who gave up a lucrative medical business in order to join the Wayne Company, and was Charles Palm's uncle); E. A. Skae, Roger L. Sullivan, and **William Kelly**. The company and car were named for **General Anthony Wayne**, who was decorated for his heroic efforts during the Revolutionary War, and Indian wars in the Northwest Territory.

The Wayne automobile was designed by **William Kelly**, who was born in Ohio 1861. He became an apprentice in a machine shop in 1879 and designed his first gasoline engine, in Detroit, about 1895. It was a 4 x 4 single cylinder, vertical, water cooled with hammer spark. He designed a second engine the following year for marine use. It had a double cylinder water cooled, with hammer spark. This engine had opposed cranks, and gave two working strokes in one shaft revolution, and none during the next revolution. **Kelly** placed his first car on the road in 1901. It had a 5 x 6 single cylinder engine with ignition jump spark. The car weighed 2000 lbs and, according to **Kelly**, vibrated too much. **Kelly** began with the Wayne Company by building his second car, a single horizontal 5.5 x 5.5, with planetary gear change, chain drive and tonneau body. It was on the road in 1902.

The "Kelly" was shown at the Chicago and New York exhibitions in January and February 1904. The price was $1000, and the weight was 1400 lbs. More orders were taken for the car than were ever filled. **Kelly** felt *the impulses of the single cylinder engine were too obvious, and that two cylinders were the fewest number suitable for a road wagon motor*. The first "Wayne" had a two cylinder motor with chain drive, and was on the road in February 1904. The motor and flywheel were hung under the chassis frame to bring the center of gravity as low as the road surfaces would permit. This design provided stability when turning corners, and reduced skidding on curves and the possibility of overturning. The 1905 Wayne had an 80 inch wb and a 56 inch tread, and came as a touring car with a rear entrance detachable rear tonneau. It sold for $1200 without a canopy.

For 1907 the Wayne had all tonneau bodies with a four cylinder water cooled engine for the seven passenger R-Model. It had 50 hp, weighed 2800 lbs, and cost $3500. It came with five lamps and a generator. Internal rear brakes were toggle expanded, and external rear brakes were the Raymond type, with camel hair felt linings. The body was of the new straight line type, made with metal (steel and aluminum) covering wooden shapes, and finished by japanning. The Wayne Company built all parts except the bodies, which were supplied by the Everitt Carriage Trimming Company. The Wayne factory was of brick construction that measured 400 x 60 ft and was three stories high. It was situated on five acres located at Piquette and Brush.

By 1907 Byron Everitt became president of the Wayne Automobile Company, and sold the Everitt Carriage Trimming Company to Walter O. Briggs who had been the manager. This was the beginning of the Briggs Manufacturing Company that supplied automobile bodies to many different companies, along with many other products. Briggs was a baseball fan and personal friend of Frank Navin. Briggs took over the Detroit Tigers after Navin died.

In early 1908 Walter Flanders left the Ford Motor Company and became manager of the Wayne Company. Everitt and Flanders soon formed a partnership with Bill Metzger, who had a hand in the Northern Manufacturing Company. The Wayne and Northern marques were soon ended when the two companies merged, and the E.M.F. Company was organized.

1905- The motor and flywheel were hung under the chassis frame to bring the center of gravity as low to the road surface as possible. This decreased skidding and overturning on curves.

Wayne asembly plant at Piquette and Brush

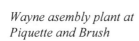

C. H. BLOMSTROM MOTOR CAR COMPANY
DETROIT, MICHIGAN
1904-1906

DELUXE MOTOR CAR COMPANY
DETROIT, MICHIGAN
1906-1909

DELUXE MOTOR CAR COMPANY
TOLEDO, OHIO
1906-1907

Carl H. Blomstrom was born in 1867 in Grand Rapids, Mich. In 1884 he took a course in engineering in Grand Rapids, then studied through the International Correspondence Schools in Scranton, Penn., and received an M.E. degree. In 1889 he married Anna Berglond and began his active career as a mechanic for the F.C. Wells Pump Company in Chicago. Starting in 1891 he designed and built marine gas engines at the Lake Shore Engine Works in Marquette, Mich. He placed his first car on the road in 1897, with a vertical two cylinder air cooled engine. In 1899 he built a second car with a different engine using a single horizontal cylinder.

He moved to Detroit in 1901, and began to design and manufacture gasoline powered marine engines at 64 Second Street. In 1902 he was manufacturing a small single cylinder automobile called the "Blomstrom." It had a single cylinder of eight horsepower with a 5.5 inch bore and a 6 inch stroke developing eight hp. By the end of 1902 he had made two dozen Blomstroms. In 1903 a new type was built with a patented two cylinder opposed engine with a 4.5 inch bore and a 4.5 inch stroke having 12 hp.

In 1904 the C. H. Blomstrom Motor Car Company was formed and located at 75 Clark Avenue near River Road, embracing 19 acres. The name of the car was changed to "Queen," which was a two passenger runabout available with a single or double cylinder engine.

In 1905 there were three body styles using the two cylinder engine, and a planetary transmission with anti-friction bearings used for the axles and wheels. Control was simple and based on a natural or "instinctive" scheme, founded on the common impulse to draw back in the face of danger. One vertical lever at the car's right controlled the car's movements. In the extreme rear position the car ran backwards; in lever mid-position the car ran forward at slow speed and forward for high speed. The footboard had one small lever by the driver's foot for controlling the throttle and one vertical pedal by which the two rear axle brakes were applied by wire cable to a triangular veneer which assured equal effect on the "Raymond" brakes. The bronze steering post contained the steering shaft pinned to the lower end. A 17 tooth steel pinion engaged an 80 tooth sector with a reduction of approximately 5 to 1. This gave the steering section ample power, and offered so much friction to hand wheel movement that the car would keep its direction "hands off." The rear wheels were driven with a Whitney roller chain using a one inch pitch with five-eights inch diameter rolls on a small 10 tooth sprocket of hardened steel and a 32 tooth rear wheel sprocket of soft steel casting.

The slogan for 1906 was "Don't be towed home at the end of a rope." The model came with running boards instead of steps. Engine bearing lengths were increased, and Parsons white brass instead of babbit was used. The Model K was fitted underneath with a formed steel shield extending the whole length of the motor. A flywheel carried out on the side to the full width of the chassis frame so as to protect the working parts from sand, mud and road dirt of all kinds.

During the summer of 1906 an application was made by the attorney general of Michigan to have the privilege of the C. H. Blomstrom Motor Company to do business, taken away from it, on the grounds that it was not properly incorporated. This action was taken at the instance of C. H. Blomstrom and the Wayne County circuit court issued an injunction. The petitioners stated that the reason for asking for the injunction was that Samuel R. Kaufman, the secretary and treasurer of the concern, resided at Marquette and therefore was not eligible to the secretaryship.

While the Blomstrom Company was going out of business, the De Luxe Motor Car Company was just starting up. The De Luxe Motor Car Company was organized in May 1906 by a number of men from upstate New York and Cincinnati, Ohio, and capitalized at $750,000. They had the services of scores of experts from other automobile manufacturing companies. George Verity, the president of the American Rolling Mill Company, in Zanesville and Middletown, was the nominal president of the De Luxe Motor Car Company, **Forest M. Keeton**, former general sales manager for Pope-Toledo was treasurer, R.C. Phillips was assistant treasurer, and F. A. Shepler was superintendant.

They leased the factory of the "former" Consolidated Manufacturing Company, maker of the Yale automobile, in Toledo, Ohio. They began with about 25 men to build models for the new car, to be called the "Toledo De Luxe" for display at the automobile shows in January and February 1907. There were two touring models, one with chain drive and one with shaft drive, priced to sell at $4000. Options were secured to buy land on the east side of the Maumee River for construction of a new factory, but capital was not forthcoming and the firm decided to move to Detroit, where there were better inducements.

The disturbances in the internal affairs of the C. H. Blomstrom Motor Co. led to an unexpected result. In October 1906, after building about 1500 Queens, the Blomstrom Company amalgamated with the De Luxe Motor Car Company, in which the Blomstrom institution lost its identity. About 200 workers from the De Luxe factory in Ohio moved to the former Blomstrom factory in Detroit. N. M. Kaufman, of the Blomstrom Company became president. George M. Verity, of the De Luxe Co., was VP, **F. M. Keeton** of the De Luxe remained secretary, and D. W. Kaufman of the Blomstrom Company was treasurer. **C. H. Blomstrom**, the mechanical expert of the Queen retained his stock in the allied concern, but was not otherwise identified with it. He left and organized another company to manufacture a car again, called the "Blomstrom."

Less than 100 "De Luxes" were made up to August 1909, when the De Luxe Motor Car Company went out of business and sold all assets, including land, buildings, and machinery to **Everett-Metzger-Flanders.**

next page (95)-
top left, *1906 Queen.* **middle left**, *1906 Queen.* **bottom left**, *Blomstrom factory 1905-courtesy NAHC*
top right, *1908 Deluxe* **middle right**, *Deluxe factory in Detroit, 1907*

DETROIT AUTO VEHICLE COMPANY
DETROIT, MICHIGAN
1904-1907

The Detroit Auto Vehicle Company was organized in the late summer of 1904 with a capital stock of $150,000. F. H. Blackman was president, **J. L. Hudson** was VP, H. H. Lind was secretary, and B. Wuryburger was treasurer. A prototype vehicle was on the road in early 1905, but production didn't commence until the Detroit Novelty Machine Company, with a factory on 65-71 Catherine Street was acquired. After the "Crown" did go into production, there was a disagreement regarding the purchase price for the Detroit Novelty Company. **Alfred O. Dunk**, Edith W. Dunk, Robert C. Yates and Louise M. Yates, stockholders in the Novelty Company, brought suit against the Detroit Auto Vehicle Company. Stock offerings and payment of indebtedness were the major points of contention.

The Crown runabout had an opposed two cylinder engine with 12 horsepower. There was also a four cylinder with 24 hp used in the touring car and delivery wagon. A Browne and Lipe sliding gear transmission with three forward and one reverse speed with bevel gear drive was used for the passenger cars, and a chain was used for the delivery wagon.

Edward T. Ross, former assistant superintendent at Cadillac for several years, was hired to re-design the Crown for the 1907 season. Ross had considerable experience with the Cadillac one-lunger and to keep the price down, he developed a two cylinder, and to provide power, the pistons were large. The first chassis was on the road in August 1907 and racked up 8000 miles in Detroit and vicinity. Making sure to have passengers and cargo representing case conditions, *the car developed no weaknesses at any point.* The wheelbase was 96 inches with a gauge of 56 inches, and 32 x 3.5 inch tires. The side entrance tonneau had doors swinging to the rear, equipped with a coat-rail, mats, horn, tools, and five lamps, all ready for extended touring. The engine was crosswise under the hood with 5.1 x 4.5 bore and stroke, *very accurately finished, showing 24 hp at 900 rpm.* It used a bevel gear drive to live axles. It weighed 1800 lbs, and the tonneau seat was 48 inches. The price for the car was $1500, plus $100 for a cape top. The name of the car was changed to "Detroit," to divorce the company from the poor image of the Crown.

John North Willys had a sporting goods shop in Elmira, N. Y., when in 1900 he saw the first Pierce Motorette, owned by a doctor in town. It was produced by George M. Pierce, a bicycle manufacturer, in Buffalo. **Willys** visited the plant and purchased the second vehicle produced by Pierce, and opened an agency in Elmira. Pierce could not produce cars fast enough for **Willys**, and in 1902 he visited the Thomas B. Jeffrey Company in Kenosha, Wisc., and obtained an agency for Rambler cars. In 1907 **Willys** contracted the entire year's output of the Detroit Auto Vehicle Company, and a second plant was opened in Wayne, Mich., to supply as many as possible. Although it appears that with a good product and a good salesman the Detroit Auto Vehicle Company should have fared well, litigation brought upon by the Detroit Novelty Company was still in court. In October 1907 the Detroit Auto Vehicle Company was formally adjucated bankrupt.

At the end of 1906 John North Willys found a factory in Indianapolis making a car called the Overland and contracted for its entire season's output, which was 47 cars. He discovered it was in financial trouble and rescued the company. He also became president, and had the output increased to 400 cars for 1909 and 4000 for 1910. He changed the name of the car to Willys Overland, and purchased the factory complex in Toledo, Ohio, of the bankrupt Pope-Toledo Company. The Willys Overland became the number three seller in the United States and Toledo, Ohio was called the "second" Detroit. Willys Overland later became part of American Motors, which then became part of Renault, and then the Chrysler Corporation.

next page (97)-
top left, *1905 Crown delivery car*   **top right**, *Crown runabout*
middle left, *1907 Detroit 3-passenger roadster*   **middle right**, *The Detroit at the Grand Central Palace, N.Y.*
bottom, *Advertisement*

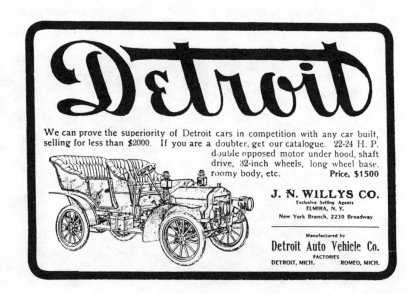

MASSNICK-MANUFACTURING COMPANY
DETROIT, MICHIGAN
1904-1908

MASSNICK-PHIPPS MANUFACTURING COMPANY
DETROIT, MICHIGAN
1908-1914

Frederic Carl Massnick was born in Flint, Mich. in 1873, and moved to Detroit when he was very young. He began learning the machinist trade when he was ten years old, and became a journeyman. He organized the Horton-Massnick Company with Bryson Horton in 1903, and by 1904 took over the entire business as the Massnick Manufacturing Company, the maker of auto parts for several leading companies. The factory was located at Lafayette and Meldrum, where the "American Electro-Mobile" was produced from 1906 to 1907 for the American Electromobile Company. (A factory was built in the summer of 1906 at 1571 River Street for the American Electromobile Company to build it's own cars, but it never went beyond the prototype stage.)

Walter Phipps was born in Birmingham, England, in 1870. His family moved to Brantford, Ontario, where he attended public schools and learned the machinist trade. He moved to Detroit and worked as a machinist with **Leland**, **Faulkoner** and **Norton** from 1888 to1904. He became a superindendant for the Brush Runabout Company, and then a master mechanic for General Motors.

In 1908 Carl and Walter formed the Massnick-Phipps Manufacturing Company, as contracting machinists. Phipps served as president while Massnick was secretary, treasurer and director. A factory was aqcuired on Lafayette and Canton Avenue. Although the Massnick-Phipps Company did not produce a car with it's name on it, in 1914 it produced the "Robie" for the Robie Motor Car Company. Fred G. Robie was in the automobile accesory business in Chicago, and contracted the Massnick-Phipps Company to produce a cyclecar of his design. The Robie first came on the market in January 1914, with a two cylinder air cooled engine. It had a wheelbase of 108 inches and a tread of 36 inches, and two passenger tandem seating. Fred Robie redesigned his car in the summer and planned to have it made by the Pullman Company in York, Penn., but his money ran out.

Massnick-Phipps also made the Perkins engine, for Robert Perkins, which was used by several automobile companies, including Remington. The **LaVigne** steering gear was also manufactured by Massnick-Phipps.

Carl Massnick sold his share of the business in 1913 due to ill health, and died in December 1914. Phipps sold his interest in the company to the J. S. Bretz Company of New York in September 1913, and stayed on as superintendant.

1911- Massnick-Phipps factory courtesy NAHC

RIED MANUFACTURING COMPANY
DETROIT, MICHIGAN
1904-1905

WOLVERINE AUTOMOBILE COMPANY
DETROIT, MICHIGAN
1905

WOLVERINE AUTO & COMMERCIAL VEHICLE COMPANY
DUNDEE, MICHIGAN
1905-1906

"THE CLEVEREST AUTOMOBILE BUILT"

The Ried Manufacturing Company was organized to manufacture bookcases, refrigerators, and automobiles. In late 1903 a prototype automobile was designed by **Walter L. Marr** for the Ried Company. In January of 1904 **Marr** rejoined **David Buick,** and was succeeded as chief engineer (at the Ried Company) by Gilbert R. Albough. Albough had experience at Olsmobile, Rambler, Peerless, and Star.

Albough made significant changes to the original Marr design, and production began in 1904 with a car called the "Wolverine." The engine was placed horizontally under the hood parallel with the axles. The drive to the rear was by axle shaft and bevel gear. The body was luxuriantly upholstered and the tonneau seats were high backed. The wheelbase was 82 inches with a tread of 56 inches. Artillery wood wheels 30 inches in diameter were shod with 3.5 inch Diamond Clincher tires. A solid 1.5 inch rear axle was housed in a 2.5 inch forged casing, automatically oiled every few weeks. A two cylinder motor rated at 15 hp ran between 150 to 1400 rpm which could propel the car 20 mph. A sliding gear transmission and water cooling were also featured. To control the car, there was a clutch pedal, brake pedal, and a hand lever for motor speed. It used a **Hussey** sliding steering wheel.

The Ried Company was in trouble in 1905 and it was re-organized in the fall as the Wolverine Automobile Company and soon after as the Wolverine Auto & Commercial Vehicle Company, in Dundee, Mich. Production resumed in 1906, but was short lived.

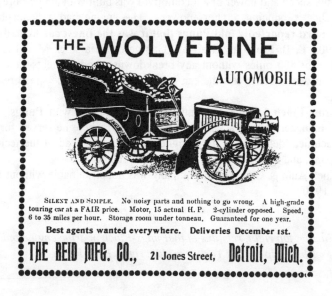

PUNGS - FINCH AUTO AND GAS ENGINE COMPANY
DETROIT, MICHIGAN
1904-1910

The Sintz Gas Engine Company in Grand Rapids, Mich., was owned by **Clark Sintz**, who had automobiles on the road as far back as 1892. A two cycle gasoline powered Sintz engine was featured at the Columbian Exposition in 1893, and a car was almost entered in the 1895 Chicago Times-Herald race by his son, Claude**.**

In 1901 the Michigan Yacht and Power Company, founded by **Ora J. Mulford**, purchased the Sintz Gas Engine Company and moved it to 1524 E. Jefferson and Baldwin, in Detroit. The Yacht Company built launches, yachts and both two and four-cycle motors.

In February, 1903, The Sintz Gas Engine Company extended operations and their stock offering. At the annual election of the company directors, E. E. Barber and James Bane of Grand Rapids, who were active in the company before its removal, retired. **Ora J. Mulford** was succeeded as secretary by **E. B. Finch**.

Edward B. Finch had an engineering degree from the University of Michigan and wanted to get into the automobile business. He formed a partnership which his father-in-law, **W. A. Pungs**, in 1902 to develop an automobile. **William A. Pungs** was part owner of the Pungs-Anderson Carriage Company, and funded **Finch** who designed and built a car in January 1903. It had a water cooled "Otto" four cylinder engine with two vertical pairs of pistons with a 4 x 5 bore and stroke, and had chain drive.

A second car, with a smaller engine, was built in March 1903, and taken up by the Sintz Company in June of 1903, for production. After driving it several thousand miles during the summer of 1903, it was decided to change from chains to a propeller shaft and bevel gear drive. The modified car was on the road in November, 1903. This car was shown at the "Garden" in January 1904. A total of 150 of these cars were sold in 1904.

Ora Mulford then left the Yacht Company to design a marine engine using the experience he gained from **Sintz**. The Gray Marine Motor Company was formed in 1905 by Mulford. In October 1904 the Pungs-Finch Automobile & Gas Engine Company was incorporated with $200,000 by **W. A. Pungs, E. B. Finch**, Ida M. P. Finch and Dwight C. Rexford.

For 1905 the motor had a 4 x 4 bore and stroke four cylinder, each single casted. The car weighed 1700 lbs and cost $1700. The chassis was pressed steel construction with cross sections of 3.5 x 1.5 x 5/32, with steel casting goose necks, *and the uniting was by hot rivet throughout.*

Whereas 1905 was distinctly known as the four cylinder year, 1906 was indisputably indicated as the year of the big four cylinder motor. Late in the summer of 1905, a one-off dubbed the "Finch Limited" (it was identified with the speed and power of a locomotive) was built with a 5.75 bore and 6.5 stroke totaling 132 cubic inches, producing 55 horsepower. The engine had a hemispherical head with an overhead cam. **Henry Ford reportedly told Pungs that it was the finest car he had ever seen.**

In October 1905 **E. B. Finch** and Harry F. George drove a stock Model F Pungs-Finch to New York. In 48 hours, it went 880 miles without any breakdowns or adjustments except those of tires and broken springs. The trip was made to test the car, but a new long distance speed record was set between Detroit and New York.

Soon afterwards, **Finch** left the company due to differences with **Pungs**. **Finch** then became assistant to the factory manager of the Packard Motor Company, where he served for a year and a half as head of the Packard technical department. He resigned and became head of the technical division of the Chalmers Motor Company, and later a Chalmers dealer in Cleveland.

The Pungs-Finch Auto & Gas Engine Company was not sustainable without **Finch**. The last year of its operation was in 1910.

next page (101)-
top left, *1904 Pungs-Finch*   **top right**, *remains of thePungs-Finch marine engine building after a fire.*
middle left, *1905 Pungs-Finch trip to New York*   **middle right**, *E. B. Finch*
bottom, *Advertisement*

Pungs-Finch Touring Car

Stands alone in its class. We are the only concern building a 4-cylinder bevel gear drive car at a moderate price who built and delivered a like model last season. This model has been thoroughly tested out by over a year's use on all kinds of roads in the hands of customers, and is in no sense an experiment.

4-Cylinder—22 H.P.   Sliding Gear Transmission—Bevel Drive

MODEL D (Rear Entrance), $1,700   MODEL F (Double Side Entrance), $1,850

Write to-day for full description

THE PUNGS-FINCH AUTO CO., Detroit

| NEW YORK | CHICAGO | PHILADELPHIA |
| C. A. Duerr Co. | The Wilson Motor Co. | The Motor Shop |

ELECTRIC POWER

James Delaney became Detroit's first lamplighter when, in 1835, the common council appointed him to service. He started by maintaining 20 lamp posts using sperm oil on Jefferson Avenue, between Randolph Street and Cass Avenue.

In 1851, Detroit's streets were lighted with gas by the Detroit Gas Light Company. Soon afterwards, homes and businesses also began receiving gas service. A second company, the Mutual Gas Light Company formed and soon after the two companies competed even in the same neighborhoods. The common council later offered the eastside of Woodward Avenue to the Mutual Gas Light Company and the westside to the Detroit Gas Light Company. Subsequently, the two companies merged in 1893 to form the Michigan Consolidated Gas Company.

The first electric light seen in Detroit was from the middle altar of The Most Holy Trinity church, in Corktown, during a mass in 1875. The Brush Electric Lighting Company formed in 1880, starting with 32 arc lamps along Woodward Avenue. Early lighting was on a moonlight basis, such that when the Almanac called for the moon to shine, the lights were not turned on that night. In 1882, a tower system using arc lamps for public lighting was inaugurated. The tallest towers were at Campus Martius, at 180 feet tall, with 3 to 6 lights each. Being so high, they seemed to light up the heavens rather than the streets. For the curious, it was exciting to watch the light trimmers who had to ascend the tall towers in a Bosun's chair elevator to adjust and occasionally clean the lamps. The towers were perilous in the winter, when they were slippery and the guy wires would rust and break. The city eventually had over 3000 towers, which stood until 1918 when the last one at Ferry and St. Aubin streets was dismantled.

Following the invention of the incandescent light bulb in 1879, it was first used in Detroit for street lighting. The arc lamp was far too dangerous for home use, but it became practical to use the incandescent safely.

In 1893, the Michigan legislature passed a bill allowing Detroit to establish an electric generating plant for public lighting. A public lighting commission was created, with **J. L. Hudson** and **Charles H. Ritter**, as members. Electricity for home and business lighting began that year. The first electric sign in Detroit was located at the foot of Woodward, advertising Queen Anne soap.

Detroit was one of the first cities to have electric power, which proved to be a factor in the automobile industry. Early factories relied on a stationary steam engine that powered overhead line shafts with leather belts coupled to grinders and lathes. The breaking belts contributed to physical harm and the production efficiency was low. Electric power allowed the use of electric motors that could be strategically located along work stations so the workers didn't have to walk back and forth continuously.

Brush Street
courtesy MSHC

LITTLE FOUR AUTOMOBILE MANUFACTURING COMPANY
DETROIT, MICHIGAN
1904

The "Little Four" steam car was exhibited at the February 1904 Detroit Automobile Show. By late March 1904 the Little Four Automobile Manufacturing Company was organized by Wyatt L. Brown, Fred L. Brown, and J. D. McLachlan with $25,000 capital. The Little Four was a runabout that weighed 400 lbs and cost $400. It had a three cylinder single acting type steam engine of six to eight hp. The boiler was made of seamless tubular steel. The speed of the engine could be doubled, using the same amount of steam, by using the reverse lever instead of the throttle. There was limited production and the firm didn't last out the year.

EUREKA MANUFACTURING COMPANY
DETROIT, MICHIGAN
1905

William Egle was born in Baden Germany in 1867. After his education, he immigrated to the United States in 1881 and joined the Army. He served as a musician at Fort Custer in Montana and became a sergeant. After leaving the army in 1891, he went to Detroit where he invented and patented a China Kiln. With his brother Henry, he organized the Egle Bros. Manufacturing Company in 1897 to manufacture China Kilns.

William was also president of the Eureka Manufacturing Company located at 169-171 Sylvester. The company manufactured manhole doors. In February 1905 the company advertised that it would produce an automobile, but didn't have a completed model for the automobile show in Detroit. The "Eureka" was a King of Belgium side-entrance tonneau on a 96 inch wb. It had a four cylinder engine and a sliding gear transmission. Production was discontinued before the year ended.

1905 Eureka

EARLY GASOLINE ENGINES

Engine speed on early cars was controlled by a centrifugal governor that prevented the exhaust valves from opening if a certain speed was exceeded, usually 750 rpm. The driver could only override the device and allow the engine to go faster. By the turn of the century, some cars had a foot pedal to increase the engine speed, called an "accelerator." The designers still believed that the best mileage was derived at optimum speed only and the car's speed was regulated by gear changing. The single cylinder, low rpm, engines were sometimes spoken of as "one chug per telegraph pole."

Atmospheric valves were used, but tended to operate best at only one speed. The inlet valves were operated by atmospheric pressure when the piston descended, and opened by suction. They required very weak springs and were a constant source of trouble from soot on the valve stem, or mis-adjustment. Also, the atmospheric valves could not be held open beyond the end of the intake stroke to fill the cylinder as full as possible with gas vapor.

Next, the intake valves were driven off a cam like the exhaust valves, using two cams, with a T-head design, and by 1906 only one cam was used, called an L-head. Other valve designs were also used such as the rotary, cuff, split-ring, and sleeve. Early engines did not have high temperature steel for exhaust valves up to 1903, which required valve grinding every 1000 miles. Up to 1914, engines had exposed timing gears, cams, tappets and valve springs, resulting in stuck valves.

The early engines suffered from severe torsional crankshaft vibration, and more often than not, oversquare cylinders were used with a bore larger than the throw. Large flywheels, and later the warner vibration damper was used before the development of the crankshaft with integral counterweights.

MIXERS

Early carburetors were like lamp wicks with one end in gasoline, except they were bigger with more surface area. The other end was in the air stream. The air/fuel ratio could not be controlled. A simple jet in a tube followed with no control of the air/fuel mixture either. By the turn of the century the crude devices were replaced with jet-in-venturi systems that required constant adjustments depending on engine speed. Then, multiple jet units with chokes and multiple metering systems were developed as up draft, side draft, and down draft types. **Henry Ford** liked to call them mixers.

THE WALKER MOTOR CAR COMPANY
DETROIT, MICHIGAN
1905-1906

The Walker Motor Car Company was organized with capital stock of $10,000 in October 1905 by **Charles L. Walker**. (**Walker** was a founding member of **Homer Warren & Company**, organized in Detroit in 1892.) The factory was located at 107 Fort Street. The "Walker" was a two passenger runabout with the appearance of a touring car. When the Walker Company designed the car, it was meant for practical use by the average person. The ads said "the runabout is the candy, absolutely noiseless." It's construction was very simple, making it desirable for the person who could spare little time to clean and repair their own machine. The use of three inch clincher tires, large enough to carry more than the weight of the light car, reduced the danger of punctures to a minimum.

A long 78 inch wheelbase gave enough room for an *artistic* body, and with a *turtle* back it was a *strikingly handsome car*. With a standard tread, one could travel through the country and follow the well-worn paths and not suffer the inconveniences because of too wide or too narrow of a tread.

The engine and transmission were located under the hood and accessible at all times. The engine was a two cylinder double opposed type 4 x 4 inch bore and stroke that developed 10 hp. A jump spark ignition was used with the spark coil fastened in front of the operator. It used a planetary transmission with two forward and one reverse gear. Final drive was by direct shaft. The Walker was capable of speeds between 4 and 25 mph. Equipment included two brass side lamps that used oil, one horn attached to the steering post, and a repair kit. The price of the 1906 Model B was $600.

The Walker Company was in financial trouble from the start, and in late spring of 1906, with judgments of $35,000 against it, authorities closed the factory and removed the machinery.

DETROIT AUTOMOBILE MANUFACTURING COMPANY
DETROIT, MICHIGAN
1905-1907

In 1890 **Joseph P. La Vigne** was awarded a gold medal from the Paris Academy of Inventions and Industry for creating the automobile *line shaft* and for positioning the engine under the hood. He immigrated to America and constructed his first automobile in 1898 while living in New Haven, Conn. The three-wheeled runabout held two passengers, weighed 1800 lbs, and could travel up to 45 mph. The car went in reverse by reversing the motor. Unfortunately it burned up from a gasoline leak three months after it was on the road.

La Vigne then moved to Detroit and organized the La Vigne Manufacturing Company. After getting the business started, part of the factory was used to construct his second automobile, which was placed on the road in August 1901. It was a three wheeler like the first, but it was larger and had a 110 inch wb, with a **55** inch tread and 35 x 3.5 inch tires. The motor was a water cooled two cycle, with three cylinders. It had the same general transmission scheme as the first car, with a reversing engine and a spiral gear to the divided rear axle. It weighed 2300 lbs and could travel 55 mph. It was never fully completed due to capital shortage.

La Vigne drove to work in a steam car. He would often go home for lunch and leave the boiler running while he ate because it took so much time to get up steam. One day in 1904 **La Vigne** went home and hurried into the house while his seven year old daughter, Olive, jumped into the car and drove away. By the time **J. P.** got to the front porch it was too late to do anything, but as he watched, she returned with the steamer intact. Olive had learned how to operate the steamer by watching her father as they went for rides around the city. That Christmas he gave Olive a car constructed just for her. She could often be seen driving around the neighborhood, and Olive always believed that she was the first child of the automobile era to drive a car. (The youngest "licensed" operator in Michigan was Dexter Brigham, who at 11 years old in 1906 drove his father's Cadillac.)

In 1905 **La Vigne** exhibited a four-wheeler at the Detroit Automobile Show priced at $375. It was placed under a banner that read "within the reach of all." The wheelbase was 51 inches, with an optional tread of 40 or 56 inches. The car weighed 350 lbs and was capable of 20 mph. The engine had an air cooled single cylinder with a two inch bore, producing three hp. The car was named "La Petite," (which was **La Vigne's** pet name for his daughter Olive). According to **La Vigne, Henry Ford** approached him at the show and said "**You have my idea, and that is what I am going to build**."

The La Petite was well accepted at the automobile show, and soon afterwards **La Vigne** secured financial backing and formed the Detroit Automobile Manufacturing Company with a factory on 177-179 Larned Street. Sales of the La Petite were almost 200 for 1905. The engines were purchased but did not meet **La Vigne's** specifications and the quality was poor. The 1906 La Petite was to have a larger engine with more horsepower, but **La Vigne** abruptly left the Detroit Automobile Company. The company then was quickly re-formed with **J. C. Forester** named as manager. They moved into a factory at 253-259 Willis Avenue East to continue production for 1906. A car was exhibited at the Detroit Automobile Show in February 1906, and was called the "Paragon" which was a fitting name, judging by the public acceptance. The price was $375, and was called "the wonder of the automobile public." Reports indicated that the engine was so muffled that it was practically noiseless.

While most automobile manufacturers were bringing out higher power and longer wheelbase machines, the Paragon was very small, light, and low powered. At first glance it looked like a small buggy, although it had two more horsepower than the La Petite. The wheelbase was 68 inches and the tread was 42 inches. It used bicycle sized tires, 28 inches in diameter. It had two forward speeds and one reverse, on one lever, which reduced the chance for stripping gears off the cogs.

Before the year was up the Detroit Automobile Manufacturing Company was out of business, and the factory and equipment were sold to the Marvel Motor Car Company with **J. C. Forester** staying on as manager.

"LA PETITE"

The Wonder of the Age — *Within the Reach of All*

Price, $375

We can now make prompt deliveries of this little runabout, which has a 2 cylinder, 5 h. p., air cooled engine, with two speeds forward and reverse, direct shaft drive; will run up a 20% grade carrying 2 persons. This car is made of the best material throughout. It has no equal. It is a strong and durable machine. Made by

Detroit Automobile Manufacturing Co.
Sales Agents: 253-255 Jefferson Avenue
Office and Factory: 177-179 Larned Street

AGENTS WANTED   DETROIT, MICH.

above top left, Olive La Vigne and her Christmas present *above top right*, Olive in a La Petite
middle left, La Petite advertisement *middle right*, Paragon advertisement
bottom, 1905 Paragon runabout

AEROCAR COMPANY
DETROIT, MICHIGAN
1905-1907

Alexander Young Malcomson was born in Scotland and immigrated to Detroit when he was fifteen years old. He worked in a grocery store, but quit to start his own grocery business. By the early 1890s he was in the coal business with his first coal yard. His slogan was "Hotter than Sunshine." By 1902 he had six coal yards with thousands of commercial and residential customers, including the Edison Illuminating Company.

Malcomson bought a Winton and must have followed the big race in October 1901 where **Ford** beat Winton at Grosse Pointe. He realized the potential in Ford's reputation for building dependable fast cars. He seized the opportunity to become a founder of the Ford Motor Company. As the Ford business blossomed with small engine, low price cars, **Malcomson** pushed for making large expensive cars, such as the Ford Model K that sold for $2800 and was a commercial disaster.

The Aerocar Company was formed late in 1905 by **Malcomson**. The Ford Motor Company stockholders viewed this as a threat to their sales and asked Malcomson to resign as treasurer and director. In July 1906 **Malcomson** sold his 255 shares of stock to Henry Ford. (By mutual agreement, stock could not be sold outside of the original stockholders.)

The "Aerocar" sold for the same price as the Ford Model K. It had a four cylinder air cooled engine with 24 horsepower, designed by **Milton O. Reeves** of the Reeves Pulley Company. John L. Poole stated that the air cooled cars were recognized as the coming thing in Europe. There were three German companies already in production, all of the leading French cars were developing air cooled engines, but the United States was now in the lead. There was a great demand for them in Norway, Sweden, Denmark, Russia and Germany.

In March 1907 C. Arthur Benjamin was named director and vice president of Aerocar Company, while **Malcomson** was President. In the days of bicycles, Benjamin was given the name of "boy wonder" and upon his early evolution into the automobile game, he was dubbed "The Original Live Wire." And so they were off, with the Model D air cooled model, and a water cooled Model F having a 4 cylinder 45 hp designed by **Leo Melanowski** who had been the chief engineer for the Dragon Automobile Company. The Aerocar Company could not make ends meet and went out of business by 1908.

The assets were purchased by **Alfred O. Dunk**. The factory was used next to build the very first Hudson car. **Malcomson** lost $90,000 out of his pocket. He continued in the coal business, merging with the C. H. Little Company in 1914 to form the United Fuel and Supply Company. He also had commercial interests in Kentucky. **Malcomson** had five children by his first wife and two by his second. He died in 1923.

The Aerocar Company, Detroit, Mich.
Beyster-Thorpe Co.
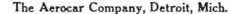
1329-1331 WOODWARD AVE.   DISTRIBUTORS.

"Built for Service."

The Aerocar Baseball Team

MOTORCAR COMPANY
JACKSON, MICHIGAN
1905

MOTORCAR COMPANY
DETROIT, MICHIGAN
1905-1908

CARTERCAR COMPANY
PONTIAC, MICHIGAN
1908-1915

"NO CLUTCH TO SLIP, NO GEARS TO STRIP"

By some accounts, **Byron J. Carter** began working with friction drives as early as 1887, first using a steam engine, then a two cylinder engine designed by Charles Brady King. He began experimenting with a model in 1901, and soon became manufacturing superintendant for the Jackson Automobile Company. A small light-weight vehicle was completed in May of 1903 with a new friction-drive, and a two cylinder Brennen engine. The runabout was later sold to Ralph H. Miller of Ann Arbor, Mich. A transmission patent, U. S. 761,146, was issued to **Carter** in 1904. The Jackson Automobile Company was not interested in the new transmission, and **Carter** continued with efforts to market his invention. He organized the Motorcar Company in Jackson, and by 1905 sold 10 automobiles with the friction drive transmission. Additional financial backing was secured and the company moved to Detroit at 21st and Baker Streets.

The Cartercar was shown at the New York Automobile Show in January 1906, along with a Lambert and a Simplicity that also had friction drive transmissions. In February 1906 three "Cartercar" models were shown at the big show in the drill hall of the Detroit Light Guard Armory. Model A had the famous King of the Belgians body, with a horizontally opposed two cylinder motor, with 20 hp. The motor was over-square with a 5 x 4.5 bore and stroke. There were not many over-square motors in use at the time. The Model B was equiped with a Victoria body, and a delivery body was also shown. The J. L. Hudson Company of Detroit purchased a number of the delivery wagons and put the first one in service starting on March 7, 1906. It provided an interesting low-cost record covering its first three weeks of operation by delivering 1595 customer parcels.

The 1907 models were essentially carryover from 1906 with a slight increase in engine size. There were 264 cars sold that year with the City Fire Department as a customer, suggesting they were solidly built. There were 600 units scheduled for the 1908 model year, including a two passenger runabout with a "beetle-back" locker which could be removed and replaced by a detachable tonneau so as to make it a touring car.

A major change in the Motorcar Company was brought about by the sudden death of VP and general manager, **Byron J. Carter**. He died of pnuemonia at his home on April 6, 1908. He was replaced by **R. A. Palmer** who began development of the 1909 models. By October 1908 there was a shake-up in management and policy with capital increased from $150,000 to $350,000. The company name was changed to the Cartercar Company and the factory operation was moved to the factory of the Pontiac Spring and Wagon Works, with two acres of floor space in a five story building, on 40 Franklin Road, Pontiac, Mich. In preparing for the 1909 models, the company announced that Models K, G, and H would be built along with a high-wheeled motor buggy continuing the name Pontiac. The model K line included a special taxicab which was apparently developed to coincide with the adoption of a new regulation in Chicago under which taxicabs could carry standard meters for the first time.

On November 1, 1908 **William C. Durant** purchased control of the Cartercar Company and installed George E. Daniels as general manager. Daniels had participated in buying the Oakland Car Company with **Durant**. In 1910 the original 1903 model was reported to be in good condition with over 25,000 miles on the road. However, not all owners were that fortunate because there were numerous problems with Cartercar transmissions. By 1912 Cartercars had a four cylinder engine and a self-starter using the explosion method.

In 1912 **R. A. Palmer** was replaced by former sales manager, **Harry R. Radford**, as general manager. This change was directed by GM president, Thomas Neal. The factory was used for manufacturing the Oakland automobile, and the Cartercar was moved to a smaller plant. Sales never reached the levels envisioned by **Durant**. When **Durant** returned to GM, it was too late for the Cartercar, and it was taken out of production on May 22, 1915. **Durant** later lamented: "How was anyone to know that the Cartercar wasn't the thing."

View of the Cartercar chassis, showing the friction drive.

Harry Radford in a 1910 Cartercar courtesy AAMA

The Motorcar factory. It was located on 230 21st Street, near Porter. It had previously been the Frederick Stearns and Company, M'FG Pharmacists.

General Machine Dept.- lathes, drill presses, grinders, a cut-off saw, and a special clamping fixture at the lower right of the picture.

Frame Assy. Dept.- Note the flat belts leading to the electric motor at the center left of the picture.

Eng. Building Dept.- The machine in the foreground is an internal grinder with a cylinder on it. Directly behind is a boring mill. Note the dust collector on the I. D. grinder and fanned out the windows.

Motor Testing Dept.- All the early Cartercar engines were of the two cylinder opposed type. Note the welded test stands.

Chassis Assy. Dept.- Note the overhead trolley, cast iron columns, and electric light drop cords.

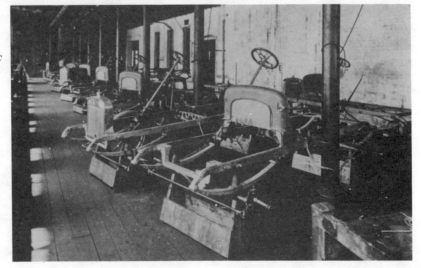

Body Dept.- The absence of woodworking machinery leads one to believe the bodies were purchased and not made by the company.

Paint and Trim Dept.- Note the overhead gas lamps and how the bodies looked like sleigh bodies of that time.

Varnish Room- Note how this was made into a clean room with pipes wrapped and the ceiling somewhat protected. The bodies were varnished after the upholstery work was completed.

Final Assy Dept.- Note the rear fenders, or splash guards, over the rear wheels and on the shelf behind the car. The Cartercar emblem is affixed to the radiator.

Chassis Painting Dept.- Note the canvass apron below the car to protect the open drive mechanism.

*Testing Room- Note the leather Michigan plate 1170 on the car to the right. This is before the 1909 Michigan porcelain plates. Also note the quantity of axles in the foreground.
All factory sequence pictures courtesy MSHC*

LOZIER MOTOR COMPANY
PLATTSBURG, NEW YORK
1905-1915

LOZIER MOTOR COMPANY
DETROIT, MICHIGAN
1910-1918

"THE CHOICE OF MEN WHO KNOW"

The Lozier Manufacturing Company was established in the early 1880s with Harry A. Lozier Sr. as president. Among the products produced by Lozier was the world renowned Cleveland bicycle. It was built in four different factories; the headquarters were in Toledo, Ohio. In 1897, just before the bottom fell out of the market, the bicycle business was sold to the American Bicycle Company for $4,000,000.

In 1900 the Lozier Motor Company was established in Plattsburgh, N. Y., to manufacture marine engines and launches. John Perrin, an engineer, and George Burwell, a mechanical genius and inventor who was responsible for the success of the Cleveland bicycle, convinced Harry Lozier to manufacture automobiles.

In 1902 J. M. Whitbeck, an engineer, was dispatched to Europe to study the automobile trade. He made drawings, purchased engines, Krupp steel, ignition systems and carburetors. Harry Lozier Sr. died in May 1903, and was replaced by **Harry Lozier Jr.** Eventually, John Perrin was given the go-ahead to study the best European designs and to build a model automobile meeting Lozier's high standards.

In less than two years the first "Lozier" automobile was driven to New York for **H. A. Lozier Jr**. and his sales group to review. It was approved and 25 more were produced at $5000 per chassis. John Perrin had designed a "better Mercedes." The first Lozier car was exhibited at the automobile show in the Madison Square Garden, and sales began to escalate.

The Plattsburg plant capacity was limited to 600 automobiles annually. In 1909 **Harry Lozier Jr.** was approached by Detroit businessmen to relocate to Detroit and build a rival to Packard. The benefits were the availability of expert craftsmen, special purpose machine tools and materials such as high strength steel alloys that were increasingly being specified by quality-oriented automobile makers. The company was re-organized and moved into a new "Kahn" factory on Mack Avenue in February 1910. **Henry Ford** and **James Couzens** visited the Lozier plant frequently and enjoyed watching the assemblers working on the chassis line. **Edsel Ford** would often visit on Saturdays while other rich-men's sons were spending their free time in teenage sports.

Lozier entered racing with driver Ralph Mulford in 1907, and by 1910 won the famous Elgin Road Race. The Lozier excelled mainly in the 24 hour events, because even though it wasn't the fastest, it had endurance. In 1911 Lozier finished second to Marmon in the Indianapolis 500, losing by 43 seconds. The Lozier used eight tires to the Marmon's three, causing delays which allowed Ray Harroun to win. (In this race, Ray Harroun drove without a mechanic and used a rear view mirror to compensate for the first time.)

In 1912 Lozier introduced crowned fenders and had 24-carat gold striping, which by all accounts most people presumed was brass. It was one of the most expensive cars in the United States. There was a disagreement among the company directors, including Stair, **Homer Warren**, Buhl, Gilbert, and Lee, some of whom thought there would be more profit in a lower price car such as the Paige. **Harry Lozier** resigned the presidency, and in August 1912, **Harry M. Jewett** was elected to replace him. **Jewett** was also the president of the successful Paige Motor Company. Several Lozier personnel soon left the company including Frederick C. Chandler, who had worked for the Cleveland Bicycle Company and became the general manager of the Lozier factory in Detroit. He went to Cleveland and established the Chandler Motor Car Company and made a light weight, four cylinder, car with a side view identical to the Lozier. The Chandler Company lasted until 1928 when the sale of its plant to the Hupmobile Company was announced.

Harry Jewett resigned in May 1913 and was replaced by Joseph M. Gilbert, former general manager of the American Tire Company. Gilbert was a merchandiser and was proud of the fact that he cut the cost of conducting his business by 8.5 percent for the month of March. (An example of how "keen" he was on factory control was the use of a low price car procured for light freight work which was cheaper to operate than a Lozier. Another example was that some of the office departments expected to have a Lozier at their disposal, but Gilbert pointed out that the trolley line was intended for that purpose.)

By May 1914 Lozier was ready to introduce a four cylinder car priced at $2100 that could outperform the popular four cylinder Cadillac. Six months after the Lozier took to the streets, Cadillac introduced its first V-8 engine, and Lozier soon went into receivership.

In 1915 a new concern called the Associated Lozier Purchasers took over Lozier. They slashed prices, moved to the **Warren Motor Company** plant at Holden and Lincoln, and sold the Plattsburg plant. By mid-1917 three Detroit businessmen re-organized the Lozier Motor Company and leased the plant of the former Standard Auto Truck Company, but production ended in September 1918. After the war the original Mack Avenue plant was occupied by the Metal Products Company, maker of automobile accessories.

After leaving the Lozier Motor Company, **Harry A. Lozier** formed the H. A. Lozier Company in Cleveland, Ohio, in June 1915. The first model had a twelve cylinder engine, and was shown at the New York Automobile Show in January 1916. The car was called the " HAL," using his initials. Lozier left the company due to ill health in September 1916. In October 1917 there were rumors of a merger with the Abbott Company in Cleveland, but they went bankrupt in January 1918, followed by Lozier in February 1918.

1900-
Harry Lozier Jr.
Lozier Motor Co.
in Toledo, Ohio.
This was the largest
bicycle factory the
company had owned.
It was used for designing
the first Lozier. A new
factory for production
was built in New York.

1905-
A Lozier is seen with
a suspension called
the platform spring.
Ample clearance was
allowed above ground
for the American roads.

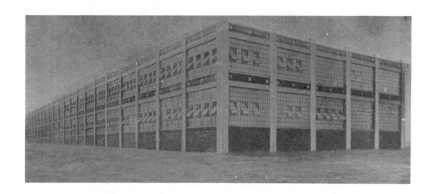

The Lozier factory, located on Mack east of St. Jean, on the outer belt of the Detroit Terminal RR. It was built in 1910 by **Albert Kahn**.

A demonstration in a Lozier showroom.
courtesy AAMA

A Lozier being pulled by a second Lozier in front of the Hotel Ponchartrain. Note the single headlight on the front car. The sign on the cart to the extreme right of the picture reads "French-American Ice Cream Co."
courtesy AAMA

*Mr. and Mrs. H. Lozier Jr.
courtesy NAHC*

RANDS MANUFACTURING COMPANY
DETROIT, MICHIGAN
1906-1907

The W.C. Rands Manufacturing Company was one of the early automobile dealerships, selling the Oldsmobile runabout in 1903. Rands was also in the automotive parts business including leather upholstery for a number of Detroit automobile companies. Initially, parts were sourced to the Wheeler Manufacturing Company by Rands. Eventually, Rands bought the Wheeler Company and made his own parts. In 1906 **W.C. Rands** built a five passenger touring car on a 106 inch wb with a 30 hp air cooled four cylinder engine, and a sliding gear transmission. Using his own shops, he built the first Rands automobile for himself, then continued building into 1907.

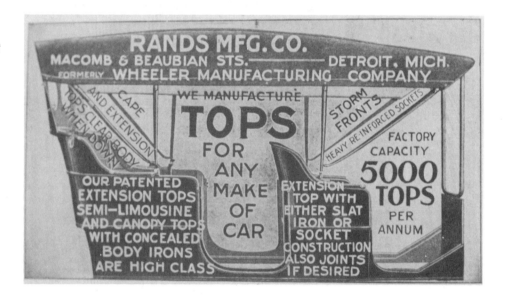

Rands advertisement

1906- Rands Special

MARVEL MOTOR CAR COMPANY
DETROIT, MICHIGAN
1906-1908

The Marvel Motor Car Company was formed in late 1906 and occupied a well-equipped factory at Rivard and Mullet in Detroit, the former home of **Paragon** Motor Car Company. Not only was the "Marvel" assembled at this plant, but the vehicle's chassis, motor, and transmission were manufactured there as well. **J. C. Forester** was the general manager and William A. Phister was the superintendent. Although he had no previous experience as a motor car designer, Phister decided to design the Marvel. **Forester** and Phister planned the Marvel's debut for February 1907, with expected sales of 325 cars.

The Marvel was a two passenger roadster that weighed 1200 lbs and sold for $800, including mat, horn, tools, and oil lamps. It had an 84 inch wheelbase, with a **55** inch tread, and an integrally casted two cylinder motor with a nominal 12 hp. The cylinders were horizontally opposed and set crosswise in front, with thermo-syphon cooling. The Marvel used a planetary transmission, cone clutch, and shaft drive. Its rear hub brakes were operated by pedal.

In August of 1907, the Marvel Motor Car Company petitioned for bankruptcy, with assets of $4000 and liabilities of $20,000. Subsequently, the Crescent Motor Car Company took over the Marvel Motor Car Company and moved its production to the former Buick factory at Meldrum and Champlain (Lafayette).

Crescent was formed in May 1907; when it purchased the **Reliance** automobile business, it faltered and, by September 1908, was taken over by investors, who were associated with the Hawley Automobile Company, in Constantine, Mich. They formed the Constantine Motor Car Company, planning to use the former Hawley factory. (The Hawley Automobile Company relocated to Mendon, Michigan; it was formed in 1906 and did not last through 1908.) **J. C. Forester** had designed a new runabout and was considering building it in Elkhart, Indiana, but the Marvel Company never resumed production, and the Constantine Motor Car Company never got started.

DRAGON AUTOMOBILE COMPANY
KITTERY, MAINE
1906

DRAGON AUTOMOBILE COMPANY
DETROIT, MICHIGAN
1906

DRAGON AUTOMOBILE COMPANY
PHILADELPHIA, PENNSYLVANIA
1907-1908

The Dragon Automobile Company was organized in Kittery, Maine in 1906. It was capitalized at $500,000 but had only $300 paid in. In July 1906, Frank S. Corlew gained control of the company and announced the manufacture of Dragon Automobiles would be in Detroit. The papers in the east were filled with "doings" of the new corporation. (The Dragon was designed by **Leo Melanowski** in 1905. He left the firm and joined **A. Y. Malcomson** to design a water cooled engine for the Aerocar.)

There probably never was as much mystery in automobile circles as that occasioned by the Dragon Automobile Company. No one in **Detroit** knew where they were located. Actually, the first six cars were constructed in a **Detroit machine shop** so they could be ready for the New York auto show.

The Dragon Automobile Company had opened offices in the **Campau Building in Detroit**, but in October announced that the car would be manufactured in Philadelphia! They purchased the factory formerly occupied by the Brill Car Company. There was nothing radically new with the Dragon. It was considered to have many well tried and proven features which had been used on other automobiles. It lasted until February 1908 when the factory shut down.

1907 Dragon   courtesy AAMA

E. R. THOMAS-DETROIT COMPANY
DETROIT, MICHIGAN
1906-1908

CHALMERS-DETROIT COMPANY
DETROIT, MICHIGAN
1908

CHALMERS MOTOR COMPANY
DETROIT, MICHIGAN
1908-1916

CHALMERS MOTOR CORPORATION
DETROIT, MICHIGAN
1916-1922

The E. R. Thomas Motor Company was formed in 1898 in Buffalo, N. Y., by Edwin Ross Thomas. The "Thomas Flyer" won an endurance race in 1908, traveling from New York to Paris with an approximate distance of 22,000 miles in 170 days. Even though low gear had to be used most of the time due to poor roads, it demonstrated to the world that the automobile was reliable. It could be argued that it opened the market for the Ford Model T.

Roy Dikeman Chapin was born in Lansing in 1880, and entered the University of Michigan in 1899. He soon became a close friend of fellow student **Howard E. Coffin**. In the spring of 1901, just weeks after the factory burned to the ground, 21-year-old **Chapin** was offered a job as a tester with the Olds Motor Works. **Chapin** was an adept photographer and helped design the new Olds catalog. He eventually became chief tester and sales manager at the Olds Motor Works.

Howard E. Coffin graduated from the University of Michigan in 1893. He constructed a steam car while at U of M in 1898-1899. He started as an engineer at the Olds Motor Works in Detroit in 1902, when it was the number one automobile producer in the United States.

Coffin was dissatisfied with the larger Olds models that were under development and asked management to let him develop a medium priced car. **Fred Bezner** was authorized to line up suppliers for the new car. Ten weeks before the New York Automobile Show in January, the company decided not to launch the Coffin designed car. A quasi rebellion occurred amongst the younger executives, and after the New York Show, they began to quit. **Chapin** left and suggested to **Coffin**, **Bezner** and **James Brady** that they should enter into a contract together.

Chapin went to San Francisco where he met E. P. Brinegan, a dealer for the Olds Motor Works and the E. R. Thomas Company. **Chapin** shared his hope to build a car designed by **Coffin**. Brinegar suggested contacting **E. R. Thomas**, who was expected on the West Coast within three weeks. **Chapin** met **Thomas** in April and struck a deal to build a car in Detroit and sell it to the Thomas Company in Buffalo for distribution through its dealers. **Thomas** agreed to pay $100,000 and the Detroit group would pay $50,000. They pooled $10,000 and borrowed $40,000 from a Detroit bank. The result was the E. R. Thomas-Detroit Company incorporated May 12, 1906.

An order for 503 four cylinder cars was received from the E. R. Thomas Company of Buffalo the same day. **E. R. Thomas** became president, **Coffin** was VP of engineering, and **Chapin** was treasurer and general manager. The Modern Match Company factory at the railroad spur at Harper and Dequindre in Detroit was leased for production, and operations began in June 1906. J. B. Philips, a former head of inspection at Oldsmobile, was made plant superintendent. A force of draftsmen was hired and immediately began work on the drawings for a new chassis.

The first car was on the road on August 20, 1906, and was tested satisfactorily with very few changes. Two more were shipped to the E. R. Thomas Company on September 20, 1906, and production began to the limit of the factory capacity, selling for $2700 each. No "first model" was ever built as the first car was part of the 503 ordered by the Buffalo Thomas Company. Though built in Detroit, the "Thomas-Detroit" was bought and marketed by the Thomas Company in New York. After delivering the first order, a second order for 750 cars was received for the 1908 season. A new factory was built and occupied in October, 1907, by the Thomas-Detroit Company.

Hugh Chalmers was born in Dayton, Ohio in 1873, to Thomas Chalmers and Jeanette Bell. Thomas was a stone cutter who immigrated from Scotland in 1865. **Hugh Chalmers** left school when he was 14 to become an office boy at the National Cash Register Company in Dayton. When he joined NCR, it was in a period of growth that was mainly due to the sales practices that were innovated by John H. Patterson. Chalmers advanced to the position of salesmen when he was eighteen by taking shorthand and bookkeeping at a night-school. He was Patterson's sales manager for Ohio in 1897, with 24 agents and salesmen under his supervision. In 1900 he was VP and general manager, with 5500 employees. **Roy Chapin** was no stranger to NCR, as he made frequent trips to Dayton to study the sales set-up. In the automotive business, there were no established sales methods or policies. There were no statistics, no territorial research, and no science of selling. As a result of these trips, **Chapin** wrote what was probably the first sales manual in the automotive field.

In 1907 **Bezner** read a newspaper account of Chalmer's resignation from NCR. **Chapin** and **Coffin** induced **Hugh Chalmers**, who at age 34 had been making $72,000 per year. **Chalmers** was offered $50,000 per year and could buy half of the interest of the principal stockholder, E. R. Thomas. The company was re-organized into the Chalmers-Detroit Company. **Chalmers** became president and **Chapin** was treasurer and general manager. The 1908 models had minor chassis revisions, and incorporated a **Herreshoff** motor, made by the American & British Manufacturing Company, of Bridgeport, Conn. It was also a four cylinder with 40 hp, but had a larger bore and a shorter stroke than the original Thomas engine. The engine was water cooled and had pistons made of a special soft gray iron, machined and ground to micrometer dimensions. The camshaft operated both, inlet and exhaust valves. The engine had a cast iron exhaust manifold with a tapered construction to insure an easy flow of exhaust gas. Known as the "Chalmers New Detroit," it was advertised for $1500, and several thousand were sold in the first year.

The car was renamed the "Chalmers 30" for the 1909 season. The engine was designed by **Coffin** with four cylinders and a one piece block, making possible a short stiff crankshaft. The 30 hp car could easily reach 50 mph. By June 30, 1909, Chalmers-Detroit had invoiced 2476 Chalmer 30s and 611 Chalmer 40s, for a profit of over $1,000,000. A 100 percent cash dividend was declared and the mortgage on the factory was paid. Under the new organization, **Thomas** and **Chalmers** each received a third of the dividends, and the remaining third was split among the rest of the stockholders.

In keeping with their plan, **Coffin** and **Chapin** conceived a new light car design in 1909, but **Chalmers** was reluctant to enter the lower price market. Detroit merchant **Joseph L. Hudson** was enlisted to finance a separate company for the manufacture of the car which became the "Hudson." **Hugh Chalmers** was involved in the initial stages, but sold his stock to **Chapin**, **Coffin**, and **Bezner** for $80,000. At the same time, he bought their holdings in Chalmers-Detroit for $788,000, and renamed the company Chalmers Motor Company.

The Chalmers models that followed got bigger and more expensive. **Hugh** wanted to build the big car like he himself would like to drive. However, **Chalmers** was well aware that bigger cars were more profitable, and by 1911 he may have felt the smaller market was reaching the saturation point. He decided to concentrate on the wealthier classes who could afford cash when credit was rarely available for an automobile. The Chalmers "Thirty-six" and "Six" were introduced for the 1912 season. The Thirty-six had a cast *en bloc* long stroke motor, with a two bearing crank shaft. A new crank case construction protected the exposed working parts of the motor from all road dirt. All new piston rings were used which entirely eliminated motor smoking. The Chalmers "Six" was rated at 54 hp, and had a wheelbase of 130 inches. The bodies were constructed of sheet aluminum, with integral cast-aluminum dashes. Each body received 21 coats of paint and varnish. The upholstery was 10 inches thick. The most radical feature was a compressed air self-starter that was demonstrated by pulling seven passengers from a standstill. Running boards were carpeted with gray battleship linoleum. A novel improvement was the elimination of polished wood dash and heel boards. They were always being scratched up because some people simply could not keep their feet on the floor. To improve the condition, all bodies were lined with leather that could not be scratched, and with a little rubbing of oiled cloth could be made as good as new. The bodies had the straight line effect, and were finished in a variety of colors including slate gray, napier green, English vermillion, brewster green, maroon, Chalmers blue, and royal blue.

In 1913 **Chalmers** backed the Saxon Motor Company, the cheapest car in the market, but sold his stock within two years, realizing it couldn't compete with the Model T. By 1915 all Chalmers came with a six cylinder, and the peak year came in 1916 with sales of 21,408 cars produced. The company was re-organized in 1916 as the Chalmers Motor Corporation, with **Hugh** moving up to the chairman of the board, and **Walter E. Flanders** named as president. The Chalmers Motor Corporation and the Maxwell Motor Company entered a business alliance that fell *just short* of a merger. Chalmers was overextended and space was leased to the Maxwell Company for automobile production. During the First World War, Maxwell produced army tractors. The two companies tried to formalize a merger after the war, but internal bickering on both sides precluded a solution. The Maxwell Company was particularly hard hit since it had gone heavily in debt to finance postwar expansion programs. In a complicated series of moves, the financial institutions to whom the two companies owed money brought in **Walter P. Chrysler**. He succeeded in guiding Maxwell and Chalmers through liquidation processes that ended with Chalmers being taken over by Maxwell in 1922.

During the First World War, **Hugh Chalmers** was head of the Chalkis Manufacturing Company, which produced anti-aircraft guns. He was also active in Washington helping to coordinate the military and naval work of the automotive industry. By 1921, as prospects for the Chalmers Company grew dim, he talked with **Roy Chapin** about a position with the National Chamber of Congress, but it didn't work out. He retired to the East with his wife, Frances Houser Chalmers, whom he married on August 22, 1901, and had five children. He died on June 2, 1932, in Beacon, N. Y.

First E. R. Thomas built in Detroit. Mr. and Mrs. George Grant drove it to the Vanderbuilt Cup Race in 1906.

The first Chalmers "30" produced was turned over to the Detroit Police and was in use for 20 hrs per day.

above- air view of the Chalmers factory, built by Albert Kahn. courtesy Kahn

Chassis assembling courtesy AAMA

Chalmers factory picture taken 1994

next page(128)- **top left**, Cliford Ely in a Chalmers   courtesy AAMA   **Top right**, tool kits were necessary   courtesy AAMA   **middle left**, 1910- Ty Cobb in his Chalmers for winning the American League batting championship award--Hugh became furious when he heard Cobb took it to Georgia and sold it.   courtesy AAMA   **middle right**, 1915 Thomas Flyer built in Buffalo, N.Y.--the last year of production   courtesy AAMA   **bottom left**, Hugh Chalmers

OAKLAND MOTOR CAR COMPANY
PONTIAC, MICHIGAN
1907-1931

PONTIAC MOTOR COMPANY
PONTIAC, MICHIGAN
1932-

"THE CAR WITH A CONSCIENCE"

In 1906, Edward M. Murphy wanted to expand his Pontiac Buggy Company and manufacture automobiles. He met **Alanson Brush** who was a consulting engineer in Detroit at the time. **Brush** had a two-cylinder engine with a CW rotation (for safety while self cranking) that Murphey was interested in. The Oakland Motor Car Company was organized in the summer of 1907 to produce the "Oakland" automobile. The Oakland was ready for the automobile shows in January 1908, although **Alanson** had already left and was working with **Frank Briscoe** manufacturing the "Brush" automobile. Sales of the Oakland in its first year was less than 300. Murphy added a four cylinder in 1909, and joined Durant's General Motors in April. Murphy died unexpectedly at the age of 44 in September 1909. The four cylinder was the only offering for 1910, and sales were up to 3000. The Oakland did very well on reliability runs and hill climbs, and it lived up to its slogan: The Car With A Conscience.

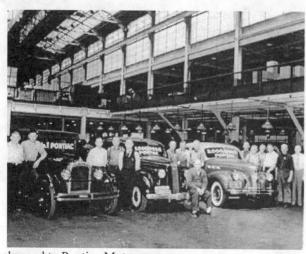

In 1926 a companion car to the Oakland was brought out called the "Pontiac," and marketed as a quality six-cylinder at a four-cylinder price. It was first shown at the New York Automobile Show in January. A sales meeting was held at the Commodore Hotel, which was renamed the Wigwam that day, and the conference was called the Pow Wow. The success of the Pontiac led GM to offer a companion car for Buick called the "Marquette," the "Viking" by Oldsmobile, and the "LaSalle" by Cadillac. Pontiac sold 188,000 cars compared to Oakland's 24,000 in 1930. The trend continued and the Oakland was discontinued in 1931. During 1932 the name of the Oakland Motor Company was changed to Pontiac Motor company.

The Oakland was named for the county in which it was built.

The Pontiac was named for the city in which it was built.

GRISWOLD MOTOR COMPANY
DETROIT, MICHIGAN
1907

J. P. La Vigne left the Detroit Automobile Manufacturing Company in late 1905. He started to design the "Griswold" in January 1906 with the support of a group of Detroit capitalists, who were favorably impressed with his scheme of vertical crankshaft and friction drive. The first car was on the road in May 1907. The Griswold Motor Company was organized in mid-summer 1907, headquartered at 521 Lincoln Avenue having a capital of $200,000. He leased space in the same factory, at Leib and Wight streets, where the **C. H. Blomstrom Manufacturing Company** was producing the Blomstrom "30."

The first models were driven by a pair of horizontal cylinders, set lengthwise of the car under the middle of the hood, with a vertical crankshaft and an aluminum alloy friction disc on the under side of the horizontal flywheel. The vertical placements of the crankshaft permitted the location of the traversing friction wheel in the middle, under the flywheel, so that the *line shaft* could be placed in the middle of the chassis frame, with a bevel gear to the divided rear axles, resulting in a symmetrical friction drive.

The Griswold had a tread of **55** inches and was offered as a 10 hp, 3-passenger runabout on a 90 inch wb; a 15 hp runabout on a 100 inch wb; and a 20 hp runabout on a 110 inch wb. They were priced around $1500.

The Griswold Company went out of business by the end of 1907 and was sold to the **C. H. Blomstrom Manufacturing Company**.

Griswold 3-passenger runabout

C. H. BLOMSTROM MANUFACTURING COMPANY
DETROIT, MICHIGAN
1906-1909

C. H. Blomstrom had severed his connection with the original C. H. Blomstrom Motor Company in August 1906, and began designing a five passenger touring car. He then formed the C. H. Blomstrom Manufacturing Company and moved to a commodious brick factory with three floors and a basement. It was located on the corner of Leib and Wight streets. Enough parts were initially ordered from outside factories to build 100 cars called the Blomstrom "30."

The engine was a water cooled four cylinder with a bore and stroke of 4.2 x 4.5 inches with 85 lbs compression, showing 30 hp at 1000 rpm. This gave enough horsepower to reach 50 mph easily on a fair road. It had Weston Mott axles with bevel gear drive. The rear wheels were on Hyatt rollers and the front wheels were on two point ball-bearings. A sliding gear was a selective type by Warner. The chassis frame was pressed steel four inches in depth supplied by the A. O. Smith Company.

In 1908 they planned to produce 200 units split between 125 touring models and 75 runabouts. Another car called the "Griswold," headed up by **J. P. LaVigne**, was being assembled in the same factory with the Blomstrom "30."

Blomstrom purchased the defunct Griswold Motor Company in late 1907. With that, he acquired the patent rights for an engine designed by **J. P. LaVigne** using a vertical crankshaft which was employed in the Griswold. On the basis that the vertical arrangement gave a gyroscopic effect, he soon renamed the car "Gryroscope." It was advertised as "the simplest and safest car on earth." The engine was started by a small friction clamp on the rim of the flywheel connected with a regular side lever, eliminating cranking. The driver stepped directly into the car, took his seat and pulled the lever to start the motor. The danger of *back kick* was removed by an interlocking device between the lever and the spark control members, preventing the operation of the lever when the spark was advanced.

The Blomstrom "30" was discontinued in 1908 with very few Gyroscopes produced. **Blomstrom** decided to sell the patented engines instead of automobiles. In mid 1909 he sold "gyroscope" engines to the Page Gas Engine Company in Adrian, Mich. After acquiring engines from **Blomstrom**, Page re-organized and became the Lion Motor Car Company and Blomstrom began shipping engines, but the Lion Company soon abandoned the gyroscope engine for a conventional four cylinder engine. The Lion automobile quickly was in production and sold seven cars in one week. They planned to build 20 cars per week using 175 men (mostly skilled workers). Every department was *up on its toes*. Then, in June of 1912, a major fire razed the factory. The town of Adrian tried to help but the Lion Motor Car Company was inadequately insured and went into receivership in December 1912. **A. O. Dunk** bought the assets saying he would build the Lion in Detroit. In 1913 **Blomstrom** sued for breach of contract and won since he had shipped a number of engines before Lion decided to use a conventional four cylinder.

The Blomstrom "30" was advertised as "The most for the money ever offered."

Chassis and a Gyroscope motor

Lion "40" in Adrian, Michigan

courtesy AAMA

Lion "40" front view

STREETCARS, BUSSES, AND SUBWAYS

In the mid-1800s, horses were used in cities to pull trolley cars. In 1872 a horse epidemic called the "Great Epizootic" spread from New York and worked its way westward. Hundreds of horses were dying each day, and thousands more were immobilized. Since steam propelled trains were commonplace, people began to think about mechanical propulsion for trolley use.

By 1890 horse cars were rapidly being replaced by electric trolley cars. The last horse drawn trolley car in Detroit ended in 1895. The last horse trolley in the United States was taken out of service in 1917 in New York. The last Detroit electric streetcar made its run on Woodward Avenue April 7, 1956. Detroit's streetcars were then shipped for use in Mexico City.

In 1907 Detroit's first municipally owned bus started service, to go across the Belle Isle Bridge. It was a "pay-as-you-enter." Two months later a bus with doors on one side only was in service and was "pay-after-you-get-in." Also that year, Detroit's **Mayor Codd** proposed an underground subway. With rock at least 100 feet below the surface or lower, it would be easy to build and would take about two years to complete. The system would have been built by the New York subway experts. Besides running strategically on and about Woodward Avenue, there was to be a belt line circling the entire city. An economic slump ended the effort.

1895- (September 15) The last horse-drawn streetcar in Detroit is seen at Brooklyn and Congress. Note the horse, train and bike. courtesy Bentley

1905- The pulling power of an automobile was demonstrated by the experimental department of the Olds Motor Works in Detroit. A 10 hp Oldsmobile pulled an electric street car with its trolley tied down. The route of the trolley was followed until the car was filled with passengers. There were 51 totaling 6885 lbs, and the trolley weighed 17,200 lbs, with extra friction from the drag of the electric motors.

undated-
Woodward Avenue
courtesy AAMA

1914-
Woodward Avenue

courtesy MSHC

1916-
Woodward Avenue

courtesy MSHC

TRAFFIC CONTROL

Even in the early 1890s, before the horseless carriage made its debut in Detroit, there were traffic problems downtown. People complained about traffic as draymen didn't seem to care who was in their way and bicyclists scurried back and forth at breakneck speed of 15 mph, sending pedestrians running. They were called "Scorchers." The police department organized a bicycle squad with handlebar mustaches, coal scuttle helmets, and frock coats, to pursue the scorchers and enforce the traffic laws. They were stationed at the main streets to help women and children cross the streets safely. Tom Reardon became Detroit's first traffic policeman and could be found at Woodward Avenue and Fort Street, where he watched City Hall and the Russel House, and controlled the flow of carriage traffic.

Because of traffic enforcement, the first noisy, smoke billowing horseless carriages in Detroit were usually driven on Belle Isle in the early morning hours to avoid scaring the horses. Mr. George Bissel, a prosperous lumberman, was Detroit's first automobile fatality on September 2, 1902, when his carriage was hit by a car at Brooklyn and Lysander. However, automobile accidents were still infrequent at the turn of the century, and next to the railroad train, the streetcar still was the biggest foe to life.

The first traffic ticket was claimed to have been issued in 1907, at the corner of Woodward Avenue and Adams Street by policemen on bicycles. The first traffic signal of the automobile era was installed in Detroit in 1910. It was a semaphore with signs saying stop and go, and operated by hand. The first stop sign was used in Detroit in 1914; also in 1914 Detroit passed new traffic regulations against crossing from one side of the street to the other in a diagonal course. This was called Jay-Walking, using the slang word "Jay" meaning newcomer, or inexperienced person. Violators were subject to arrest. The first lighted traffic signal was used in Detroit in 1915 at the intersection of W. Grand Blvd. and Second Street. Three-color lights, worked manually from a crow's nest, were set up in New York in 1918, and in Detroit in 1919. Detroit patrolman William Potts developed the first automatic traffic signal in 1920 at Woodward and Michigan Avenue, which was the busiest intersection in the United States. By 1933 Detroit had 933 automatic traffic signals.

The "Scorcher Patrol"
They were often stationed at the main streets to help women and children cross the street safely.
courtesy AAMA

*The first Detroit Traffic Division with their coal scuttle helmets and frock coats.
courtesy AAMA*

*1915-
The first traffic light. It was located on 2nd Avenue and Grand Boulavard in Detroit. A kerosene RR switch lantern was borrowed from the Michigan Central RR. It was on a stand, with two green and two red slides. It was turned by the police officer to direct and stop traffic.
courtesy AAMA*

*1916-
Traffic crows nest on Woodward
courtesy MSHC*

FEE & BOCK AUTOMOBILE COMPANY
DETROIT, MICHIGAN
1907-1908

Robert La Gora Fee was an eminently successful businessman in Detroit. He was born in New York in 1863 and went to public school in Binghamton. After learning the cigar trade, he moved to Detroit and went to work for the Detroit Cigar Company and later for the Banner Cigar Company as a traveling salesman.

In 1897 he began manufacturing cigars under the name of La Gora Fee Company. In 1901 he married Eva Austin Warren in Fowlerville, Mich.

The Fee & Bock Automobile Company was formed to sell Elmore cars that were produced in Clyde, Oh. Robert L. Fee was president, Walter W. Bock was secretary and treasurer, and the firm was located at 254 Jefferson. A member of the staff named Edward Zolle convinced Fee that they could produce a two cycle that could compete with the Elmore. As a result, the "Fee-American" was produced and sold for $1500. It had a two cylinder engine with 20 hp and used shaft drive.

A prototype was made in secret until it was shown at the February 1907 Automobile Show in Detroit. Limited production followed through 1908. The company was re-organized as the Fee-Vincent Electric Car Company with a garage at 344-346 Jefferson for battery charging. The company sold the Woods Electric, along with the Elmore and REO. Robert L. Fee died in 1917 of pneumonia.

1907- Fee-American
French Gray body with red running gear   courtesy NEWS

Robert L. Fee

REGAL MOTOR CAR COMPANY
DETROIT, MICHIGAN
1907-1918

In the fall of 1907 the Regal Motor Car Company was incorporated with a capital stock of $100,000. It was formed by Fred W. Haines, a prominent Detroit engineer and president of Regal, and the Lambert brothers: Jacob E., Charles R. and Bert, who were associated with the Clayton Lambert Manufacturer Company, producer of fire pots. The factory was located at the corner of Trombley and Beaubien. Nelson Clayton was president, Charles Lambert was VP, and Bert Lambert was treasurer. They hired Paul Arthur to design their car. The Clayton Lambert factory was secured at Beaubien and Trombley streets for construction of several dozen 1907 models. The medium priced car was well advertised and started off succesfully.

A new plant was built in 1908 at 201 Piquette and Woodward Ave. The Regal was advertised with a price of $1250 with a water-cooled four cylinder, with sliding gear and shaft drive, that could carry five passengers.

In 1913 the Regal Motor Car Company increased its capitalization to $3,000,000, with the management staying the same. The Regal shifted to left drive and came in three body types with underslung springs and one body type with overslung springs. The big feature was the Rushmore electric cranking and lighting. **Because the electric lighting could be dimmed for city driving, there were no side lights on the body.** The bodies were *brought up to the minute* with the cowls sloping into the body proper. The doors fit snugly to the body outline, and none of the hinges or handles were in view, creating a flush side effect. Regal was annually exporting over 1000 cars per year, including an English model called the Seabrook R.M.C. A new factory was built near Toronto, in Berlin, Ont., with a capacity of 1000 cars annually.

In 1914, 8250 square feet of floorspace was added to the Detroit building for new machinery to help cut production costs and add a high degree of mechanical accuracy. A Foote-Burt cylinder boring mill was added. It could bore four cylinders simultaneously, and was capable of turning out thirty castings per day. A Baush multiple drilling machine with twenty spindles could complete 50 crankcases per day. Piston work was accelerated by the use of a Potter & Johnson automatic chucking machine turning out pistons at a capacity of 100 per day. Four gear hobbers could turn out twenty sets of gears per day. In addition, radial and upright drills, turret lathes, engine lathes and cylinder grinders were procured along with an extensive tool room.

A single chassis of a new design was offered in 1915, selling at a reduced price of $1085. The new chassis marked the passage of the underslung construction by obtaining the same low center of gravity by the use of springs hung from the underside of the axle. **With the use of an Atwater Kent ignition system with automatic spark advance, the steering wheel mounted spark control was rendered unnecessary, while the control of the throttle was by foot pedal.** Both hands were free to steer and change gears. Another feature was the **radiator filler inside the hood**, becoming an integral part of the radiator outlet connection. **This prevented unsightly rusting of the radiator surface caused by overflowing water when filling.** With the same 3.75 x 5 bore and stroke as the previous year, the weight of the motor was reduced by 151 lbs. This was accomplished by better balance, lighter reciprocating parts, and the use of a steel stamped lower crankcase instead of a cast iron type. A V-8 engine was added to the light weight four cylinder, but sales volume was only 200 the first year.

Due to the onset of World War I, material shortage caused financial problems and the company went into receivership until the creditors voted to liquidate. By the summer of 1918 Maurice Rothschild purchased the Regal property and made spare parts.

In the summer of 1908 the Regal Plugger, a stock model 30 with 30 hp traveled across the country from New York to San Fransisco, then repeated it five more times before returning to Detroit with 22,000 miles.

courtesy Tinder

Offices of the Regal Motor Company.

courtesy Tinder

A portion of the million dollar Regal factory in Detroit.

ANDERSON CARRIAGE COMPANY
DETROIT, MICHIGAN
1907-1911

ANDERSON ELECTRIC CAR COMPANY
DETROIT, MICHIGAN
1911-1919

DETROIT ELECTRIC CAR COMPANY
DETROIT, MICHIGAN
1919-1939

Motors used for early electric automobiles were an outgrowth of street railways and other applications. Battery improvements were concomitant with advances in the vehicles. Tire problems were acute because of the heavy mass of the early batteries, but gradual improvements eliminated what was termed "impossibilities" in connection with electric vehicles. They sold well in the United States because wealth was concentrated in the cities. Because of the tall glass area, they were referred to as mobile china cabinets.

William C. Anderson was born in Milton, Ontario, in 1853. His family moved to Lexington, Mich. when he was a boy, and he attended public schools and business college. In 1884 he established the Anderson Carriage Company in Port Huron, Mich., where he manufactured carriages and farm implements. The Anderson Carriage Company moved to Detroit in 1895, and received financial help from department store magnate, Cyrenius A. Newcomb, who became VP until 1905. Other officers were William Locke, treasurer, and W. P. McFarlane, secretary. The factory was located at Riopelle and Clay streets, with a warehouse on 81-83 Jefferson Avenue. **Anderson** also formed the Anderson Manufacturing Company, and merged it with **William Pung's** Michigan Railway Supply Company to form the Pungs Anderson Carriage Company. It was eventually dissolved due to patent litigations against **Pungs** in 1906.

The Anderson Carriage Company produced 160,000 carriages before a decision was made to make automobiles. Most of the employees were discharged, machinery was thrown out, and carriage production was dropped altogether. The company officers deliberated on the type of car to produce, and felt the market was saturated with gasoline cars, and that an electric car had never been built in Detroit. They sought the best electric vehicle designer available and hired George M. Bacon. The first car was tested on average roads and achieved 140 miles on one charge. The first "Detroit Electric" was delivered in June 1907, and was followed by 125 more that year.

The heart of the Detroit Electric was a series wound, four-pole motor manufactured by the Elwell Parker Electric Company, in Cleveland, Ohio. It was mounted in the center of the car, and rated at four to five hp. The car was controlled by a pair of operating levers mounted on the side of the seat in parallel position. The longer lever was a tiller for steering, operated with the right hand. The shorter lever controlled the motor and was operated by the left hand. The drum-type, continuous torque, series controller provided five forward and five reverse speeds. A slight backward movement of the speed controller operated an external contracting brake at the front of the electric motor. A foot pedal operated internal expanding brakes on each rear wheel. A pressed steel frame was used with a conventional I-beam front axle, and a floating live rear axle with a 5 to 1 gear ratio.

Electric cars were among the first automobiles to use an enclosed body. All Detroit Electrics used an all-aluminum exterior body formed over white ash. Noise and vibration were more perceptible in the quiet running electrics. All moveable body panels were hinged the entire width like the lid of a piano to reduce noise. The doors opened foward instead of backward, offering the advantage that if the door were accidentally left open while the car was in motion, it could not be torn from the hinges. Every joint was shouldered, mortised, glued, and anchored with screws. A crimping process was used in its construction to reduce the number of rivets and bolts, and thereby reduced their inevitable noise.

In 1919 the Anderson Electric Car Company became the Detroit Electric Car Company, and moved its offices to a new building on Cass and Antoinette. In 1929 the stock market crashed in October, and Anderson died in November at age 76. Only 25 Detroit Electrics were built that year.

A. O. Dunk took over after Anderson's death, but sales continued to slide downward. By 1938 A. F. Renz was the head of the Detroit Electric Vehicle Manufacturing Company, making ten cars per year in Detroit's smallest automobile factory at 731 Tenth Street. He predicted that the electric automobile might stage a comeback in 20 to 50 years. He said it would be due to a depletion of gasoline reserves and increased development of cheap electricity.

Interior of the Anderson carriage factory.

courtesy Burton

Anderson Electric vehicle factory under construction.

courtesy Burton

above- Anderson factory in 1912.

right- Anderson factory picture taken in 1995

Dr. Charles P. Steinmetz, chief consulting engineer of the General Electric Company (man on the right, with his hand on the car) purchased a 1915 Detroit Electric Broughham Pullman. (He produced the Stienmetz Electric delivery truck from 1920 to 1927.)
courtesy AAMA

Charging station "A" (B and C were in other parts of the city) was located on Woodward Avenue. It was 192 feet x 90 feet, with two floors. Over 100 cars could be charged at once. After washing, polishing, and checking every mechanical device, a "chaser" (as he was known in the vernacular of the electric garage) donned a pair of white gloves so that any grease or perspiration on his hands would not dirty the controller or steering lever, and delivered vehicles to their owners.
courtesy Burton

*1913-
Randolph Street, between Gratiot and Monroe. A Detroit Electric can be seen to the right.*
courtesy AAMA

John W. Anderson

Thomas Edison gives a Detroit Electric the "once-over"

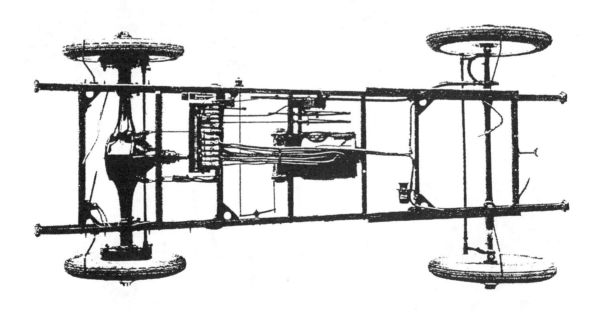

The Detroit Electric Society's Town Car

Clear Vision Brougham   Model 42

Why Not Buy a DETROIT ELECTRIC?

That is a fair question, is it not? Let us fairly answer it together by analyzing your automobile needs.

Won't at least 90% of your driving be done in the city and its environs? Admit that you may (?) do some touring, you can well afford to hire a car for this purpose and then be money ahead at the end of the year, but that is not the big point.

For *all* occasions—business and formal—in all kinds of weather, the Detroit Electric Clear Vision Brougham will meet all of your requirements *economically* with *every luxury* of the most expensive limousine, minus the necessity and inconvenience of a chauffeur. It has all the mileage you can possibly need. It can climb any hill that any other car of any type will climb.

If you wish to install your own charging set, the current will cost you less than any other form of energy. Oil is not a big item as there is no complex mechanism requiring it. If you wish to have your car garaged—called for and delivered, washed, adjusted and charged—you can have this done for a nominal charge of from $35 to $40 per month, which *includes the cost of the current used in charging*.

What other type of car can offer you any such luxurious service at anywhere near the price?

Then there is the tire question. In the Detroit Electric you get cushion tires *guaranteed* for 10,000 *miles*. They will always bring you home.

Another point—when you buy a Detroit Electric there is no *extra* equipment problem. It is electrically started and electrically lighted. It has *all* the advantages of electricity—not one or two of them.

You will appreciate the Detroit Electric in congested traffic. All speeds are instantly available with *one* lever. There is no possible chance for your motor to stop when changing speeds.

Quick "get-away" is a *pleasure*.

Our new 1913 catalog has just arrived. It will be sent upon request. Our new model 42 Clear Vision Brougham for 1913 is now here. Let us take you for a ride in this car over any roads or hills that you may select.

ANDERSON ELECTRIC CAR COMPANY
416 Clay Avenue   DETROIT, U. S. A.

BRANCHES:
New York—Broadway at 80th St.
Chicago—2416 Michigan Ave.

Boston
Brooklyn
Buffalo

Cleveland
Evanston
Kansas City
Minneapolis

THE HOTEL PONCHARTRAIN

The National Hotel was built on the corner of Cadillac Square in 1836. It was replaced by the Russel House in 1857, which became the scene of almost every social event of any importance in the city's life. There were statesmen, theatrical figures, and businessmen among the guests. It even entertained royalty; the Prince of Wales and later King Edward VII in 1859. The Russell House was replaced by the Ponchartrain in 1907, probably the most famous of all Detroit hotels.

The history of the automobile industry might almost be said to have been cradled in the taproom of the Hotel Ponchartrain. It had a high dark ceiling, gleaming mahogany, leather cushioned cozy corners and booths and a long bar with an oil painting of Count Ponchartrain on the south wall. It was a room of pleasant and comforting shadows and quiet lights. The main entrance was through a narrow hallway that contained telephone booths and led from the main lobby. It was the center for business men and women. Investors, inventors, speculators, promoters, engineers, writers, advertising men, capitalists, and near capitalists. It became known throughout the world that one who wished to get into the automobile game must go to Detroit, and once in Detroit must stop at or at least hang around the Ponchartrain. Sooner or later everyone came to the Ponchartrain, whether they were teetotalers such as **Henry Ford**, occasional drinkers, or old hands at the rail like **Louis Cheverolet**, **Walter Flanders** and the **Dodge brothers**. They came to bring their own gadgets, to see what the others were making, and to swap ideas. No one seemed to think of money except as something that was used to build an automobile. It was not an uncommon sight to see four or five men carry a piece of heavy machinery into the room, place it on the floor or table and set it in motion. **Albert Champion**, newly arrived from France, walked in with an elaborate electric set and showed a group of strangers his "supérieur" porcelain for spark plugs. Tire vulcanizers, rims, valves, brakes, carburetors, magnetos, were shown off in the Ponchartrain bar. There, people began to speak a strange new language. It became the world's first automotive demonstration room.

One day **E. LeRoy Pelletier**, the leading publicist and Ford's first advertising manager, told William Chittenden, the manager of the Ponchartrain: "If I were you, Bill, I would take up that Turkish rug which covers the floor of your lobby. Then I would buy myself some red paint, the reddest I could buy: then I would have a big circle drawn in red paint around the whole lobby, and within that circle I would paint this legend: "The Heart of the Auto Industry!" Chittenden thanked **Pelletier** and walked away, quickly.

The world will never know how many giant corporations were formed in the smoke clouds of the old tap room, how many millions were passed between hands that rested on the tables amid the damp rings of beer or highball glasses, how many fortunes were made and how many failures were born. It was a stock exchange and the spirit of the bonanza was in the air. Many of the transactions were made with all the careless disregard of the great events that once marked the trading in gold mine expectations, and stocks worth millions were traded between drinks. Very often as the evening advanced and the glow of many glasses warmed their spirit, bets were made on various things, the turn of the dice, the date of a patent, the record of a racer, and many of these bets were made in stock. It was not uncommon to see a man with a glum and serious face then an hour later half mad with delight as some unexpected turn had given him a "closed deal."

This stock for that, a dollar one way or another on the quotations meant nothing. One man threw a bundle of stock certificates on the floor under his safe following a trade and it lay for a year until its original price of $12 had leaped to over $100, and later the bundle became the body of a great fortune.

Another man who had agreed to buy certain stock could not find his fountain pen and was too lazy to go to the hotel desk. He agreed to take the stock "the next time I see you." He didn't see him before the stock was sold to someone else and it rose in value from $700 to $89,000.

Another man went to sleep and let a bundle of stock certificates fall on the floor. It was found by the sweeper and was laid in the porter's room on a shelf for three days. Their value was enough to make a person independent for life.

The Ponchartrain was razed in 1920 and replaced by the First National Bank. Its demise was due to the opening of the Detroit Athletic Club in 1915, which became the new center for the automobile industry, and prohibition.

The Ponchartrain Hotel-Note soldiers monument is to the left. An arc-light tower is in the foreground. At one time there were 1800 lighting towers from 75ft to 150 ft tall in Detroit.

The Bagley monument is seen in the foreground, to the right, in this view.

The lobby of the Ponchartrain.

MANUFACTURED VERSUS ASSEMBLED CARS

When the industry first started, automobile companies followed the tradition of the buggy companies and manufactured the complete car themselves, using artisans (craftsmen) to custom make all of the parts. By 1906 there were typically more than 30 different trades pursued in each automobile factory, and a number of skilled help in clerical and office positions. The trades included designers, draughtsmen, chemists, pattern makers, foundrymen, molders, coremakers, sheetmetal workers, carpenters, tool makers, general machinists, blacksmiths, brass finishers, assemblers, platers, grinders, upholsters, wheelwrights, painters, decorators, furnishers, electricians, fireman, and about a dozen others.

Automobile companies continued to make the public think they manufactured every part, and yet they recognized that quantity production would lead to less cost and higher quality. Thus many standard parts were purchased that were *cast* with the car companies' name to help in the deception. Usually, the more expensive cars maintained the in-house manufactured image to the greatest extent, Packard being a good example.

After about 1908, most automobile manufacturers made no secret that they used parts from well known parts suppliers, and by 1916 "assembled" was no longer a stigma.

MASS PRODUCTION

Manufacturing is the operation of making a standardized product for a general market. It is derived from the Latin "manus," hand, and "factus," made.

Mass Production is a productive organization that delivers in continuous quantities a useful commodity of standard material, workmanship and design at minimum cost.

Henry Ford developed a system of **progressive assembly** of automobiles which made **mass production** possible. **Progressive assembly** was used for making railroad cars, bicycles, and sewing machines in the 1800s. What really made the manufacturing of the automobile different was that it contained thousands of parts instead of hundreds, and that it had many reciprocating parts that travel at a higher speed compared to other products. Consequently, many of the cars parts required both precision and tensile strength. These requirements were achieved with new and innovative machine tools and new materials.

A key factor for **mass production** was the use of **interchangeable parts**, as first developed by Eli Whitney. Simeon North was the first producer of guns to take advantage of the concept in 1813. Later, factories making railroad cars, locomotives, watches, sewing machines, agricultural implements, and bicycles used interchangeable parts. In the automotive industry, **Henry Leland** was recognized for his work with interchangeable parts used to manufacture Cadillac automobiles. His training came from the small arms industry. At Ford, **Walter Flanders** brought similar technology from his work manufacturing Singer Sewing Machines. Perhaps the most noteworthy contribution to the practice of making **interchangeable parts** was developed by **Carl Edward Johansson**, who was called the "Edison of Sweden." In 1901 he pondered over the inaccurate methods of measuring and gauging then in practice. At the time the micrometer was used, but it was only accurate down to a thousandth of an inch, mainly because its accuracy depended on a screw and nobody could make a perfect screw. He spent 10 years to develop a set of 81 standard gauge blocks made in sizes marked to ten-thousandths of an inch that were so smooth and parallel, that when held together, they acted like they were magnetized (they were later called Jo-blocks). Two things came out of this. First, a system of producing gauge blocks so accurate that 34 of them totaling four inches could be stacked together and compared to a solid four inch block without any discernable difference. And second, a steel so seasoned that it would hold its accuracy. Plain steel, without seasoning, would change much in the same manner as wood; expanding, contracting, and twisting.

The origin of the **conveyor line** dates back to the early 1800s in stockyards where animals were strung up on an overhead trolley to be dressed. Although the animals were disassembled and the reverse procedure was used for assembly of automobiles.

ALBERT KAHN

Mills used for splitting grain using human and animal power date back 5000 years and are considered to be the forerunners of today's factories. Gradually, mills were developed with strong timber frames clad with weather boarding, and were in proximity to running water, which meant that they were often built on marshes using timber pile foundations. These mills were subject to wetrot, and that eventually led to brick construction.

Before the 18th century, large buildings with a substantial number of workers began to appear, and were called factories. **The term factory implies a certain level of organization.** The first evidence of this can be identified with the monastic institutions of rigidly structured monks being allocated specific tasks. Factories were usually multi-story, with an internal structure of wooden beams, joists, columns and trusses. Spinning machinery would be positioned on the top floor, and the lower floors contained other manufacturing activities requiring mechanical power from horizontal shafting driven by a water wheel, which was limited to about 20 hp. Steam power was revolutionary since the factory was no longer tied to a site near the water. Since these factories were lighted by naked flames, flammable lubricants used on the machinery created a great fire hazard.

At the beginning of the 20th century the "mill-type" factory was in common usage. Most New England textile mills were based on a rectangular structural grid of between 9 and 12 feet, and were four to seven stories high. The need for line shafts, natural lighting, and ventilation, kept the proportions long and narrow. These buildings had brick bearing walls with timber columns, girders, and hardwood flooring over joists. Hardwood was economical and readily obtainable. After several serious fires, the so called "slow-burning mill" construction was developed, which called for all timber to be planed and free from sharp edges, and substituted timber beams that were four to six feet apart with planed and splined plank underfloor and a hardwood wearing floor.

Next, the wooden columns were substituted by iron "Lally" columns, then steel was substituted for the columns, girders, and beams. The use of steel made it possible to construct higher buildings, and the use of the bearing wall gradually gave way to the skeleton type.

The early factories were usually designed around the powerhouse and machine shop that were flanked by one or two story wooden buildings. These were constructed to protect the machinery which they contained from the elements, rather than the workers at the machines. No thought was given to expansion, and if the business increased, lean-to after lean-to was added to the original factory. This resulted in long rows of straggling wooden buildings built to no plan, without hope for efficient work processing. Above all, the threat of fire was so great, that it became an impetus for better construction. Concrete was tried for building construction, but **would often crumble**, and was considered a **poor building material**.

Detroit, along with the rest of the United States, was plagued by fires, and the loss of life among workers on upper floors was enormous. In 1895 the explosion of a steam boiler in downtown Detroit set fire to two adjacent buildings, and trapped 50 people. Thirty workers lost their lives in a factory explosion in 1901. Some burned to death, while others drowned in the basement during the firefighters all-night battle against the flames. Part of the Oldsmobile factory was destroyed only two years after it was constructed. Insurance companies constantly increased their rates.

Albert Kahn was born in 1869 in the Rhineland, the eldest of six children. His father was a rabbi as well as a peddler. **Albert** immigrated to the United States at age 11 and with a sound basic education from Germany, never returned to school. In Detroit, Albert's father became a peddler of fruits and vegetables, while his mother cooked in a small restaurant near the Michigan Central railroad station. **Albert** worked as a busboy in the restaurant at night, and afterwards helped his father groom the family horse. He was also an office boy for an architect; filing prints, grinding India ink, and running errands for no pay.

The Kahn family lived in a house without running water, and he was limited to Saturday night baths. The combination of the kitchen and horse work caught up to him at the office, and in 1883 he was fired for being offensively malodorous. Albert's mother arranged for him to take drawing lessons from Julius Melchers, an artist and sculptor. In turn, Melcher introduced **Albert** to his good friend, **George D. Mason**, who years later designed the Detroit Masonic Temple. Mason hired **Albert** as a draftsman, but the head draftsman soon detected that **Albert** was color blind. He was marched into Mason's office and asked to identify colors on a rug. He identified reds, yellows, and guessed right on green. He later joked that "If I had guessed brown, I might be a butcher today." He received his first significant assignment in 1887 to design the world's longest wooden porch, measuring 660 feet, for the Grand Hotel on Mackinac Island. In 1890 **Kahn** won the American Architect's traveling scholarship to Europe. His companion was Henry Bacon, Jr., who later designed the Lincoln Memorial.

Kahn set up his own office in Detroit in 1895 and was carried upward in the industrial boom. **Albert** married Ernestine Krolik, the daughter of a client, in 1896. He accepted an offbeat assignment and designed his first industrial work in 1900, a small mill construction on Second Avenue in Detroit for **Joseph Boyer**, manufacturer of pneumatic hammers. His younger brother **Julius** was studying engineering at the University of Michigan and asked **Albert** how he calculated the strength of concrete. **Albert** said that he did not and that it was a matter of knack and intuition. **Julius** was bemused and began to devise a means of calculating the shear force and settling of concrete. He decided that concrete could be improved by reinforcing it, and developed the "Kahn bar." This was a continuous bar to resist tension along the bottom of a beam or girder with frequent projections upward and outward toward the ends to counter diagonal tension stresses. It had a square cross section, placed on edge, with continuous shear reinforcing wings to either side. They could be bent upward at 45 degrees as needed, and acted approximately parallel to the forces tending to cause concrete to fracture.

In 1902 the brothers were commissioned to design a new engineering building for the University of Michigan in which they used the Kahn bar reinforcement. Another key event in Albert's career occurred in 1902 when **Joseph Boyer** introduced him to **Henry B. Joy**. When the Packard Motor Company moved from Warren, Ohio to Detroit, Mich., **Joy** called upon **Albert Kahn** to design Packard factories. Up to that time engineers designed factories, not architects. **Kahn** took a career gamble in 1903 and started on "building number one" by designing a big wood-framed mill, but with better light than most. It was reported as one of the new style factories. By 1905 **Kahn** was designing the 10th Packard factory. By then, enormous and heavy machinery vibrated timber frames, and fires were a constant threat. Crammed assembly halls were dangerous and inefficient. Later in his career he said, "where the problem is new, a new solution must be found to fit." Building number 10 became the world's first reinforced concrete factory. Steel reinforcing bars were supplied by the Truscon Steel Company, owned by **Julius Kahn.** Concrete was used for floors, columns and beams, and the outside was dressed up with paving brick. To allow more natural light, large steel sashes were used. **Henry Ford** took note that they required minimal upkeep and were durable. Building 10 became a tourist attraction and propelled Albert's career.

Kahn's next reinforced concrete factory was designed for the George N. Pierce Company, in Buffalo, N. Y., in 1906. The structural work covered 280,000 square feet and was completed in six months, exactly half the time it would have taken contractors using the traditional forms of mill-type construction. The plant consisted of eight mostly single-story structures that were ideal for continuous-flow processing. Raw materials could be made into parts, and parts assembled into cars all on one level. The layout's potential escaped George Pierce and his workers continued to build the old way with excess motion, crowded assembly rooms, resulting in high costs.

The first concrete office building in Detroit was under construction in 1907 at Lafayette Avenue and Wayne Street, called the "Trussed Concrete Building," using the Kahn trussed bar. The Packard and Pierce work led to the selection of Kahn by **Henry Ford** in mid-1908, following the introduction of the Model T. **Henry** could see that the Piquette plant was too small and too inefficient to produce the volumes that would be required. "Mr. Kahn, can you build factories?" Ford asked. "I can build anything" was Kahn's reply. When **Albert** went with **Henry** to the site of an old racetrack on Woodward Avenue, he had doubts. **Ford** wanted to build a massive factory to consolidate the entire manufacture of the Model T under one roof. Later **Albert** said, "I thought he was crazy. I didn't think it was possible, but I didn't want to tell him that. I wanted to see what could be done." **Ford** made some rough drawings and said, "you've got only part of the idea." **Albert** suggested centering the building in the middle of the 52 acres, but **Henry wanted to** keep it near the street to allow room for expansion. It was then that **Albert** began to grasp the magnitude of Ford's conception.

Besides erecting virtually all of Ford's buildings, he designed for Hudson, Chalmers, Paige-Detroit, Cadillac, Maxwell, Anderson, Chrysler, and over 150 for General Motors. **Kahn** also designed the Warren Tank Plant, and the Willow Run bomber plant. He was involved with over 2000 factories worldwide. **Albert Kahn** had a third heart attack on December 8, 1942, and died at the age of 73.

KAHN SYSTEM
of Reinforced Concrete

Specify Kahn System. You will if you investigate. The United States Government has given substantial endorsement to the Kahn System in over 20 buildings erected at West Point, Annapolis, and Washington.

The Main Building of the E. R. Thomas Detroit Co. Factory. Large Power Plant not shown. The Entire Plant was Built Complete in 65 Days.

The E. R. Thomas Detroit Plant
BUILT COMPLETE IN 65 DAYS

Kahn at 71 years

FORD HIGHLAND PARK FACTORY

"IF THE MODEL T DOESN'T RUN, CHECK THE GAS TANK"
-Henry Ford

The site for the new factory was served by the Detroit Terminal Railway, a beltline connected to every railway entering Detroit. **Ford** was a charter member of the Wayne County roads commision, established in 1906. He got the county to pave Woodward Avenue from the Ford plant to Pontiac, Mich. The construction of the plant, as well as the roadway, owed a great deal to **Ford's** friend and idol, **Thomas Edison**. For an iron mining venture, **Edison** had devised giant rock crushers to make "little ones out of big ones" and other machinery. Although the venture was a disaster, the equipment was perfectly adaptable to making cement, and in 1907 **Edison** incorporated the Edison Portland Cement Company. Adopting Edison's machinery, **Albert Kahn** and the road builders began exploiting Michigan's vast deposits of marl, gypsum, and lime, and a new industry was born. **Ford** persuaded the water commision to run a pipeline to the site, but it was beyond the reaches of the Detroit Edison Company to supply electric power. **Ford** used his own electrical experience and erected a power plant capable of supplying a city of a quarter million people, rivaling Detroit Edison as the largest producer of electricity in the state. A five-stacked power plant was connected to **Ford's** office by an overhead walkway, and he would often kibitz with the watchmen and engineers.

The original unit of the plant fronted Woodward Avenue and Manchester Street. **Kahn** wanted to position the plant in the middle of the property, but **Ford** insisted to start at one end in order to accomodate expansion. As it turned out, production roughly doubled each year from 1908 to 1913, which required continual expansion through 1914. The first building had four stories, and was 860 feet long and 75 feet in depth. Columns were spaced at 20-foot intervals along the entire length. Three 25-foot bays made up the 75-foot depth. The sash was made of steel, imported from England, and of maximal size uninterrupted by any major mullion subdivisions, so that it gave the impression of being a taut skin or screen. The expansive glasswork gave rise to the nickname of "The Crystal Palace."

Ford wanted to consolidate all operations under one roof, and all major operations on one floor. At the Piquette plant, **Ford** began to use gravity chutes to transfer some parts and materials from one operation to onother. He was counting on a highly systematized and organized work process with maximum utilization of gravity for transport of raw materials, parts, and subassemblies, and created the need for the multi-storied Highland Park factory.

In 1912 **Ford** reduced the work week from ten down to nine hours per day, and five hours on Saturday. (In April, the Farrand Company of Detroit was first to follow **Ford's** lead with a work week of 50 hours. "The change was made to convince the employee that we are not determined to screw him down to the last notch.") **Ford** made 75,000 cars in the 1912 season that required 150,000 animals skinned to make seat upholstery. There was already thought being given to a substitute for leather. As production demands increased, **Ford** struggled to keep up with the sales demand for the Model T. In the summer of 1913 **Charles Sorensen** had a chassis pulled slowly by a windlass across 250 feet, timing the process all the while. Behind him walked six workers, picking up parts from carefully spaced piles on the floor and fitting them to the chassis. From this experiment grew what would be termed "mass production."

At this time period there was a backlog of 100,000 orders. With the steadily increasing production volume, **Wills**, **Couzens**, and other executives met with **Ford** to discuss scrapping the two nine-hour shift day and operate around the clock. Actually, workers had been fighting unsuccesfully for an eight-hour day for over a half century, and now **Ford** came to the only logical conclusion, to divide the day into three eight-hour shifts. On Monday, January 5, 1914, The Ford Motor Company announced the working day was cut from nine hours per day to eight hours per day, and a minimum wage of $5.00 per day at a time when factory workers were getting $2.34 per hour.

As production volume increased, the factory was always oversold. No car was ever stored for more than a day, simply because there was no place to store that many cars. Eventually, the railroads could not handle the volume. The solution was to ship knockdowns to satellite plants and let them assemble the complete car. The first plant of this nature was in Kansas City, Mo.

Henry Ford is quoted as saying that you could have any color as long as it was black. The reason was that the typical varnishes took weeks to finish, and **Ford** was forced to use the jappaning process that only came in black.

*Nearly completed-
The Highland Park
Assembly Plant is seen
along Woodward Ave.
Note the two story
office building
to the left.*
   courtesy KAHN

*Four gas/steam generators
at 5000 hp each were installed
in the power house. Heat waste
from the gas cylinders made
boiling water to drive steam
pistons. When the Dodge Bros.
terminated on July 1, 1914 all
machine work in connection
with building the Model T was
done in the Ford Highland Park
Plant. The gas/steam generators
were replaced with 7000 hp units
of 3750 kw each.*
   courtesy Lodge

*The power plant,
with five smoke stacks can be seen
to the left in this view. Note the office
building has four floors.
courtesy Tinder*

This was the largest automobile plant in the world. In the foreground, people are lined up to ride the street cars. Although many tried, being late for work because of a street car was never acceptable.
courtesy MSHC

The Highland Park State Bank is seen to the right facing Woodward Avenue and Manchester Street. It was founded in 1909 as the richest institution in Michigan in proportion to the size of the community. James Couzens was president, F. L. Klingensmith was VP and Henry Ford was also associated. In 1911, Flanders, Rapid, Oakland, Cartercar and Welch began paying employees by check. Henry wasn't interested in it and continued paying cash. When the payroll grew to extreme proportions, he had a tunnel dug between the bank and his paymaster at the plant. A hand-operated winch and a cash box was used to transport the money.
courtesy MSHC

There were also buildings perpendicular to Woodward Avenue with spacing in between them. A craneway between each pair of buildingss brought supplies from RR cars below and delivered them to galleries on either side of the craneway. The design of the building was such that raw material was hoisted as near to the roof as possible to let work down in the proccess of manufacture. Thousands of holes were cut through the floors so the parts gravitated down, through chutes, conveyors, or tubes.
courtesy KAHN

The completed buildings, including the expansion as Ford explained to Albert Kahn. The newest section can be identified by the windowed wall, parallel to Woodward, and behind the original structure. The factory was three times the original size, 65 acres larger, with an output three times greater. (In between the old and new buildings was John R. Street, where the bodies were taken outside onto a ramp and dropped onto a chassis.) *courtesy KAHN*

(The first "assembly line" used by Ford was used for making magnetoes, similar to the Ed Huff design.) For chassis production, conveyor number "one" is seen here.

circa 1914-
Chassis production in the new building. Note that Kahn innovated and provided heating ducts and registers inside of the structural cement columns.
         courtesy AAMA

John R. Street looking southeast- underneath the body chute.

Body Chute on John R Street looking Northwest- an endless-belt controlled by the man at the extreme right of the platform. Two handlers at the top of platform to left of lever-man ready to start next body down chute. Near the foot of the chute, under the rocking gallows frame, two men make slings hung from the gallows cross-bar fast to the body. The man to the right controls the movement of the gallows frame. The man to the left directs the body when launched. courtesy KAHN

John R Street looking northwest. Same side of building where body chute was located (beyond the bridge.)
    picture taken 1994

EVERITT - METZGER - FLANDERS COMPANY
DETROIT, MICHIGAN
1908-1912

FLANDERS MOTOR CAR COMPANY
DETROIT, MICHIGAN
1909-1912

STUDEBAKER AUTOMOBILE COMPANY
SOUTH BEND, INDIANA
1902-1912

STUDEBAKER CORPORATION
SOUTHBEND, INDIANA
1912-1954

E. Byron F. Everitt was born in Ridgetown, Ontario, Canada, in May of 1872. He quit school early and worked as a carriage maker until he was 19. He moved to Detroit and worked on carriages for Hugh Johnson for two years, before accepting a job as manager at the C. R.. Wilson Carriage Company. After seven years with Wilson, **"Barney"** formed his own buggy and wagon business at Brush and Woodbridge. Later that year, the Olds Motor Works factory was erected just a few miles away from **Everitt**. When fire struck the Oldsmobile factory in 1901, and the Olds Company quickly began purchasing components from local suppliers, **Everitt** supplied bodies for the curved dash runabout. Sales of the runabout bodies exceeded **Everitt's** resources, and he enlisted **Fred** and **Charles Fisher**, from Norwalk, Ohio to manage the Oldsmobile runabout body production. When the Oldsmobile Company shifted production of the runabouts to Lansing, contracts with Detroit body suppliers were terminated to save shipping costs.

In 1903, **Everitt** helped in the formation of the Wayne Automobile Company, serving as general manager, while at the same time, he was supplying bodies to the Ford Motor Company. In 1904, **Everitt** hired **Walter O. Briggs**, and made him general manager of the Everitt Company. **Briggs** had been a switchman for the Michigan Central Railroad, then became a foreman of the body operations at the C. H. Little Company. In 1907, **Everitt** moved his operations to Clay and Dequindre, near the Ford factory on Piquette. In 1908, the management of the Wayne Automobile Company went through a change and Everitt became president.

M. William E. Metzger was born in Peru, Illinois, and moved to Michigan in 1878 when he was ten years old. He became a successful bicycle merchant with a shop in Detroit on Woodward Avenue. **His automobile career began when he traveled to London, England in 1895 to attend the world's first automobile show. He helped organize the first automobile show in the United States in 1900 at Madison Square Garden.** In 1901, Metzger established an *automobile store in the Biddle House*, in Detroit, and sold a batch of electric automobiles, followed with steam powered automobiles. **This was probably the first automobile dealership in the United States,** making "Smiling Bill" a pioneer in automobile retail industry. He also helped organize the first automobile race in the midwest, at **Grosse Pointe** in 1901, and the first automobile show in Detroit, although it was combined with a dog show.

In 1902, Automobile Row, as it was called, consisted of four dealerships, which was enough for the Tri-state association to hold a show exclusively for motor vehicles. **It was from this show that Detroit automobile dealers count their beginnings.** Of the cars on display, **Metzger** was the Detroit representative for the Waverly, Baker and Columbia electric's, the Mobile and Pope Toledo steam cars, and the Winton, Olds, Knox, and Silent Northern gasoline cars. He had the largest dealership in the United States. Among his staff were William Nuemann, William Rush, **James J. Brady**, Frank Riggs, William

States. Among his staff were William Nuemann, William Rush, **James J. Brady**, Frank Riggs, William Hurlburt, Joseph Schulte, and Walter Bemb. Walter was in the service department and his principal duty was to go around the town and retrieve the electric's which failed because their owners let the batteries run down. A dead battery was a substantial alibi when the owner failed to arrive home on time.

Business outgrew Metzger's quarters, and in 1902 he entered into a lease with the **Joy** estate for a new two story building on Jefferson Avenue. In 1903, **Albert Kahn**, of Mason & Kahn designed a four story addition of slow burning construction with paving bricks used for the outside walls, with stone trimmings and copper cornice. He sold Baker and Waverly electrics, and Pope Toledo, Orient, Northern, and Cadillac (he became the sales manager of Cadillac Motor Company) gasoline machines. He paid cash for his cars and was paid only after they were purchased.

F. Walter E. Flanders was born in 1871 in Rutland, Vermont, the son of a country doctor. He left school at age 15 to work on machinery, and served an apprenticeship at the Singer Sewing Machine Company before leaving for Cleveland, Ohio, to work for Thomas S. Walburn in general machining. An order came from **Henry Ford** for a thousand crankshafts that **Flanders** filled with high quality and delivered on time. **Ford** hired **Flanders** as his first production manager. **Flanders** was the best when it came to arranging production machinery, and setting up procedures and time saving methods. He developed various multiple drills, a vertial boring mill, and valve grinding machines. **Walt "Bullneck" Flanders** was huge at 275 lbs, with a loud voice that **Charlie Sorensen** said, "could be heard in a drop-forge plant."

Flanders was admired by the men working for him, but his personality was in conflict with **Henry Ford**. Besides, he wanted to have a larger role in the fledgling automobile industry. He left the Ford Motor Company in March 1908, with the position of general manager at the Wayne Automobile Company waiting for him. Wayne spokesmen were quoted as saying that plans were proceeding for production on a large scale. **William Metzger** was retired as sales manager of the Cadillac Motor Car Company, and was invited to join **Everitt** and **Flanders** at the Wayne Motor Company. **Flanders** wanted to produce a mid-priced automobile, and said survival demanded a capital investment of two million dollars. **Everitt, Metzger, William Barbour**, J. B. Gunderson, Charles Palm, and **J. B. Book** provided the capital. Metzger was involved with the Northern Motor Car Company and quickly gained control and merged it with the Wayne Automobile Company.

The formation of the E.M.F. Company was formally announced on June 2, 1908, with three ready made plants: the Wayne and Northern factories in Detroit, and the Northern factory in Port Huron, Michigan. **Everitt** was made president, Palms was the treasurer, **Metzger** was the sales manager, and **Flanders** was the general manager. The chief engineer was **William E. Kelly**, who had built his first automobile in 1895, and designed the Wayne.

An alliance was formed with the Studebaker Company, of South Bend, Indiana, to use the 4000 Studebaker agencies for distribution of the E.M.F. Each company would distribute half of the production. Studebaker sold under the name "E.M.F. Studebaker" and took the south and west regions of the United States, and overseas, markets. The E.M.F. Company, under **Bill Metzger**, took the east and north sales regions of the United States.

The E.M.F. "30" had a 30 hp four cylinder engine with a 4 x 4.5 bore and stroke, with 226 cubic inches. The cylinders were cast in pairs and had integral water jackets and large mechanically operated valves. The valve guides were made of steel, and press fit to make it easy to service when worn. A splash lubrication system was governed by an automatic vacuum feed. One filling of the oil reservoir was good for 300 to 500 miles, depending on the road conditions. A single float feed carburetor was adjustable from the driver's seat, and a dual jump spark ignition system consisted of a quadruple coil, commentator and magneto built into the engine. Thermo-syphon cooling was initially used, but after experiencing overheating, **Kelly** personally recalled the vehicles and installed water pumps. (The Ford Model T was introduced at the same time as the E.M.F., with a water pump, but Ford soon changed to thermo-syphon cooling.) The transmission was a three speed sliding gear type, housed in the rear axle. The frame was U-channel pressed steel construction, with semi-elliptic front springs and full elliptic in the rear. The steering was worm and sector and the clutch was an expanding ring.

In early 1909, the E.M.F. provided the "pathfinding" for the Detroit-to-Denver Glidden tour. The roads were in extremely poor condition, however the car made it and received good publicity. When the trip was over in April, E.M.F.'s advertising manager said it wasn't necessary to fake hardships, because there were plenty. *The steering never failed, so she wasn't ditched. The brakes never failed to hold, so she always stopped when necessary and didn't bump the lamps or the radiator. The radiator was properly suspended, so it didn't spring a leak. Fenders were properly made and attached, so they were still in place as on the day she started out. And the car was washed every night so it would look classy like an E.M.F. should.* The AAA contest boards "official pathfinder," Dai Lewis, had been skeptical about starting on such a trip in a car that was new to the market, but afterwards he said the E.M.F. "30" was a wonderful car!

The "big three" began quarreling over Studebaker: **Metzger** never thought the Studebaker alliance was a good idea from the start, and he got **Everitt** to agree. **Metzger** always expected to manage the entire sales and distribution. In May 1909, **Metzger** and **Everitt** sold their shares for $360,000 to Studebaker, which took over all of the distribution. They took **William Kelly** and proceeded to form the **Metzger Motor Company** to produce an automobile called the "Everitt."

After **Barney Everitt** and **Bill Metzger** left the E.M.F. Company, **Walter Flanders** decided that a low price car was needed to compete with the Ford Model T. Since **Flanders** had taught **Henry Ford** much of his manufacturing skills, it was reasonable to expect that E.M.F. could be competitive. With the help of the the Studebaker Company, **Flanders** acquired the former factory of the **Deluxe Motor Company** at 1000 Woodward Avenue for the purpose of building a four cylinder car to compete with the Model T Ford.

The Flanders Motor Car Company was organized to manufacture the "Flanders 20." It was a scaled down version of the E.M.F. 30, with 20 hp, a 100 wb, and at a price of $750, which sold for less than the Model T. Soon after its introduction in late 1909, there was trouble between **Walter Flanders** and the Studebaker Company regarding payment to the E.M.F. Company for cars and advertising, which cleverly gave the impression that E.M.F. cars were built by Studebaker. The Flanders 20 was unable to outsell Ford, but the combination of the Flanders 20 and the E.M.F. 30 gave the E.M.F company the number two spot in sales for 1911.

(In January 1911, **Walter Flanders** also formed the **Flanders Manufacturing Company** in Pontiac, Michigan as a separate entity from the E.M.F. Company with a capital of $2,500,000. Most of the capital was raised from former E.M.F. backers. The company was originally set up to produce motorcycles in Chelsea, Michigan. **Leroy Pelletier**, the advertising manager for E.M.F., convinced **Flanders** to produce an electric car, called the "Flanders," featuring worm drive, and a cradle spring suspension, making it a foot lower than other electric cars. The company relocated to Pontiac, Michigan and received orders for 3000 cars priced at $1775. After 100 cars were built, the Flanders Manufacturing Company went into receivership. **Pelletier** bought the Flanders business in 1913 and moved to Flint, Michigan. He renamed the car "Tiffany," but it didn't sell. He received permission from **Flanders** to move back to Chelsea and rename the car the "Flanders Electric," but it was too late for success with electric vehicles and he went out of business.)

Studebaker owned one-third of the stock in E.M.F., and decided to acquire the entire company in 1912. After a bitter court battle, Studebaker took over the E.M.F. Company and all subsequent automobiles were called "Studebakers." **Flanders** rejoined **Everitt** and **Metzger** later in 1912.

1910- E.M.F.
courtesy AAMA

*E.M.F. factory
on Piquette Ave.
courtesy AAMA*

*E.M.F. factory
on Piquette Ave.
courtesy Tinder*

*The E. M. F. factory
on Piquette.
picture taken 1994*

*1909-
E.M.f. pathfinder
tour*

*1912-
Flanders "20"*

Walter E. Flanders

HERRESHOFF AUTOMOBILE COMPANY
DETROIT, MICHIGAN
1908-1909

HERRESHOFF MOTOR COMPANY
DETROIT, MICHIGAN
1909-1914

The Herreshoff family was known for boat making in the early 1880s, and even made a coal burning steam buggy, but did not pursue production. In August 1908 Charles Frederick Herreshoff, the Bridgeport, Conn. naval architect, inventor, designer and builder of the Den, the speedy motor boat, severed his connection with the American and British Company, of Bridgeport, Conn. He moved to Detroit and formed the Herreshoff Automobile Company with Louis Mendelson, formerly of the Modern Match Company, as president. Herreshoff designed a marine motor that would be used for an automobile, *and anticipated that it would create something of a furore.* It had 40 percent less weight than a conventional automobile motor, which saved gas, oil, tires, and allowed better manipulation of the vehicle.

The first "Herreshoffs" were 1909 models with 24 horsepower four cylinder motors. They had a unit power plant incorporating the engine, flywheel, clutch, and a three speed progressive gear change all in one unit. A three point suspension was featured and a full floating rear axle by tubular shaft enclosed in a torque tube. Weight was 1600 lbs and the price for the standard type was $1500. The total production for the first two years was to be sold in the east by Harry S. Houpt of New York City. Afterwards, the cars were also to be marketed in the West. Herreshoff and Houpt were at odds from the start and the Houpt Company of New York gained control of the Herreshoff Automobile Company and re-organized as the Herreshoff Motor Car Company. By October 1909 a Detroit attorney, Harry Helfman, bought back the Herreshoff Company on behalf of his friend Charles F. Herreshoff, who then became president of the company.

(The Harry S. Houpt Company in New York then passed out of existence and was succeeded by the Harry S. Houpt Mfg. Company which would market a new automobile called the Houpt, built in Bristol, Conn., by the New Departure Company in 1909. From 1910 to 1912 the name was changed to Houpt-Rockwell. The New Departure Company also built a taxicab called the Rockwell from 1909 to 1912.)

In 1913 a small six cylinder motor was introduced with a stoke-bore ratio of 1.33 to 1, putting it in the long stroke field. The valves and springs were completely enclosed with cover plates, two to a side. Thermo syphon cooling was used with a large fan at the front of the motor. Cranking motors and electic lights were also offered.

In the fall of 1913 the Herreshoff Motor Company found itself in financial difficulties and called together a number of the larger creditors to ask for an extension of time. The *situation was canvassed* and it was decided to grant such an extension provided the creditors could be represented on the board of directors, and provided further that the Lycoming Foundry and Machine Company, which had been supplying engines, could send a representitive to Detroit to see that the proper motors were secured and installed in the cars after that date. In March 1914 an involuntary petition in bankruptcy was filed against the Herreshoff Motor Co. Considerable dissatisfaction was created among the creditors owing to the fact that Lycoming was said to have endeavored to grab off money and cars to pay its claims in preference over other creditors subscribing to the extension agreement. Within days, Charles Herreshoff left the company and went to Troy N.Y. to organize a cycle car company.

On May 8 the Herreshoff Motor Company real estate was sold to the Jacob-Nicol Realty Company and the personal property went to Harris Bros. Company. In June the entire belongings of the plant consisting of all patterns, tools, records, blue prints, jigs, stock, etc., was purchased by the American Motors Company in Indianapolis, Ind. Herreshoff owners could obtain spare parts from that point after July 1914.

(Charles Herreshoff organized the Herreshoff Light Car Co. in Troy, N.Y., in April 1914 to build a cycle car with a four cylinder motor built by the City Pattern Works on 93-95 Catherine Street in Detroit. The car's name was changed to Harvard in 1915 and was produced until 1921.)

Just off the "row," on Seventh Avenue, in New York, Harry Houpt had a two story showroom and garage.

1910- Herreshoff courtesy AAMA

The Herreshoff Company originally used the Modern Match Company factory and the E. R.. Thomas factory. Seen here is the new building on Woodward Avenue.

HUPP MOTOR CAR COMPANY
DETROIT, MICHIGAN
1908-1915

HUPP MOTOR CAR CORPORATION
DETROIT, MICHIGAN
1915-1941

"THE SMARTEST AND BEST LITTLE CAR EVER MARKETED IN AMERICA AT ANYTHING LIKE THE MONEY"

Robert Craig Hupp was born on June 2, 1887 to Charles J. Hupp and Anna Klinger Hupp of Grand Rapids, Mich. The surname is Huguenot, originally spelled Houppe. The family moved to Detroit in 1884 where Charles became assistant general freight agent for the Michigan Central Railroad from 1884 to 1907, when he became general freight agent. **Robert Hupp** attended Detroit public schools including Central High, which was located in the Biddle House on Jefferson Avenue. **Robert** played quarterback for the Detroit Athletic Club football team in 1896. His working career started with one of the railroads in Detroit, but he saw no opportunity to get ahead. It was apparent that the way to get ahead was to hire on with a company, start at the bottom and work his way up. The main idea was to learn a business, no matter what the hours were. He joined the Olds Motor Works in Detroit in 1902, unloading coal from freight cars, sweeping floors, and loading castings. He worked to the best of his ability, and when a position in the assembly department opened up, he got transferred. He advanced to assistant in the engine testing room, inspector in the final testing room, and occasionally drove cars on the road. He finally became a regular tester, then a year later was in the repair department and became manager. In 1905 the Olds Motor Works moved to Lansing and Hupp went along, but after several months he listened to the siren song of a Chicago manufacturer of soda fountains, who made him an exceptional offer.

After nine months in Chicago the automobile fever came back and he returned to Detroit to work for the Ford Motor Company. **Hupp** was in charge of the repair, claim, and accessory departments, and set out to acquire a thorough mastery of every department possible. He became an assistant in the purchasing department and to production superintendent **John Dodge** during the six cylinder Model K period. At the same time, Ford announced the first four cylinder runabout at a low price with manufacturing on a large scale. The experience **Hupp** gained at Ford and the Olds Motor Works gave him insight into factory and cost conditions. In early 1908 he left Ford and worked with his father at the Michigan Central Railroad for a brief time, but spent most of his time and energy trying to build a medium priced automobile.

Hupp gained financial support from Joseph R. Drake, J. Walter Drake, and John E. Baker with the objective of organizing an automobile company. A small machine shop on Tower Court in Detroit was used to build a prototype car. Development and testing progressed until winter when the group was satisfied. On November 8, 1908, a week after a prototype automobile was completed, the Hupp Motor Car Company was organized, with a capital stock of $25,000, and $3500 paid in. A factory was rented at 345 Bellevue Avenue to begin production. A few months later, Edwin Denby, who later became secretary of the Navy, bought one-fifth interest and paid in $7500, to bring working capital up to $11,000. A number of former Olds Motor Works employees joined **Hupp,** including Charles D. Hastings, C. H. Dunlap, J. H. Peterson, and **Emil A. Nelson**, who had also been an engineer for the Packard Motor Car Company. **Nelson** was made chief engineer for the Hupp company, and was most responsible for designing the first car with **Robert Hupp**. The company's presidency was given to J. Walter Drake, who at 33 years old gave up a 12 year law practice. Charles Hastings was assistant general manager in charge of sales, service and accounting. In addition to his work, he frequently worked 18-hour days, helping workmen assemble and wash cars, making him a key to the initial success of the Hupp Company. Hastings was from Hillsdale, Mich., and began his career as a traveling salesman for a hardware company. Then he worked as a clerk for the Michigan Central Railroad, then became a partner in a wholesale liquor agency in 1898. In 1901 he joined the Olds Motor Works and became a sales manager for the Detroit and Lansing offices, and also served as the foreign sales manager. He developed a real interest in the export trade saying that it had the "greatest broadening influence a man can have." In 1907 he joined the Thomas-Detroit Company as office manager and sales executive. He left the Thomas Company when **Hugh Chalmers** took over and brought in his own cadre of people.

After final touches the first "Hupmobile" was shown at the Detroit Automobile Show in February 1909. It had the seat positioned as far rearward as possible and had a patented three point suspension, making it comfortable for two large people. Years later **Henry Ford** noted, "I recall looking at **Bobby Hupp's** roadster at the first show where it was exhibited and wondering whether we could ever build as good a small car for as little money." The Model 20 two-passenger Roadster had an 86 inch wb and a 56 inch tread, weighed 1100 lbs, and was priced at $750. It had an L-head, four cylinder, four stroke engine producing 16.9 hp. The cylinders were cast in pairs with a 3.25 inch bore and 3.2 inch stroke. Cooling was thermo-syphon with a front mounted fan-bladed flywheel, and a Mercedes-type radiator with vertical tubes and straight horizontal fins. A sliding gear transmission and Bosch high tension magneto was above average for cars in its class. Over 1500 Hupmobiles were sold in 1909, and over 5000 in 1910. With Hastings as an export trade enthusiast, 200 Hupmobiles were shipped to New Zealand, and eventually a large percentage of production would be committed to overseas markets.

Hupp was determined to procure better gray iron, drop forgings, and machine work for the manufacture of the car of which he was in charge. He invested his personal capital in the Hupp-Guyman Foundry Company, Hupp-Turner Machine Company, Hupp-Johnson Forge Company, Hupp-Detloff Patern Company, Hupp-Ellis-Rutley Construction Company, R. C. Hupp Sales Company, and the Rotary Valve Motor company, in the new Fairview district east of Detroit. **Hupp's** companies were consolidated in May 1910 as the Hupp Corporation, with **Robert** as president, Charles Hastings as VP, and Robert's brother **Louis G. Hupp** as secretary and treasurer. In June of 1910 the Hupp-Yeats Electric Car Company was formed, which was reported to have marked the most epochal development in construction, design, and price the industry had ever seen. **Hupp** wanted to expand the operations of the Hupp Motor Car Company, similar to General Motors, United States Motors, the Studebaker-E. M. F. combine, and Albert Pope's organization. Drake and Hastings were vehemently opposed to over-extending the newly formed company. In August 1911 **Robert Hupp** sold his stock to the company's officers with the intention of producing another Hupmobile through the Hupp Corporation. Hastings and Drake quickly filed a lawsuit in the Wayne County Circuit Court to prevent **Robert** and **Louis Hupp** from using the **Hupp** name associated with a gasoline automobile. The suit was upheld and the name of the Hupp Corporation was changed to the R. C. H. Corporation.

After **Robert Hupp** resigned, the Hupp company continued to expand, and was capitalized in November 1915 at $8,000,000, and re-named the Hupp Motor Car Corporation. In September 1919 the Hupp Corporation and Mitchell Motors of Racine, Wisc., organized the H & M Body Corporation to supply bodies to both companies. Detroit Auto Specialty Company, another Hupp subsidiary, supplied fenders and sheet metal parts. The main Detroit Hupp plant was enlarged in 1924 to 1,570,000 square feet of floor space, and over 250,000,000 square feet with subsidiaries counted. Employment was over 5000. In January 1925 the Hupp Corporation brought out model E-1 with an eight cylinder, that also had Lockheed hydraulic brakes. Unfortunately, there were minute pin-holes in the castings, and the brake cylinders tended to leak, making the eight cylinder hard to stop. In April 1925 the Murry Body Corporation purchased the Hupp Company's Racine body plant for $1,500,000, and received a five year contract to supply bodies. The Hupmobile final body inspection was managed by **Calista Conwell**, who was in charge of 400 men and 50 inspectors. She was reported to be the first woman in the automotive industry with such significant authority.

Sales in 1928 reached over 65,000 and the defunct Chandler-Cleveland Motors Corporation of Ohio was purchased for expanded production. Sales in 1929 fell to 50,000, and the company was in trouble even before the stock market crash. For 1932, **Raymond Loewy** introduced form-fitted fenders, which most people called cycle fenders, and chrome plated wheel discs. Hupmobile entered the 1932 Indianapolis 500 race with the Hupp Comet and finished fifth. **Loewy** designed aerodynamic Hupmobiles for 1934, with faired-in headlamps and three piece windshields. At the same time, a furious battle for control of the Hupp Corporation ensued. It began when Archie Andrews sent a letter to all of the stockholders charging gross mismanagement. This forced DuBois Young to step down as president, with outsiders taking control. J.

Walter Drake and other original members of the firm eventually regained control, but the company was in shambles. Production was suspended in late 1935. A Federal reserve loan request was denied, and the Hupp Corporation was forced to sell off some of its plants.

A conservative approach was taken while the Hupp Corporation tried to get back on its feet. There was no new 1937 Hupmobile, but a new six cylinder and eight cylinder without the aerodynamic styling was introduced in 1938. In August 1938 Samuel L. Davis was brought in as president and he hired Norman DeVaux as general manager. His DeVaux cars failed in the early 1930s, but he believed that the Hupp could be rescued with another recently defunct car. The tooling for the front drive Cord 810/812 was purchased to use for the rear wheel drive Hupmobile for $45,000. John Tjaarda, who designed the Lincoln Zephyr, was brought in to revise the design for the Hupmobile. In 1938 four cars were built, and renamed "Skylarks." The Cord was a low volume, expensive car, and to keep tooling costs down, there were many individual pieces of sheetmetal, especially in the roof, that required lead and sanding. In addition, the trunk floor pan was flat because the Cord had front wheel drive.

To make the Skylark cost competitive, it needed new sheet metal tooling. Hupp was dangerously close to bankruptcy, and was forced to close down again in 1939. J. Walter Drake was brought in again to preside over the company, and a call was made to **Joe Graham**. In September 1939 the Hupp Corporation entered into an agreement with the Graham-Paige Company, which would build cars for both companies using the former Cord tooling. Graham-Paige was to have the roof made with one piece of sheet metal and the trunk floor pan revised to clear the differential, besides creating a new front end treatment for its version called the "Hollywood."

Although the Skylark was produced at the Graham-Paige Dearborn plant, the Hupmobile Corporation continued to manufacture its own engines and chassis parts in its Detroit plant. While the Hupp Corporation continued to sell off plants to meet expenses, the body dies were being shipped from Auburn's Connersville factory to the Graham-Paige plant in Dearborn.

The first Skylarks rolled off the assembly line in May 1940, eighteen months after the official announcement was made. Most of the 6000 orders had been canceled due to the extreme waiting period. The Hupp Corporation was exhausted by then, with less than 30 cars manufactured per week, and no money to pay the Graham-Paige company for building them. The sheet metal was never revised, and the trunk floor pans had to be hammered out with ball-peen hammers. Of the 319 Skylarks built that year, Hupp only paid for eighty. The remaining cars were sold to Hupp distributors by the Graham-Paige Company. After production of over 500,000 Hupmobiles, the last Skylark rolled off the line during the week of July 8, 1940, using another company's obsolete body, and assembled in a competitor's plant.

*1908-
Hupp Bellevue
Avenue plant.
courtesy Tinder*

Hupp Bellevue Street plant courtesy Tinder

1911- Hupp Jefferson Avenue plant courtesy NAHC

Hupp Jefferson Avenue plant courtesy AAMA

Hupmobile starting on a tour around the world courtesy AAMA

D. E. Meyer Co. courtesy MSHC

advertisement

Hupp factory on Milwaukee Ave.

1925- Hupmobile 8 cylinder courtesy AAMA

1938- Hupmobile courtesy AAMA

1939-
Hupp 6 cylinder
(Cord body)
courtesy AAMA

Hupp factory on
Milwaukee Ave.
picture taken 1981
courtesy MSHC

Robert Craig Hupp

GENERAL MOTORS COMPANY
1908-1916

GENERAL MOTORS CORPORATION
1916-

"THEY CLAIM THIS MOTOR CAR BUSINESS IS A MYSTERY, IT HAS AN ELECTRIC PLANT, HAS A WATER PLANT AND A WHOLE LOT OF GADGETS, BUT AFTER ALL IT IS NOTHING BUT ORGANI-ZA-TION. BRICKS AND MORTAR, MEN AND MACHINERY, AND ORGANI-ZA-TION. I'M GOING TO BE ONE OF THE BIG MOTOR CAR MANUFACTURERS OF THIS COUNTRY."
-as Billy Durant said to a salesmen for Timken, Mr. Lewis

William Crapo Durant was born on December 8, 1861, in Boston, to William Clark Durant and Rebecca Crapo. William's grandfather, Henry Crapo, had made his fortune in Flint with lumber, and in 1864 he became the governor of Michigan. Crapo died in 1869, and the Durants visited Flint for the funeral. Shortly afterwards, Rebecca and William separated and Rebecca moved to Flint with her two children. After the death of Mrs. Henry Crapo in 1875, Rebecca received over $500,000, and purchased a substantial home. In 1879 **Billy** came home from school and announced that he quit and was going to work. He took a job at the Crapo lumberyard piling lumber, and at night he clerked at a drugstore. He became more interested in selling and quit both companies to travel around selling patent medicines.

By the 1880s the lumbering boom was ending with the depletion of the forests, and Flint was just a pleasant place to live with 8000 inhabitants. Begole, Fox & Company, which had operated one of the largest lumber mills chose to make wagons in its idle buildings. It became known as the Flint Wagon Works in 1882. **Durant** started selling cigars and in no time replaced three salesmen, then turned to the city's Waterworks collecting overdue bills. Waterworks were so poor, people were refusing to pay. He visited each household and interviewed customers and in eight months had the Waterworks back on its feet. In 1885 he married Clara Miller Pitt, and settled in a house at Garland and Fourth Avenue in Flint. They had two children, Margery and **Clifford**.

In 1886 **Billy** was working for an insurance company and for the utilities. While making a trip to read a meter, Johnny Alger passed by in his **road cart**, which resulted in the start of **Billy's** new career. As **Billy** was at the hardware store chatting with his friend **Josiah Dallas Dort**, Alger offered to give **Billy** a ride. **Billy** looked the cart over and it seemed quite flimsy. Alger pointed to the springs, which were held with stirrup shaped mounts under the shafts. This was the secret, he said, a unique seat suspension. **Billy** had to get to the gas plant so he hopped in and waved goodbye to **Dort**.

Billy was told the cart was purchased in Coldwater, Mich., from the Coldwater Road Cart Company. In the morning, **Billy** went directly to Coldwater and found it to be a small old-fashioned carriage shop. **Durant** and Thomas O'Brien exchanged introductions and **Billy** asked if he could buy an interest in the shop. O'Brien asked, "why not buy it all?" **Durant** and **J. Dallas Dort** became partners and founded the Flint Road Cart Company on September 28, 1886, and leased a one story mill built in the early 1880s. They became the largest manufacturer of horse drawn wagons in the United States until they stopped production in 1917 to use the factory for building Dort automobiles.

In 1890 **Durant** met **Charles W. Nash** at the W. C. Pierce hardware store in Flint. He offered him a job at the Road Cart factory, in the blacksmith shop, and then did piece work stuffing cushions for buggy seats. Nash worked so fast that he made the other laborers look bad. "I need the money," he explained. (Charles Nash was born in Illinois in 1864. He had been "bound out" by his father at age seven to a farmer in Flint to work for room and board until he reached adulthood. At age 12, he ran away and worked as a laborer on other farms and formed a successful hay proccessing firm, Adams & Nash, in 1882, then worked in the hardware store in Flint.) **Nash** quickly moved up through the ranks into positions of authority, as a leader of men and a man who had an instinct for saving money in the factory and working hard, and became general manager and VP. **Durant** changed the name of the firm to the Durant-Dort Carriage Company in 1895. **Dort** left the company in 1898 and went to Arizona because of his wife's failing health.

A. B. C. Hardy took over as president, but became burdened with too many details and tried to go out of town to relax only to get constant telephone calls from **Durant**. Finally, **Durant** ordered him to go to Paris for a rest. He enjoyed the trip so much that he returned and took his family back for an 11 month tour in 1901. He learned that the automobile was beyond the primitive stages as seen in Flint, and surmised it could even replace the horse and carriage. He studied cars in detail and when he returned, he warned the directors of the Durant-Dort Carriage Company to "get out of the carriage business before the automobile ruins you."

In late August 1904, **Billy Durant** visited **James Whiting** about the Buick Car Company he had purchased from **Ben Briscoe**. On November 1, 1904, **Durant** was elected to the Buick board of directors. As usual, he declined the presidency and Charles M. Begole was elected president, and George L. Walker became VP. A short time later, Arthur C. Mason, Buick's engine superintendent from the beginning in Flint, was developing an engine capable of 4000 rpms, compared to the average 1800 rpm. While walking home one day, **Durant** was stopped by one of Flint's leading citizens and was asked about the new motor. He said that Durant was gambling away his whole carriage business for a passing fancy. **Durant** hired a motor expert named Simmons to study the Mason engine. He reported that "This thing is basically unsound and extremely dangerous. It's quite likely to explode, and in fact, I would suggest the purchase of a bushel basket with every one sold in order to be able to pick up the pieces." Mason listened quietly with mounting anger and finally stepped up to the motor, started it, and placed his head alongside the engine block: "If it explodes I might as well go with it."

1908-Billy Durant

In 1909 **Durant** was approached by **Benjamin Briscoe** to develop an automotive combine. Although they did not come to terms, **Durant** dug in his heels. Almost nobody noticed when articles of incorporation were filed on September 16, 1908, with a capital stock of $2000, forming the General Motors Company. The incorporators were Benjamin Marcuse, Arthur W. Britton and George E. Daniels, acting president. **Durant** increased the capitalization to $12,500,000 on September 28. General Motors then purchased Buick from **Durant** for $3,750,000 in stock. On October 10, **William Eaton** became president and Durant was VP. On November 12, General Motors purchased the entire outstanding stock of the Olds Motor works in Lansing which was having serious sales problems. GM paid just over $3,000,000 in GM stock, except for $17,279 cash. **Durant** worked with the Oldsmobile engineers and showed them what kind of a car to make. He had the wooden body taken off his Buick and sawed into four equal parts. Then he ordered the sections moved apart from each other. "Make a bigger Buick," he said "put an Oldsmobile radiator and hood on it, and raise the price $200 above the Buick Model 10.

Within two years, General Motors Cormpany included: Buick in Flint, Cadillac in Detroit, Cartercar in Pontiac, Champion Ignition Company in Flint, Dow Rim Company in New York City, Elmore Manufacturing Company in Clyde, Ohio, Ewing Automobile Company in Geneva, Ohio, Jackson-Church-Wilcox Company in Jackson, Mich., Auto Parts Company in Detroit, Michigan Motor Castings Company in Flint, National Motor Cab Company and Northway Motor & Manufacturing Company in Detroit, Oakland Motor Car company in Pontiac, Olds Motor Works in Lansing, Rainier Motor Company in Saginaw, Rapid Motor Vehicle Company in Pontiac, Reliance Motor Truck company in Owosso, Welch-Detroit Company in Detroit, Welch Motor Car Company in Pontiac, Seager Engine Works in Lansing, Marquette Motor Company in Saginaw, Randolph Truck Company in Flint, Bedford Motors Company in London, Ontario, Novelty Incandescent Lamp Company, Heany Lamp Companies, McLaughlin Motor Car Company, Ltd., in Oshawa, Ontario, Oak Park Power Company in Flint, and the Weston-Mott Company in Flint. GM also had interests in the United Motors Company, Maxwell-Briscoe and Lansden Electric.

Billy's pace was extraordinary, holding meetings as late as 1:30 a.m. One night **Durant** and his personal secretary, W. W. Murphy were headed to Detroit by car to catch a train, with **Durant** driving. They slowed by a tractor pulling hay which obstructed the whole road. **Durant** gunned the car around the hay wagon and veered into a ditch. There was a loud snap as a leaf spring broke. "What was that?"

Murphy shouted. "Never mind," yelled **Durant** as he pulled out of the ditch in front of the wagon. "We're still running."

By the fall of 1910 it was widely known that **Durant** had expanded too quickly and far exceeded his capital reserves and borrowing power. Hopelessly in debt, **Durant** was forced to pay exhorbitant commissions and interest to banks to help get back on his feet. Although he stayed on as VP, he lost control of his company to the New York Banking Syndicate.

Durant turned to **Louis Cheverolet** to design a new car, and eventually it sold so well that **Billy** began trading Cheverolet stock for GM stock. After a monumental effort in acquiring GM stock, **Durant** announced to the directors in the middle of May 1915 that he was in control of the majority of the GM voting stock. He was back in control, and **Charlie Nash** quit as president. **Charlie** purchased the Thomas B. Jeffrey Company in Kenosha, Wisc., and renamed it Nash Motor Company. **Billy** then asked **Walter P. Chrysler** to take over the presidency of Buick. Chrysler was making $5000 per year at the time, and decided to leave GM along with **Nash**. **Durant** asked that he reconsider and to name his price. He came back with a request for $50,000, and **Durant** said fine, I'll pay you $500,000, so he accepted.

In July 1918, the **Du Ponts** and **Durant** were looking for gilt edge industrials to add to their holdings. They would have preferred railroads, but because of the war, the government was in control. Expansion continued until **Billy** lost control in 1920 for the second and last time. He went on to form another automobile corporation to produce the Durant and Star, but didn't make it through the depression.

His last effort was the creation of a family bowling chain in Flint. He died on March 18, 1947, at age 85.

GM building- completed in 1920, was located Grand Boulevard in Detroit. It was built 15 stories tall with 20,000,000 cu. ft. of office space.
courtesy AAMA

January 12, 1940- celebration of the 25,000,000 th GM car. Chairman Alfred P. Sloan (R) seen leading Billy (L) on stage before 5000 GM corporate guests in Detroit to pay him tribute.

FACTS ABOUT A FAMOUS FAMILY

A car for every purse and purpose

In the automobile industry several distinct price classes have developed.

General Motors, a family of car and truck builders, offers a choice of models in each class.

In Buick, Cadillac, Chevrolet, Oakland, Oldsmobile and GMC Trucks, there is a car for every purse and purpose—purchasable on a sound payment plan.

Back of each car are all the resources of General Motors—an assurance of scientific excellence, continuing service and satisfactory value.

GENERAL MOTORS

BUICK · CADILLAC · CHEVROLET · OAKLAND
OLDSMOBILE · GMC TRUCKS

General Motors cars, trucks and Delco-Light products may be purchased on the *GMAC* Payment Plan.
Insurance service is furnished by General Exchange Corporation

COMMERCIAL VEHICLES

The word "truck" was derived from the Greek word "trokhos" meaning "wheel." It became widely used to define heavy wagons drawn by beasts-of-burden. It was called the "horseless vehicle industry" in the early days. In the late nineteenth century, the word "motor" was placed in front of "truck" to distinguish self-propelled vehicles from horse drawn vehicles. Different words were used regionally to mean the transportation of goods by truck. In Detroit the word "cartage" was used, while in California "drayage" was used.

Traffic conditions were greatly improved with the advent of the truck. Three major advantages over the horse were found. The motor truck was less destructive on the street, and reduced the maintenance charges for street repair; there was a reduction in horse debris, due to the supplanting of the horse by the truck. It also resulted in a betterment of the public health, due to the more sanitary conditions of the public streets. The increase in tonnage of the motor truck reduced traffic congestion at rush hours. The third improvement was based on the motor truck supplanting three two-horse teams and a wagon, while occupying only 30 percent of the space. Also, because the motor truck had greater flexibility and greater control, it could follow closer to the vehicle in front of it. Estimates were given that for a given amount of traffic, four times the tonnage could be carried if all traffic was carried on motor vehicles.

above photos courtesy AAMA

GMC TRUCKS

The first gasoline powered truck to be used on the streets of Detroit and one of the first in the country was manufactured by the Grabowsky Motor Vehicle Company in 1902. **Max Grabowsky** was born in Detroit in 1874, and was educated in the Detroit public schools. He demonstrated a talent for mechanics at an early age, and while in his teens he began experimenting with the construction of commercial vehicles. He went to work for Winder & Woodward, a locksmith, gunsmith, and a bicycle repair shop. In 1900 **Max** and his brother **Morris** designed a single-cylinder truck with a horizontal engine, and formed the Grabowsky Motor Vehicle Company. They used structural steel and machined every part of the vehicle in their own shop. Their first truck was sold to the American Garment Cleaning Company in 1902.

In late 1902 the company name was changed to the Rapid Motor Vehicle Company. In 1904 the Rapid Motor Vehicle Company was incorporated with a capitalization of $13,000. The company needed more space and the first factory in the country exclusively for manufacturing gasoline trucks was built in Pontiac, Mich. By 1905 the company had produced 75 trucks, along with several passenger cars. In 1906 12 types of trucks, passenger busses, and a wide range of other styles were being produced.

In 1909 GM purchased both the Rapid Motor Vehicle Company and the Reliance Motor Truck Company to form the General Motors Truck Company. At first, GM marketed the Rapid, Reliance, and Randolph from 1909 to 1910. (**Morris Grabowsky** left the Rapid Company in 1910 to work for **US Motors**, after it purchased the Alden-Sampson Company and moved it to Detroit.) The Lansden electric was added in 1911 and was produced until 1916. In 1912 all GM trucks were marketed as GMC. A two-ton Lansden, renamed GMC, was used at the "electric farm" of the Boston Edison Company, in the outskirts of Boston, where all possible operations were electrically conducted. Electricity cut the fodder, milked the cows, washed the dishes, pumped the water, churned the butter, and took the produce to market by means of an electric powered truck. The farm was a demonstration project where thousands of visitors were entertained.

Max Grabowsky left Rapid in 1908 and formed the Grabowsky Power Wagon Company. A major feature of the "Power Wagon" was a front-end removable power plant which became widely imitated, and it proved to be the forerunner of REO's "Speedwagon." The company went out of business in January 1913, selling its belongings, except for the factory and machinery to the Seitz Automobile and Transmission Company of Wayne, Mich. The Seitz company built their own trucks from 1908 to 1913. The factory was sold to the Budd Manufacturing Company which moved in to make steel bodies. (**William J. Seitz** got his start in Grape, Mich., hauling lime by horse and wagon. In 1901 he was at a threshing bee watching the threshing machines work under its own power. He got the idea to place a gasoline engine on a lime wagon to help hauling during the spring when the roads were deep with water and mud. His first model was pulled by horses and had no steering and iron band tires.)

Max Grabowsky went on to help coordinate the manufacture of airplanes and automobile parts during World War I, then worked as a factory appraiser and for various industrial associations. He later worked for the city of Detroit until he died in 1946. He was survived by his wife, Celia, and two daughters.

Rapid Truck-
T. B. Rayl had a popular hardware store on Woodward Avenue.
courtesy AAMA

DETROIT-DEARBORN MOTOR CAR COMPANY
DETROIT, MICHIGAN
1909-1910

Arthur E. Kiefer was one of the founders of the Detroit Edge Toolworks before helping to organize the Detroit-Dearborn Motor Car Company. He was born in Detroit and attended public schools and German Seminary. He attended Polytechnic of Karlsruhe in Munich, Germany, and graduated with an engineering degree in 1880. His working career began as a surveyor and draftsman.

The Detroit-Dearborn Motor Car Company was organized in August 1909 with $50,000. Along with Kiefer as VP, Edward Bland of the Rieland & Mathews Manufacturing Company was president, and Samuel D. Lapham was treasurer. They secured a lease on a building in the village of Dearborn. Immediately afterwards they acquired 71 building lots four miles away from the Detroit City limits in Dearborn to erect a factory.

The "Detroit-Dearborn" was designed by former Regal designer Paul Arthur with the first pilot model on the road in November 1909. The "D.D.", as it was called, was powered by a four cylinder motor cast in pairs, rated at 35 hp. The bore and stroke was 4.124 x 4.75, water cooled by thermo-syphon. It had a T-head with intake and exhaust valves on opposite sides. Lubrication was by the splash system with sight feeds and a tell-tale on the dash. A jump spark ignition with a magneto was standard equipment. A three speed sliding gear change *with direct in high* was mounted in the center of pressed steel chassis. A five passenger torpedo touring body, called the Minerva, and a two passenger Nike model were offered, each weighing 2200 lbs. The front floorboards were covered with aluminum, a cocoa mat was fitted in the tonneau, and the running boards had cork covering, giving the car a little class.

A voluntary petition of bankruptcy was filed by the Detroit-Dearborn Motor Car Company in October 1910 after a total production of 110 cars. Liabilities footed up to $117,383, of which $90,860 were unsecured. According to the attorney for the company, the trouble was over the refusal of a few creditors to grant extension, to which 90 percent of the creditors agreed. The company was just beginning to market its cars, he said.

The assets were purchased by Vernon C. Fry, who marketed the Vernon 30. The venture was short lived, and in March of 1911 the assets and factory were purchased from Fry by the Huron Motor Car Company. The President was J. F. Burns, and J. F. Sughrow was VP. The remaining inventory from the Detroit-Dearborn parts and the Vernon 30 were used to produce a few Hurons before the company went out of business.

1910- Detroit-Dearborn Torpedo type roadster with runabout body and front door. Elliptical gas tank behind the front seat and a large carrying trunk at the rear.

ANHUT MOTOR CAR COMPANY
DETROIT, MICHIGAN
1909-1910

BARNES MOTOR CAR COMPANY
DETROIT, MICHIGAN
1910

The Anhut Motor Car Company was incorporated in October 1909, with William M. Walker as president. The young Michigan senator, **John N. Anhut,** was vice president, and Detroit Mayor Breitmeyer was one of its financial backers. A factory on 510 Howard Street was leased to build the "Anhut." **H.C. Barnes**, who was superindendent of the Overland plant in Indianapolis, Ind., severed his Overland connection to move to Detroit and become factory manager for the Anhut Motor Car Company. The car had a six cylinder engine and was priced at $1800 when displayed at the Detroit Auto Show in January 1910. The sales were very encouraging and the capital stock increased from $150,000 to $300,000. The Chatham Motor Company in Ontario, which produced cars from 1907 to 1908, was purchased in order to penetrate the Canadian market although the Anhut car was never produced there. During the summer of 1910, stockholders feared the worst, the Anhut Motor Company was in trouble. **Senator Anhut** could not be found as he was vacationing in Europe, getting ready to campaign for re-election.

By the time **Senator Anhut** returned from Europe, the Detroit press was exposing some financial misdeeds aimed at the young senator. Soon his name became a deterrent to the company, so the senior officers placed **H. C. Barnes** in charge with new capitalization of $375,000. The company's name was changed to the Barnes Motor Car Company. The "Barnes-Six" was to be offered for $2250, and a "Barnes-Four" for $1400.

By November 1910, the company went bankrupt without making any Barnes automobiles. In January 1911 Frank Howard paid claims of $34,000 and purchased the stock and equipment for $10,025. He planned to re-organize and resume operations under a new name, but never did. Among those who had claims against the Barnes Motor Company were the Griswold Body Company of Detroit ($6245), and R. C. Durant (son of W. C. Durant), of San Fransicso ($2812). William M. Walker, as president of the Barnes Company, was made receiver and went to the East Coast to try and straighten out the tangle.

1910 Anhut courtesy AAMA

1910 Anhut courtesy AAMA

1910- Anhut leased factory at 510 Howard (became P.O.S.S. Motor Company in 1911) courtesy AAMA

1910- Anhut Factory

DEMOTCAR COMPANY
DETROIT, MICHIGAN
1909-1911

RITTER AUTOMOBILE COMPANY
MADISON, WISCONSIN
1912

The Demotcar Company began to form in late 1908, and was organized in the fall of 1909 with $100,000 in capital. The name was a contraction of Detroit-Motor-Car. C. H. Ritter, a retired wealthy wholesale liquor dealer was president. George Anthony of the American Radiator Company was VP, Frank T. Lodge, a prominent local attorney was treasurer, and William Elsey was secretary. Guy Hamilton was chief engineer, and George T. Homeier was the factory manager. The factory was initially located at 21st Street. In December 1909 the Demotcar Company moved into the former 40,000 square foot factory of the Sun Stove Company on 1305 Bellevue Avenue in Detroit.

The "Demot," sometimes called the "Little Detroit," was a two seat runabout weighing about 1100 lbs with two average people, that sold for $550. It had an 80 inch wb and a 56 inch gauge. The engine was a two cylinder opposed with thermo-syphon water cooling. The ignition was a jump spark, and a splash lubrication system was used. The transmission was a planetary with selective control with a drive shaft. Springs were semi-elliptic front and rear.

The car was announced in September 1909 with an estimated 3,000 cars to be marketed, and dealers from Texas, Oregon, and Maine were flocking to Detroit to have a look. The Harper-Aldrich Auto Company at Woodward and Warren were the distributors for the state of Michigan. They took a Demot on a 1500 mile grind through the state which they claimed proved its stability.

The Demotcar Company went bankrupt in August 1910, and production continued under receivership. In December of 1910 C. H. Ritter acquired possession of the plant to resume operations. R. A. Skinner, who had previously been in the printing business, and A. W. Voege, one of the trustees of the bankrupt Demotcar Company, were new officers of the company. The company and the officers moved to Madison, Wisc., where they re-organized with $25,000 as the Ritter Automobile Company.

They produced a larger car with a 90 inch wb. It had a 15 hp, four cylinder engine with a 3.75 inch bore and a 3.4 inch stroke. The Ritter sold for $685 and came with a top, windshield, lamp equipment and a gas tank. Production was stopped in 1912.

1910- Demot courtesy AAMA

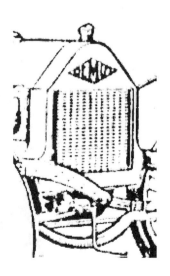

CARHARTT AUTOMOBILE CORPORATION
DETROIT, MICHIGAN
1909-1912

Hamilton Carhartt was born in Macedon Locks, N. Y., in 1855. He was educated in public schools in Jackson, Mich., and graduated from Episcopal College in Racine, Wisc. He married Anette Welling in Jackson, and had a son in 1882.

He began his business career at Young, Smythe, Field and Company in Philadelphia. He moved to Detroit in 1884, and started a wholesale furnishing business under the name of Hamilton Carhartt and Company in 1899. He changed to working men's clothing as "Hamilton Carhartt Manufacturing Incorporated" where he was the president. He was also the vice-president of the Peninsular Bank and of the Art Commission.

1901- Hamilton Carhartt & Co.
S.W. corner of Michigan and Tenth, Detroit

By 1910 Detroit was a burgeoning automobile center and **Hamilton Carhartt** grasped at the opportunity to enter the automobile business. In March 1910 the Carhartt Automobile Corporation was established and by August it was in production. There were 25 hp and 35 hp four cylinder engines offered with one chassis and several body styles.

The cars were a commercial success at first, but by the end of the year sales were falling off. The 1912 model offering was reduced although a second chassis was added and horsepower was increased. By March 1912 **Carhartt** stopped automobile manufacturing and returned to clothing with his son Hamilton Jr. The Carhartt factory was taken over by the Monarch Motor Company.

1910- First Carhartt (wasn't available until 1911) courtesy AAMA

KRIT MOTOR CAR COMPANY
DETROIT, MICHIGAN
1909-1915

The Krit Motor Car Company was organized in July 1909 with a capital stock of $100,000. **Claude S. Briggs** was president and W. S. Piggins was vice president of the company. The designer and chief engineer was Kenneth **Crittenden,** who had worked at Ford and Regal. The former C. H. Blomstrom Manufacturing Company's factory on the corner of Wight and Lieb was secured to build the first cars as 1910 models, and employed about 150 men.

Right from the start they were swamped with applications for agencies. The car was noted for simplicity and ease of control. It had a four cylinder cast *en bloc* with ball bearings throughout. The transmission was a sliding gear with two speeds forward and one reverse. The motor and transmission were one unit suspended on the frame on three points. The clutch was multiple disc immersed in oil operated by releasing a foot pedal.

Claude S. Briggs resigned in 1910 and became general manager for the Brush division of U. S. Motors, and Piggins was made president. The Krit Motor Company was having financial problems, and in 1911 Walter S. Russel of the Russel Wheel and Foundry Company led a syndicate to purchase control. **Crittenden** was made vice president and head of the engineering department, and Piggins was replaced by Lawrence Moore. Capitalization was increased from $100,000 to $250,000 and the former Owen Motor Car Company factory at 1620 East Grand Blvd. was secured giving them three times more floor space. Howard Henry Crawford was promoted from purchasing agent to director and general manager in 1912. Crawford had worked for Durant's Flint Wagon Works before that as paymaster at Buick, and at Weston Mott. Former Owen plant manager A. A. Gloetzner stayed on to work for the Krit Company.

A six cylinder was offered in 1913, but sales were down. The six cylinder was dropped in 1914 and the body was redesigned with wide doors, concealed hinges and door handles. The gasoline tank was repositioned from under the seat to the cowl which was very popular in the U. S. To carry the streamline appearance, no side lights were used, headlamps alone were featured. The wiring was such that the lights could be switched in series for dim light for city driving and in parallel for bright glow.

More changes were made to streamline the body in 1915. Crowned fenders, which nearly all makers were recognizing as *meritorious*, were offered, as well as a rounded top radiator with coped over edges and sloping hood. A new North East two-wire type electrical system including a starter motor/generator, was used with a 120 ampere hour Willard battery. The frame was 5/32 inch in thickness with a channel depth of 5/8 inches with three cross members. A new feature was the grouping on the dash instrument board. It contained a gasoline gauge, speedometer, ammeter, switches and carburetor control. This arrangement made it easy to illuminate all the devices with one dash lamp.

The Krit Company was always in financial trouble and new models, including a low priced version, but it didn't provide the help that was needed. The First World War aggravated the problem because the Krit Company relied on export trade.

In January 1915 the Krit Company was petitioned into bankruptcy. **Samuel L. Winternitz** in Chicago, **Alfred O. Dunk** of the Puritan Machine Company, and the service department of the Krit Motor Car Company all offered spare parts.

Perhaps the very last KRIT can be construed as the "Gremel" made in 1916 by **H. G. Gremel,** who was the manager of the Puritan Machine Company. It was made up from 102 different automobiles, with the body of a Krit being the largest part.

Right at the start, the K-R-I-T Motor Car Company adopted as it's symbol the ancient Swastika "good luck" emblem.

The K-R-I-T factory on E. Grand Blvd., near Mt. Elliot courtesy NAHC

frame build-up courtesy AAMA

chassis build courtesy NAHC

*1910-
KRIT with Ken Crittenden behind the wheel. (When he was at Detroit High School, his pals called him "Crit.")
courtesy AAMA*

*1911-
advertisement*

*1914-
Dealer in Brooklyn N.Y.*

ABBOTT MOTOR CO.
DETROIT, MICHIGAN
1909-1913

ABBOTT MOTOR CAR COMPANY
DETROIT, MICHIGAN
1913-1915

CONSOLIDATED MOTOR COMPANY
CLEVELAND, OHIO
1915-1916

THE ABBOTT CORPORATION
CLEVELAND, OHIO
1916-1918

Charles Stewart Abbott decided to go into the automobile game and hired **John Utz**, who was widely known as the designer of the popular Chalmers "30." **Utz** was a student of aviation and applied advanced ideas from this field to the automobile in such things as dual fans. One conventional fan was behind the radiator and another used the flywheel as a fan blade. He warped the fan blades for more thrust of the air, which Wilbur Wright verified as giving a much stronger current of air. The chassis had a pressed steel frame and three quarter elliptic rear springs.

Abbott brought John R. Phillips from Chalmers to run his factory, which initially occupied the former Northern Motor Company plant. Phillips then helped design a new factory at Beaufait and Waterloo streets. The first car left the factory in January 1910 and was originally called the UTZ, but soon the name was changed to "Abbott-Detroit." It was an assembled car using an F-head Continental engine. Later in 1910 the Abbott-Detroit was one of the first cars to offer electric headlights with tungsten bulbs. Production capacity was set at four cars per day for the first year.

To publicize the Abbott-Detroits, the editor of Health Magazine, Dr. Charles G. Percival, drove an Abbott around the U. S. border from coast to coast three times, traveling 100,000 miles. Dr. Percival had a pet bulldog with him, and when the first picture was taken of the group starting this trip, the dog was prominently placed in the picture. Although the dog was intended to be a mascot, the press picked up on the dog and dubbed the car "Abbott-Detroit Bull Dog."

A motorcade of Abbott-Detroits drove around Belle Isle in March of 1911 to celebrate the first year of production. In 1912 a Continental motor was again used and the car switched from one pedal control for clutch and brake to separate pedals. In 1913, electric starting and electric lighting was added. Four models were offered, called the 44-50 and three different 34-40 models. The 44-50 had a Continental motor with a bore of 4.5 inches and a stroke of 5 inches. It was rated as 32.4 hp using the SAE formula. It was lubricated by the constant splash system. The front and rear springs were the underslung type which lowered the center of gravity without loss of road clearance. Production capacity was up to 3500 per year.

Abbott retired in 1911, and by 1913 there were problems with the creditors. Edward Gerber of Pittsburgh, Pennsylvania bought the company and paid off the creditors at fifteen cents to the dollar. There were reports that he would move the production to Pittsburgh, but Gerber reported that the factory would not move from Detroit. The plant was operating with 140 workers. He did make changes to increase capacity and lower expenses at the plant, but he expected to add another 60 workers by the end of the year.

Gerber's experience stemmed from being a dealer. He knew what was needed in a car to make it an *easy selling proposition*. He wanted to listen to the dealers and cater to their demands, believing that they should know better than the factory what should be incorporated in a car. This policy was new to the "automobile game." The Abbott-Detroit battleship roadster was an interesting outcome of this study. It had sheet metal with rows of rivet heads showing, and the radiator was a unique V-shape.

In the first quarter of 1914 the Abbott Motor Company added departments for body trimming, top work and painting. A three story addition to the factory was made of the slow burning mill-type construction with automatic sprinklers and electric power elevators. A six cylinder chassis was added, called the "Belle Isle" model. The engine had Continental's latest oiling system, a circulating splash system with a new type of pump. The car had left hand drive and center control.

In addition to fours and sixes, a V-8 was offered in 1915 and sold for $1950. But the increased cost of production hurt the company. A former manager of Cartercar, **R. A. Palmer**, was brought in to improve the company's fortunes, and its name was changed to Consolidated Motor Company. Before long, the company leased a plant in Cleveland, Ohio, and the car was simply called the Abbott. In 1916 the company underwent another name change to become the Abbott Motor Corporation. The four cylinder was dropped and the eight cylinder was made by Hershell-Spillman with a 3.25 x 5 bore and stroke.

Due to insufficient sales in 1917, the struggling company considered a merger with **Hal Motor Car**, but that company was not doing much better. The Abbott Corporation went into bankruptcy in January 1918.

John Utz with glasses on the left and Pat O'Conner. (Pat started with Olds in Detroit, then Packard, Abbott, then Liberty.
courtesy AAMA

Abbott racer
courtesy AAMA

courtesy NAHC

*Note pet bulldog
held by Dr. Percival
in the center of the picture.
courtesy AAMA*

1911- Sunday picnic at Belle Isle. Standard colors were medium blue, deep Derby red, and Brewster green.
  courtesy AAMA

Abbott-Detroit Battleship roadster, a radical departure from conventional body design.

John Utz

PAIGE-DETROIT MOTOR COMPANY
DETROIT, MICHIGAN
1909-1927

"THE MOST BEAUTIFUL CAR IN AMERICA"

Harry Mulford Jewett was born in Elmyra, New York in 1870. He obtained a chemical engineering degree from the University of Notre Dame in 1890, and began his career as a civil engineer on the Chicago drainage canal control, and later as an assistant engineer with the Michigan Central Railroad. Next, he entered the coal mining business with the W. P. Rend Coal Company of Chicago. He moved to Detroit, and served on the U. S. S. Yosemite with his brother **Edward** during the Spanish American War. In 1903 he founded the Jewett, Bigelow and Brooks Miners & Wholesale Dealers Company in Detroit, and was made president. He was also president of J. B. B. Coal Company. In 1909 Jewett met **Fred O. Paige**, who had been president of the Reliance Motor Car Company before it was purchased by General Motors. **Paige** had enlisted an engineer named Andrew Bachie to design an automobile using a three cylinder, two cycle motor, with 25 hp. Jewett took a test drive and decided to organize an automobile company. **Jewett** was successful in the mining business, and began looking for other investors. E. D. Stair, the publisher of the Detroit Free Press joined in, followed by an attorney, Charles B. Warren, Arthur and Willis Buhl, Alex McPherson, president of the Detroit National Bank, and Sherman Depew. In the fall of 1909 **Fred Paige** was named president of the newly formed Paige-Detroit Motor Car Company, with a capitalization of $75,000. Willis Buhl was made VP, Gilbert W. Lee was treasurer and William B. Cady was secretary.

In January 1910 the Paige-Detroit was publicly displayed for the first time at the Palace in New York. It was reported to have the most perfect two-cycle engine ever built for an automobile. The engine was mounted vertically under the hood, and had a 2.75 inch bore and a four inch stroke. The motor and transmission were built as a unit, and had a three point suspension. The transmission was the semi-selective gear type with two forward speeds and one reverse. The engine was cooled by thermo-syphon, and used a magneto for ignition. The car only came as a roadster with a 90 inch wb. The gasoline tank was cylindrical, and mounted behind the seat. The body was unique, using pressed steel construction and with baked enamel the same as the gasoline tank, hood, and fenders.

By the spring of 1910 **Jewett** became dissatisfied with the design of the car and removed **Paige** from the company. **Harry** took over the presidency, and made his brother **Edward Jewett** VP. He had the car re-engineered, and shut down the assembly line in order to bring out a new car for 1911. The car's name was shortened to Paige, and it came with a conventional four stroke, four cylinder engine that developed 25 hp. A customer named W. J. Marshal said that his 1911 Paige four cylinder *has a motor which motes every dollar spent on a Paige* to give back more than that in comfort and reliability. By 1913 the Paige-Detroit Company had five factories and was constructing its sixth factory on McKinstry Avenue, while Harry Jewett was also the president of the **Lozier** Motor Company.

In May 1914 sales increased 136 percent from 504 units during the first four months of 1913 to 1223 units over the first four months of 1914. The increase was partially credited to the new factory which was used since the beginning of the year. *The factory was being worked four nights a week in the machine shop departments as well as Sunday forenoons.* General manager Bourguin, in surveying the future of the automobile industry, claimed that if crops were good there would be no limit to the sale of motor cars for next season. Conditions for 1914 crops were better than the past years, and in many sections of the country where crops were poor in 1913, such as Kansas, there was more ready money than might be expected. The Kansas farmer husbanded his resources in 1913 and was forced to sell his cattle because it was a dry season. They were sold at a good market and he saved his money. Business on the Pacific Coast was poor, along with Texas, but many of the large cities such as Chicago and Minneapolis were strong. The lake states were specially strong, but many points in Kansas and Missouri were a little slower. Sales totaled 4631 in 1914, and jumped up to 7749 in 1915, with the advent of the company's first six cylinder.

In 1916 an entirely new model called the Paige Six-36 was produced with a 3 x 5 inch *en bloc* motor carried on a three point suspension. It had a 112 inch wb, 36-40 hp, and cost $1095, compared to the 1915 model Six-46, which sold for $1395. The gasoline tank was mounted on the dash with a capacity of 14 gallons. Water was circulated with a centrifugal pump through a zigzag cellular type radiator. A three speed selective transmission was attached to the rear axle for better balance. The clutch consisted of seven driving discs with 36 cork inserts in each, running in oil. The electrical system consisted of a Gray & Davis separate generator and starter. The headlamps were furnished with dimming attachments. By regulating the headlamp intensity, separate low intensity cowl lamps for city driving were no longer a necessity. An 18 inch one-piece steering wheel with a corrugated edge and a horn button in the center were located on the left hand side of the car. Both ignition switches and lighting switches were directly in front of the driver and the speedometer, oil pressure gauge and ammeter were in direct view. The various control elements were illuminated with a dash lamp. The Paige was advertised as having an L-head type motor, pointed radiator, and a tapering hood, with the slogan "The Most Beautiful Car in America."

In 1921 Ralph Mulford drove a 6-66 roadster at Daytona Beach and set a new one-mile straight-a-way record of 102.83 mph. Following up on the successful speed run, a three passenger roadster with a seat for two and a pull-out drawer on the right side for a third passenger was offered in 1922 called the Daytona. A smaller car was also brought out in 1922 called the Jewett, which was the first car without sector for spark or throttle on the steering wheel. It was produced until January 3, 1927, then became a Paige model 6-45. In 1925 the Paige Company was in 10th place in the sales race, but soon it started to lag, and by 1927, **Harry Jewett** had lost $2,500,000. In June of 1927 the **Graham Brothers** purchased the company and re-organized it as the Graham-Paige Motors Corporation.

Model number 1 Note headlamps not installed. In front of administration building.
  courtesy AAMA

Paige administration building
  courtesy NAHC

*Jewett-
A subsidiary of Paige, introduced January 12, 1922. (Harry Jewett is standing by the car.)*
  courtesy AAMA

The Jewett factory on Warren Avenue became the Graham-Paige factory.
  courtesy KAHN

Harry M. Jewett

HUDSON MOTOR CAR COMPANY
DETROIT, MICHIGAN
1909-1954

AMERICAN MOTORS CORPORATION
DETROIT, MICHIGAN - KENOSHA, WISCONSIN
1954-1987

"LOOK FOR THE WHITE TRIANGLE"

Joseph L. Hudson had automotive experience as a VP for the Detroit Auto Vehicle Company, but was not enthusiastic about reentering the field. His niece was married to **Roscoe Jackson**, and they persuaded uncle **Hudson** to finance $90,000 for a new car company. The board of directors of the Hudson Motor Car Company was made up of **Joseph L. Hudson, Hugh Chalmers, Roscoe E. Jackson, Howard E. Coffin, Frederick O. Bezner, Roy D. Chapin, James J. Brady, Lee Councelman**, and with only one share, **George W. Dunham**. **Hudson** was elected president, **Chalmers** was VP, **Chapin** secretary, and **Jackson** was elected treasurer and general manager. In April 1909, the company operated an office in Detroit's Penobscot Building and moved into a factory at Mack and Beaufait avenues where the Aerocar had been produced. It also purchased the Selden Patent license of the former Northern Automobile Company. The first advertisement for the "Hudson 20" was published in July 1909, resulting in 4000 orders with $25 down payments. The first models built from July to October 1910 were considered as 1910 models by the factory and were priced at $900 F.O.B. Detroit.

The first Hudsons used a four cylinder L-head engine made by the Atlas Engine Works in Indianapolis. When sales demand exceeded capacity, Buda engines, made by the Buda Company in Harvey, Ill., were also used. Continental engines were used beginning in 1911, and Hudson produced its own engines starting with the 1916 six cylinder.

Hugh Chalmers accepted an offer that let him buy out the **Coffin, Chapin** and **Bezner** shares in the Chalmers-Detroit Motor Company, and he sold his shares of his Hudson stock to them. Of the original Hudson shareholders, only **James J. Brady** and George W. Dunham stayed with the Chalmers-Detroit Company. The Hudson Motor Company was totally independent as of December 20, 1909.

With a record breaking first-year production record, the company purchased land five miles from Detroit's City Hall at Jefferson and Conner avenues to construct a 172,000 square foot "Kahn" factory. The 1912 model became the first medium priced car with a 6-cylinder engine, and was the last year with right hand drive. Of seven body styles, one was a roadster with a 115 inch wb, called the "mile-a-minute" model. The 1913 model 54 had the first 4-speed overdrive transmission in the industry. Over 7000 Hudsons were sold in 1914, and when asked about the 1915 expectations, Chapin responded that the general labor situation was not good throughout the country and many mercantile lines suffered a business shrinkage because of the cut in buying capacity. **Chapin** looked for good crop conditions for the 1915 season and went under the assumption that business always follows money tendencies. He expected 1915 to be a bigger season than 1914, and his company would *cut its cloth accordingly*. In 1915 a new development by Hudson engineers was made public, called the "super-six" principle. This idea was a completely counter balanced crankshaft that raised the horsepower over 50 without a change in the bore or stroke. The success of the L-head six was so assured that the company abandoned all experiments with V-8s and V-12s. Just prior to World War I, the size of the factory was doubled, raising capacity to over 100 cars per day, and in 1916 Hudson sold 26,000 cars.

In early 1919 the Hudson Company introduced a new name in the low priced car field called the "Essex." It was a four cylinder F-head with a 179 cubic inch displacement that developed 55 hp, and by December 1919, 20,000 were produced. The Essex was manufactured in a new plant with a continuous moving production line. The plant had capacity for 150 cars per day and used two 75-car capacity conveyors to allow switching between 75 and 150 car capacity without changing the efficiency. One hundred seventy five men were required to assemble Essex chassis and bodies on a 75-car-per-day basis.

The departments were: Chassis assembly-45, Body assembly-30, Final assembly-49, Wiring assembly-3, Wheel assembly-7, Final touch-up-3, Final equipment-7, Final tuning-11, and Final repair-20 men.

The 1922 Essex enclosed "coach" was priced at $100 above the open touring car and started the public toward year round use of the automobile. The Essex was made of a composite structure with steel panels on a wood frame. It had four seats with a door by each seat, instead of a driver's door and a single door on the opposite side. The coach represented 55 precent of all Essex production in the second quarter. In 1923 when the Hudson Motor Car Company was the largest producer of closed cars in the industry, **Chapin** resigned as president and was succeeded by **Roscoe E. Jackson**. **Roy Chapin** returned temporarily in 1932, and helped raise money for an Essex derivative called the "Terraplane." The Essex Terraplane was introduced for as little as $425, and was followed in 1933 with the Terraplane 8. It had the best power-to-weight ratio of any stock car, and set many records for hill climbing, speed, and endurance during the following six years.

Hudson introduced automatic shifting at the steering column using a vacuum device called the "electric hand" in 1935. Hydraulic brakes were added for 1936, with an added safety feature of a cable brake system that was activated with further pressure on the foot pedal. The double brake system was used throughout the remaining years of the Hudson.

A. E. Barit, who started with the company in 1910, became president in 1936 and remained until 1954. The company made a small profit during the war years, and jumped back into the car business in 1945, as the fifth largest manufacturer. The 1946 and 1947 models were face-lifted pre-war designs. In October 1947 a radical change in production methods was made to produce the 1948 Hudson. The "monobilt" step-down unit-body was one of the great designs of the post war era. They were low and sleek, and hugged the ground. The body used a sturdy stub-frame, fore-structure welded to a cage that constituted the main body structure. The stub frame was mainly used due to space considerations on the assembly line. The plant could produce a greater number of shorter bodies. Like all Hudsons since 1932, it had a unit body that was extremely strong and rattle free. The nickname "step-down" came from its recessed floor, which was completely surrounded by frame girders. That gave more leverage to resist tortional stress, making it the safest package in its time. The old Detroit rule against "restyling and re-engineering" in the same year was violated, and a new 262 cubic inch Super Six engine was brought out also, making the marque one of the quickest and road-worthy cars. By 1951 the engine evolved into the 308 cubic inch Hudson Hornet engine, making Hudson the king of stock car racing. The wrap around frame became an asset when the cars hit the retaining walls by protecting the tires and suspension from damage.

In a related occurrence, the Nash-Kelvinator Corporation introduced the first modern compact car in 1950, and revived the name "Rambler" that dated back to 1902 when Thomas B. Jeffery created the first Rambler in Kenosha, Wisc. In 1953 Hudson similarly came out with a compact car called the "Hudson Jet" to compete head to head with Ford and Chevrolet.

There was a problem with the full size Hudson "step-downs" because the design was hard to change, and the Hudson Motor Car Company lacked the financial base to produce new models. With minor face lifts year after year, and the decline of the seller's market after 1950, sales went spinning into a downfall. On May 1, 1954, **A. E. Barit** stepped down from the presidency, and the Hudson Motor Car Company was merged with the Nash-Kelvinator Corporation to form the American Motors Corporation. **George W. Mason**, the president of the Nash-Kelvinator Corporation since 1937, was named chairman and president, and 47 year old **George Romney** was elected executive vice president. The last Detroit-built Hudson left the assembly line in August 1954. **Mason** died suddenly on October 8, 1954, and four days later the board of directors elected Romney as president.

Production was moved to Kenosha, Wisc., and the Hudson body became a re-badged Nash, often referred to as a "Hash." The last Hudson only came in one model, the Hornet V-8 with 255 hp that could go from 0 to 60 in 9 seconds. However, **Romney** had decided to stake everything on the Nash Rambler. He hired **Roy Abernethy**, former sales executive with Packard and Willys, as sales chief. Beginning with the 1958 model year, American Motors stopped making big cars and dropped the Nash and Hudson names.

1909-
The first Hudson
courtesy AAMA

First Hudson
factory-
Mack and Beaufait
80,000 square feet.
(former Aerocar
factory)
courtesy AAMA

First Hudson
factory
courtesy AAMA

*1910-
Hudson used for
delivering the Detroit
News
   courtesy AAMA*

*Jefferson Avenue
plant
   courtesy Tinder*

courtesy Kahn

courtesy Kahn

*1932-
Amelia Earhart next to a Terraplane (she christened the first 1931 model)
courtesy AAMA*

*1934-
Orville Wright (L) and Roy D. Chapin (R) viewing a Terraplane*

*Hudson
courtesy AAMA*

*1955-
A Hudson Hornet climbs a 28 percent grade at the American Motors proving grounds near Burlington, Wisc. The test area covered 204 acres including a 1.22 mile outside high-speed black-top oval.
courtesy AAMA*

*The Hudson Triangle-
The first side was the work of the designer who planned the car and the factory workers who built it.*

The second side was the cordial welcome dealers all over the country gave the Hudson.

The third side was of the owners.

In 1975, J. C. Long (Chapins biographer) speculated Chapin, Coffin, and Bezner were the three sides!

UNITED STATES MOTOR COMPANY
1910-1912

After the collapse of the Buick-Maxwell merger negotiations, **Durant** was successful in forming General Motors in September 1908. **Briscoe's** first attempt to find a new partner came in November 1908, when he was given an opportunity to purchase 60 percent of the Cadillac Motor Car Company stock for $150 per share, but he was unable to find financial backing. **Briscoe** went back to **Durant** and offered the Maxwell-Briscoe Company, the Briscoe Sheet Metal Company, and his brother **Frank's** Brush Runabout Company for $5,000,000. The Morgan interests vetoed the deal, insisting that **Durant** had to provide $2,000,000 in cash. **Durant** had his hands full trying to purchase Cadillac for $4,000,000, and the merger plans collapsed. **Billy** was aquiring the better companies and **Benjamin** had to make his move.

The United States Motor Company was formed when the Maxwell-Briscoe Motor Car Company combined with the Columbia Motor Car Company in February 1910. **Briscoe** knew the Columbia was not an automobile winner, but the company was owned by wealthy Easterners such as Anthony N. Brady, and it was the possesor of the **Selden Patent**. This was his answer to General Motors, with backing from Anthony N. Brady, the Morgan interests, and investment banker Eugene Meyer. Rumors abounded that the E. M. F. Company would be acquired, but it fell into the hands of Studebaker. REO would have given the combine a car with a volume second only to Maxwell's 10,000 per year, but at $7,000,000, the price was beyond reach. Brother **Frank's** Brush Runabout did not live up to expectations as it became an outmoded design. The Gray Motor Company of Detroit was added, along with Stoddard-Dayton, which also made the Courier car, and Sampson.

Eventually there were 14,000 employees building 52 models in 18 plants under seven marques distributed through 34 branches and not one car selling well. The corporation extended from New York to Michigan, Connecticut, Rhode Island, Ohio, and Massachusetts making it impossible to maintain control, and **Briscoe** ran out of money.

In mid-October 1912 the corporation was in mid-receivership, and everyone wondered what would happen next. **Briscoe** said it would happen without him and resigned.

The Briscoe brothers went to Paris to learn how to build automobiles the "French way." They built a motor in Billancourt, emulating the Sizaire motor with a small bore and a long stroke. They returned to the United States and formed the Briscoe Motor Company, settling in the former Standard Electric factory in Jackson, Mich. The first "Briscoe" was a small roadster on a 106 inch wb powered by a 157 cubic inch L-head engine. The car was easily recognizable by its cyclop's headlight at the top of the radiator.

Benjamin served in the First World War and was awarded the Navy Cross. In October 1921, Briscoe left the company and appointed Clarence A. Earl to take over. The name of the company was changed to Earl Motors, with the car name changed to Earl. The company was out of business by early 1924. Briscoe also produced the Argo cyclecar from 1914 to 1917, which became the Hackett from 1917 to 1919. Benjamin Briscoe moved to Montreal and managed the Frontenac Oil Company, and developed improvements in refining oil. He then went to Colorado and became involved in gold mining. In 1938 he suffered a stroke, and his wife, Lewis Snyder Briscoe, died. Briscoe retired to Florida and remarried to Ellen Lustig Briscoe, and experimented with tung trees and varieties of hay. Benjamin died on July 27, 1945. His brother Frank died in 1954.

43,000 Maxwells in daily use.

Maxwell

Need we say more?

MAXWELL-BRISCOE MOTOR COMPANY 61st STREET and B'WAY **NEW YORK CITY**
Division of UNITED STATES MOTOR COMPANY

Builder of Motor Cars for Seventeen Years.

THE COLUMBIA MOTOR CAR COMPANY 61st STREET and B'WAY **NEW YORK CITY**
Division of UNITED STATES MOTOR COMPANY

None can go farther.

"Stoddard=Dayton"

None can go faster.

DAYTON MOTOR CAR COMPANY 61st STREET and B'WAY **NEW YORK CITY**
Division of UNITED STATES MOTOR COMPANY

Liberty-Brush $350

Everyman's Car

Standard Brush Runabout, $450

Low enough in price to be within the reach of everyone.

THE BRUSH RUNABOUT COMPANY 61st STREET and B'WAY **NEW YORK CITY**
Division of UNITED STATES MOTOR COMPANY

Sampson
Freight and Delivery Motors

Strong as its name suggests.

ALDEN SAMPSON MANUFACTURING COMPANY 61st STREET and B'WAY **NEW YORK CITY**
Division of UNITED STATES MOTOR COMPANY

Sampson 35

The best touring car at the price.

ALDEN SAMPSON MANUFACTURING COMPANY 61st STREET and B'WAY **NEW YORK CITY**
Division of UNITED STATES MOTOR COMPANY

ALDEN SAMPSON MANUFACTURING COMPANY
PITTSFIELD, MASSACHUSETTS
1904-1911

ALDEN SAMPSON
DIVISION OF UNITED STATES MOTOR COMPANY
DETROIT, MICHIGAN
1911-1912

The first automobile hit Berkshire County, Mass., in 1900 and just as everywhere else, people regarded it as another diversion like the rest of the sports. The first automobiles in Pittsfield, Mass., were owned by doctors, who were determined to make the automobile more than a sport or fad. They were fed up with horses. In 1902 the Berkshire Automobile Club was created by a group of doctors. They mapped out and led tours of the club over Berkshire County's hazardous dirt roads. Among the officers were **Alden Sampson**, who in 1905 owned the Alden Sampson Machine Company, in Pittsfield, Mass.

In 1903 the Moyea Automobile Company was formed by Henry Cryder. He began production of the Moyea in Middletown, Ohio, while building a factory in Rye, N. Y. The Moyea Automobile Company was making a copy of the French Rochet-Schneider under license. In January 1904 the Moyea Automobile Company changed its name to Consolidated Motor Company. During its first year of production it became apparent that the Moyea needed structural improvements. Before the Rye, N. Y., factory was completed, the Alden Sampson Machine Company was asked to develop a prototype. In 1904 **Alden Sampson** bought out Consolidated Motor Company and renamed the car "Alden Sampson."

In 1905, the Alden Sampson Company absorbed the Crest Manufacturing Company located first in Cambridge and then in Dorchester, Mass. Crest had produced cars since 1901 called the Crestmobile. Soon afterwards, the Alden-Sampson car was discontinued and a truck called the Sampson was manufactured. They were hand-made using a great deal of nickel steel in their mechanism and were regarded as indestructible.

Alden Sampson died in 1909. By 1911 his widow had re-married and sold the company to the United States Motor Company for $200,000. A factory was procured in Detroit and in addition to the truck, a car called the "Sampson" was produced with a 35 hp engine, shaft drive, and like most cars, it had right hand steering. The price was $1325. Both car and truck were victims of the collapse of the United States Motor Company in 1912.

Heckers flour mill in New York demonstrated one of its Sampson trucks and 12 horses that it replaced.

1904- Sampson 3A 4 cylinder car with tonneau and hood removed.

1911- Alden-Sampson factory in Detroit courtesy NAHC

1911- Alden-Sampson factory in Detroit courtesy NAHC

THE STARTING CRANK

There was one rule to observe when cranking a gasoline engine--first retard the spark. It should have been easy to remember, but was usually forgotten until the engine "kicked." The object of cranking the engine is to draw a charge of gasoline into the cylinders, compress it, then ignite it. The spark should occur when the piston is moving outward on the stroke following the compression stroke, so that an explosion will start the engine. With the spark retarded, ignition occurs when the piston has reached the end of the compression stroke, or maybe has moved outward a little on the power stroke.

If the spark is not retarded while the crank is revolved, a charge is drawn into the cylinders and the spark ignites the fuel mixture before the pistons have reached the end of the compression stroke. As the starting crank is pulled and the pistons are forced upward, the pressure that is developed by the explosion suddenly forces the piston to move downward. This makes the crankshaft move backward, and you are trying to prevent it by your grip on the starting crank. It happens very suddenly and before you know it the starting crank is torn out of your hand, and in its swing is only too likely to hit your forearm, wrist, or back of the hand, with every possibility of doing serious injury.

Most automobile engines have a ccw rotation so that when you stand facing the front, the crank pulls up on your left side (the engine appears to rotate clockwise when viewing it from the front of the automobile). It is natural to grasp the starting crank in the right hand, pull up, and then press it down with the elbow stiff and the weight of the body bearing on it. This leads to trouble when the engine "kicks." When you are pressing down, your stiff arm can be broken. The proper procedure is to pull up; never push down. Another safety procedure is to grasp the starting crank with your thumb hooked over. Better yet, using the left hand was the safest, but very few people complied.

1911-
Cranking an Abbott
on Belle isle
  courtesy AAMA

BRUSH RUNABOUT COMPANY
DETROIT, MICHIGAN
1906-1910

BRUSH RUNABOUT COMPANY
DIVISION OF <u>UNITED STATES MOTOR COMPANY</u>
DETROIT, MICHIGAN
1910-1912

"EVERYMAN'S CAR"

Alanson Partridge Brush was born on February 10, 1878, in Detroit, Mich., to Charles and Hannah Brush. He attended school in Detroit and Birmingham, Mich., and had a natural inclination for mechanics as a boy. He enlisted in the 31st Michigan Volunteer Regiment when the Spanish-American War broke out and served in Cuba until 1898. When he returned to Detroit in 1899 he went to work for the Leland & Faulkner Manufacturing Company to gain experience with machinery. After L & F received an order for 2000 engines for the Olds runabout in 1901, **Alanson** had his first big challenge. With production problems mounting and the possibility of losing the contract, **Henry Leland** asked **Alanson**: "you are a good mechanic, can't you think of some way out of this difficulty?" **Brush** went back to the shop for several days and designed some of the gear cutting tools that got the job done. **Leland** let **Brush** experiment after that and he developed a mixer that led **Charles Strelinger** to have **Alanson** design a small engine for marine use. The mixer design was reported to have been used in 1903 on Ford's racer 999. After receiving a $1000 payment for the design, **Alanson** married his girlfriend Jane Marsh in March 1902.

Alanson improved the Olds engine with a design that produced 10 hp compared to the Olds 3 hp, and it became the first Cadillac engine in 1902. By 1905 he had designed the four cylinder engine that helped **Leland** transform Cadillac into a higher price range. He and Jane lived on Harper Avenue in Detroit, across the alley from the Ford family. **Henry Ford** and **Alanson Brush** would play ball in the alley with young **Edsel**.

Leland was not always an easy person to work for, and he seemed worse since he took over the reigns at Cadillac. In 1905 **Alanson** was unhappy about the way **Leland** tried to control and use his patents. Finally **Brush** left with a $40,000 settlement for the use of his patents and designs for two years. In turn, **Brush** would not participate in any business that was competitive with Cadillac. After two years, the Cadillac Company would pay royalties to **Brush** for any patent they still needed to use. **Leland** informed the Cadillac directors that with great confidence, "I will eliminate all the **Brush**" in our cars inside two years. **Leland** was unable to follow through and the Cadillac Company paid **Brush** for some of the patents until they expired.

For the next two years **Alanson** experimented in a shop he built near his home and designed a small car to test out new ideas such as a coil spring suspension system. After completing his car and pushing it out to test, he took a ride with **Frank Briscoe**, who said "Al, I think there is a future for this little car, just as it stands, and I want to build it!"

The formation of the Brush Runabout Company was announced in November 1906, with a capitalization of $200,000. **Frank Briscoe** was general manager, George B. Yerkes was president, F. A. Harris was VP, Emil D. Moessner was secretary, and A. C. Miller was treasurer. The Eclipse Manufacturing Company on Euclid Avenue in Detroit was purchased for building engines, and the Briscoe factory in Detroit was used to assemble the car. The first automobiles went on sale on April 1, 1907. The design of the "Brush" was very simple and stayed that way for the life of the marque, mainly due to the fact that **Alanson** was not a member of the company. The engine consisted of a single vertical cylinder, with a 4 x 4 bore and stroke, rated at 6 hp at 1000 rpm. **The engine rotated clockwise, making the hand crank rotate counter clockwise, for safety in the event of kick-back.** The cylinder head screwed on and could be removed with very little trouble to inspect or clean the cylinder walls. A Briscoe flat tube radiator was located in front of the hood for water cooling.

There was no transmission, but rather a variable speed clutch that gave eight forward speeds and one reverse. At high speed it became simply a friction clutch. The main shaft drove a counter shaft through a bevel gear and from the ends of the counter shaft there were two short chains to drive the rear wheels. The compensating gear was unique, so that when going around a curve the outside wheel overran and there was no tendency to make the two wheels revolve in opposite directions.

Steering was by wheel, located on the left side, using a new Brush-type of steering gear that was a slow powerful irreversible gear when driving straight ahead, but the further the steering wheel was turned, the faster the front wheels turned in proportion.

The wheelbase was 74 inches, with standard tread, although a 60 inch tread was available for the wider wagon ruts in the southern regions of the United States. The body was of piano box design with a helical suspension spring at each corner that was stressed in tension rather than compression as most cars used later. **Brush** was not the first to use coil springs, as the Cannstatt-Daimler and commercial vehicles used this device in the mid-90s. The Brush had wooden axles, and a steel-armored wooden frame that gave the short wb car a smoother ride. The factory made a point of telling prospective owners that the axles were made with selected maple, air dried, then kiln dried, and oil treated before painting. Mention was also made that wood was not subject to the metal weakening known as crystallization. The car's construction gave rise to the expression "wooden body, wooden axles, wooden wheels, wooden run."

The Brush runabout was a bargain at $500 and production climbed from 500 in 1907 to 10,000 in 1910, equaling the output of Benjamin Briscoe's Maxwell-Briscoe Company. Both companies were joined that year into the United States Motor consolidation. When U. S. Motors collapsed in 1912, the Brush lasted until 1913 then was discontinued.

Although **Alanson** received remuneration from the Brush company, he was not actively involved. Instead he became co-founder of the Oakland Motor Car Company with Edward W. Murphy who was the head of the Pontiac Buggy Company, founded in 1890. When Murphy decided to get into the Automobile business, he saw the Brush runabout and liked it. He asked **Alanson** to design the Oakland, since his two year agreement with the Cadillac company expired. He brought with him a two cylinder engine design that rotated clockwise, designed while he was at Cadillac. From Oakland, Brush worked as an engineer at Buick, then as a consultant.

By the end of the 1920s most automobile companies had their own engineering staffs, and **Alanson's** engineering service was declining. He started a new career as one of the best known consultants to lawyers in patent litigation. **Alanson Brush** died in March 1952.

1910-
Brush on
South Boulevard
in Detroit
courtesy AAMA

NEW HOME OF THE BRUSH RUNABOUT COMPANY, DETROIT

Brush factory courtesy NAHC

Brush factory courtesy Tinder

1910- Brush closed runabout courtesy AAMA

coil springs were stressed in tension

Alanson P. Brush

TEMPLETON-DuBRIE CAR COMPANY
DETROIT, MICHIGAN
1910

Arthur Auguste Caille was born in 1867. In 1888 he and his brother Adolph invented and patented a new type of service for conveying cash in stores. They later invented coin-slot devices which resulted in the establishment of penny arcades in the leading cities of the world.

Adolph Caille started his career making furniture in Saginaw, Mich. By 1901 the brothers had the largest coin operating machine factory in the United States. Located in Detroit at 1427 to 1457 Woodward Avenue, the factory had a capacity of 300 workers.

Arthur was president and treasurer, with an office in another factory on 1300 to 1340 2nd Street. Their slot machines weighed over 300 lbs, and were basically of *home manufacture*, with the cabinets made in Saginaw. The only part purchased was the music box which came from Switzerland.

Slot machines were considered unlawful, and raids by the police were a common occurrence. The machines were hard to conceal once the raid was in progress, and machine loss was expensive. The Caille brothers also manufactured scales and other equipment to offset the loss of their machines. In 1905 Arthur joined John Kunsky and opened the first theater in Detroit devoted exclusively for showing motion pictures at Monroe and Farmer, called the "Casino" Theater.

The automobile industry offered another opportunity for the Caille brothers, and in 1904 they met up with Stanley R. DuBrie. He was a mechanical engineer from the University of Ohio. DuBrie built his first experimental car in 1896 and established the DuBrie Motor Works in Freemont, Ohio, for the manufacture of two stroke marine engines. He sold his business to the Clauss Shear Works in 1900, and went to Canada and built his second experimental car for the Charles Beck Manufacturing Company in Penetang, Ontario. Then he designed a two stroke marine engine for Peter Payette and Company in Penetang.

DuBrie returned to the United States and received financial backing from the Caille brothers to build a two cylinder, two stroke engine and a touring car body and chassis. Testing began in July 1904. The Caille brothers did not pursue the project any further, but did produce marine engines under the name of the Caille Perfection Motor Company.

The Templeton-Dubrie Car Company was organized in the early part of 1910 to manufacture a combination pleasure-delivery car designed by Stanley DuBrie. The factory was located at 687 Mack Avenue, just a few doors away from the former Ford Mack Avenue factory. The "Templeton-DuBrie" had 20 hp, a 124 inch wb, and sold for $1200. A two cylinder, two-stroke gasoline engine, similar to a Wolfe compound steam engine that had no valves or crankcase compression, was used with a planetary transmission with shaft drive. The company did not last through 1910, although DuBrie manufactured the "DuBrie" marine engine.

The Du Brie-Caille 2-cylinder, 2-cycle car

Du Brie-Caille engine

Templeton-Du Brie

THE PURITAN MACHINE COMPANY, LTD.

The Puritan Machine Company, Ltd was formed in 1903, and was located in the Puritan Building at 10-12 Gratiot Avenue, Detroit. The company started out making novelties, with George Maitland as manager. In 1905 Maitland became chairman, and in 1906 **A. O. Dunk** became treasurer.

A. O. Dunk began manufacturing automobiles and parts in 1898, starting with Locomobile steamers. In 1907 the company began making coin machines and providing general machinists work, with Thomas Dunk as secretary. The firm moved to 72-76 Brush Street and produced electrical heating units, while Detroit was the biggest producer in the country.

By 1911 **Dunk** was president and the "Deluxe Auto Parts" business, at 51-57 10th Street, was formed to supplement Puritan. In 1913, another plant was added at 413-415 Lafayette Boulevard. In 1915 **Dunk** left the Auto Parts manufacturing and it was taken over by **William Metzger**. The Puritan Machine Company became inactive in 1924. In 1927 **Dunk** acquired control of the Detroit Electric Car Company following the death of **W. C. Anderson**.

Companies like Puritan played a strategic role in the evolution of the automobile industry. Spare "pupose" built parts and standard "aftermarket" parts helped make the automobile affordable. The automobile was much more complicated than the horse carriage and there were too many parts to have custom made by artisans.

A. O. Dunk

BEYSTER-DETROIT
DETROIT, MICHIGAN
1910-1911

Henry E. Beyster and **Tom Thorpe** headed up the Beyster-Thorpe Motor Company. The factory had a 60 foot frontage and 500 feet back located at 1329 Woodward Avenue between Antoinette and Vienna streets. Besides producing motors that were designed by Henry Beyster, The Beyster-Thorpe Motor Company had been a distributor for the **Aerocar**. Accordingly, the two men had expertise in manufacturing and sales.

Although the automobile market seemed to be saturated, they were convinced that no delivery vans were being produced, other than converted runabout automobiles. They organized the Beyster-Detroit Motor Car Company in December 1909 with a capital stock of $50,000, and the use of the Beyster-Thorpe Motor Company factory. Light delivery cars were the principal focus of the firm.

The "Beyster-Detroit" delivery car had a 105 inch wb with a 56 inch tread, and 60 inch for the southern trade. The capacity of the delivery body was 1200 lbs, powered by a 4-cylinder engine with 25 hp, using a 3.25 x 3.6 bore and stroke. The transmission was a selective shift with sliding gears, and final drive was by double chains, in oil, aluminum housings. The speed ranged from 25 to 55 mph, according to the gearing desired. They said the "lines were stylish, and the body was a masterpiece of cabinet construction."

A number of runabouts using the Hupmobile 20 engine were produced also, but competition was too strenuous. In April 1911 the company announced it was "winding up its affairs."

Tom Thorpe

Henry Beyster

Beyster-Detroit delivery car Brewster green with yellow wheels

WARREN MOTOR CAR COMPANY
DETROIT, MICHIGAN
1910-1913

Homer Warren was born in Shelby, Mich., in 1855. He was educated in a public school in Albion, and graduated from Albion College. He moved to Detroit in 1872 and began working at J. M. Arnold & Company, retailer of books and stationary. He married Susie Leach in 1879, and was appointed deputy collector of customs in Detroit that same year. He entered the real estate business from 1886 to 1892. The Homer Warren & Company was organized in 1892 with Cullen Brown and **Charles R. Walker**. **Homer Warren** was appointed postmaster of Detroit by President Roosevelt in 1906.

During the fall of 1909 the Warren Motor Car Company was organized with John G. Bayerline, former purchasing agent for the Olds Motor Works, as general manager. W. H. Radford, former engineer at Hudson, was chief engineer. The factory was 300 feet long by 60 feet wide, located on Holden Avenue and the Grand Trunk railroad. The "Warren-Detroit 30" had 30 hp, and sold for $1000. John Bayerline announced a deal with the Taylor Distributing Company of Philadelphia for the purchase of the entire 1910 output and 500 more in 1911.

The motor had four cylinders, cast *en bloc*, with a 4 x 4.5 inch bore and stroke. The transmission was a selective type sliding gear, with hardened steel gears, and liberal bearings of parson's white brass. The body trimmings included white pyramid rubber molded mats, brass scuff plates, a square mahogany dash with polished brass lamp brackets, and a polished aluminum bonnet ledge.

In January 1911 nine distinct models built on the standard Warren "30" chassis were shown at the New York Auto Show at the Palace: a four door roadster, four door touring and torpedo types, round tank roadster, a roadster with a dickey seat, a five-passenger demi-tonneou, a standard touring car, and inside drive coupe, and two types of delivery cars with 1000 lbs capacity. The "Wolverine" was in the heart of the exhibit. It was driven by "snowball" Bill Smith from Detroit to New York in deep snow and blizzards, alone and unaided. It was designated by the New York papers as the "blizzard buster."

In 1911 the Warren Company was selling cars in: Sydney N.S.W., Stourbridge, England: Guatemala, Central America, Mexico City and Monterrey, Mexico, Denmark, St. Petersburg, Russia, and Dublin, Ireland. The Taylor Distributing Company did not renew its contract with the Warren Company in 1912, and sales plummeted. The Warren Motor Company was in receivership in April 1913.

1911- Warren on Belle Isle with Vista Victoria behind the wheel. Finished in Royal Blue, striped in ivory; running gearstraw color, striped with black; fenders black.
courtesy AAMA

Homer Warren

Warren-Detroit "30"
An Uncommon Car at a Common Price
Which will serve you tomorrow as well as today.

ROADSTER (34x3½ tires) $1200. TOURING CAR (34x3½ tires) $1325. FORE-DOOR (35x4 tires) $1500.

Power, style and comfort are all incorporated in this reliable machine. The purchaser secures a 110-inch wheel base, selective type, sliding gear transmission, cone clutch, double ignition system and full equipment. There are eight models: roadster (dickey seat); roadster (gasoline tank); demi-tonneau, touring; fore-door; torpedo; inside drive and light delivery wagon. Deliveries prompt. Write for announcement.

WARREN MOTOR CAR CO.
Detroit, Mich.

Warren-Detroit "30"

SIBLEY MOTOR CAR COMPANY
DETROIT, MICHIGAN
1910-1911

The Sibley Motor Car Co. was formed in early 1910, with **Fred M. Sibley,** a Michigan lumber dealer, a principal financial backer. Clyde B. Warren was the president, C. P. Moore was VP, and Fred's son, Eugene, was secretary and treasurer. Former chief engineer of Chalmers, and designer of the Abbot-Detroit, **John G. Utz** was also involved.

Offices were opened up at 870 Woodward Avenue, and the Detroit Valve & Fitting Company factory at the N. E. corner of Solvay and Mackie Avenue was leased for manufacturing.

The "Sibley 20" sold for $850, and only came as a two-passenger roadster on a 100 inch wheelbase, with 32 inch wheels. It had a four cylinder motor with 30 horsepower, and a three speed selective shift transmission.

In January 1911 the Detroit Valve and Fitting Company sued the Sibley Company for recovery of its plant, charging default on the lease agreement, which ended the Sibley Motor Car Company. Eugene Sibley moved to Simsbury, Conn. in the fall of 1911, and formed a partnership with an automobile dealer named Joseph J. Curtiss. Together they organized the Sibley-Curtiss Motor Company, but only sold two automobiles through 1912.

In 1914 **August Fruehauf** had a blacksmith shop on Gratiot Avenue. Besides shoes and wheels, other wagon parts were made there, including complete wagons. **Fred Sibley's** lumber yard was also on Gratiot Avenue and he asked the **Fruehauf** establishment to make a wagon to pull loads, for his new Ford roadster. At that time it was well understood that a horse could pull much more than it could carry. **August Fruehauf** did not realize that he was founding a new industry when he delivered his first "trailer" to **Fred Sibley.**

Eugene Sibley behind the wheel and Fred on the right.
courtesy AAMA

Sibley factory courtesy NAHC

The Fruehauf blacksmith shop on Gratiot in Detroit

The first Fruehauf trailer courtesy Sibley Lumber

OWEN MOTOR CAR COMPANY
DETROIT, MICHIGAN
1910-1911

In 1898 **Ralph Owen** developed an experimental automobile while he owned the R. M. Owen Carpet & Rug Manufacturing Company, in Cleveland, Ohio. From 1901 through 1903 he was the head of the Owen Motor Company in Cleveland, and produced the Owen automobile. In January 1910 he organized the Owen Motor Car Company with a capital stock of $500,000, along with **Angus Smith** and Frank E. Robinson. The factory was on the N. W. corner of East Grand Boulevard and the M.C.R.R., just west of the Packard Motor Company.

The new Owen had a straight-line torpedo body made of aluminum sheet, and was offered in four body styles. The adoption of a *closed front* protected the occupants from wind and weather and was a preventative for trailing robes or garments. It had 42 inch high wheels with pneumatic tires, providing 200 percent to 300 percent greater mileage. The riding qualities were that it provided the ability to break a path in deep snow, sand, or other *heavy going*, and increased traction area on the road, consequently allowing freedom from side slips and skids. Left-hand steer and right-hand control was also featured. It had a four cylinder L-head with a 4.3 inch bore and a 6 inch stroke developing 50 hp, which identified it in the long stroke class.

Sales were slow, so **Ralph Owen** contacted his brother Raymond (whose R. M. Owen & Company was a distributor for the REO Motor Car Company), and a deal was made with the REO Company to take over the entire Owen organization. In return, **Ralph Owen, Smith**, and Robinson received REO stock. The REO organization completed 35 cars but could only sell 31, which indicated that Owen would not be a commercial success, and it was discontinued in 1911. **Ralph Owen** marketed the last cars as the "R. O.," perhaps a way of telling Ransom E. Olds that the two men had something in common.

The two brothers went to New York and in 1912 began working on refinement of an electric transmission, developed by Justis B. Entz. In 1915 the "Owen Magnetic" car was shown at the New York Automobile Show and was produced until 1921.

1910- Owen with Ty Cobb behind the wheel courtesy AAMA

Cobb, his wife and Ty Jr.

rear view of the Owen factory courtesy NAHC

METZGER MOTOR CAR COMPANY
DETROIT, MICHIGAN
1910-1912

EVERITT MOTOR CAR COMPANY
DETROIT, MICHIGAN
1912

FLANDERS MOTOR CAR COMPANY
DETROIT, MICHIGAN
1912

FLANDERS MOTOR CAR COMPANY
DIVISION OF UNITED STATES MOTOR COMPANY
DETROIT, MICHIGAN
1913

The Metzger Motor Car Company was incorporated on September 20, 1906 with capital of $500,000, of which $300,000 was paid in. Byron Everitt was president, William Kelly was VP, and William Metzger was secretary, treasurer, and general manager. All three officers had left the E.M.F. Company when it became apparent that Studebaker planned a takeover. William Kelly designed a runabout and a five passenger model that was already prototyped when the Metzger Company was incorporated. The plant of the former Jacob Meier Company, truck manufacturers, located on the corner of Dequindre and Milwaukee Avenue, by the Grand Trunk R. R., was purchased by the Metzger Company. Although the plant was several years old, it was considered ideal, and was in a good location for manufacturing automobiles.

Obtaining membership into the Association of Licensed Automobile Manufacturers was not as easy as it would seem for the Metzger Company. Although William Metzger was one of the early directors of the A.L.A.M. and all of the companies he had been connected with, Northern, Cadillac, and E.M.F., were members, it required the combined efforts of Kelly, Everitt, and Metzger. They contacted William Hewitt and purchased the Hewitt Motor Company in N.Y., N.Y. which was licensed by the A.L.A.M. The Hewitt truck was being manufactured with capacities of one, two, three, five, seven and ten tons. The Metzger Company added sight-seeing cars and taxicabs, along with the pleasure car designed by Kelly.

The Metzger Company produced a car named "Everitt" which in 1911 came as a demi-tonneau touring car that sold for $1350. It had a 110 inch wb, and came with a 56 or 60 inch gauge, and weighed 1650 lbs. One engine was available, a 30 hp, cast *en bloc* four cylinder with vacuum feed lubrication. Water was circulated with a centrifigul pump for engine cooling. A selective, sliding type change gear was used with a leather faced aluminum cone clutch, and shaft drive.

It was said that 1912 would be the biggest six cylinder year ever known, and the new Everitt six was unusual in many ways. The most striking was the cast *en bloc* type cylinder casting in which the six cylinders, intake and exhaust pipes, and upper part of the crank case were all cast in one piece, doing away with about 200 parts. Besides the six cylinder chassis, two four cylinder chassis were available, priced at $1850, $1500, and $1250.

In May 1912 plans were completed for the merger of the Metzger Motor Car Company and the manufacturing interests of Walter E. Flanders, including the Flanders Manufacturing Company. Walter Flanders was not satisfied with the gradual E.M.F. takeover by Studebaker. The big three were back together in June of 1912, with the newly formed Everitt Motor Car Company. As soon as Flanders was released by Studebaker, the Everitt Motor Car company was re-organized as the Flanders Motor Car Company. (This was the second company with the same name. The first company that bore the name was under a proviso of the

Studebaker Automobile Company. After Walter Flanders left Studebaker, the "Flanders 20" was renamed the "Studebaker 20.")

The new model produced the Everitt Company was called the "Flanders Six," and was introduced at the 1913 New York Automobile Show. In the Fall of 1912 Walter Flanders had been asked by the creditors of the United States Motors Company for help in re-organizing after it had gone into receivership in Septenber 1912. Flanders accepted under the condition that the United States Motor Company buy out the Flanders Motor Car Company for $3,750,000, which resulted in a handsome profit for the "big three."

In 1915 Everitt got back into the automobile painting and trimming business with his brothers, Roland and Gordon. The new firm was located at 669 Mack Avenue. Bill "twinkle eyes" Metzger stayed out of the automotive manufacturing business and worked for the Automobile Club of Michigan.

Annette Kellerman famous swimmer who shocked the nation wearing skin-colored tights in her diving act courtesy AAMA

Metzger factory courtesy NAHC

THE STEELY AUTO ENGINE COMPANY
DETROIT, MICHIGAN
1910

The Steely Auto Engine Company was incorporated in the summer of 1910 with a capital stock of $150,000 for the manufacture of two cycle engines, steering gears, pnuematic tires of leather and canvass, and convertible motor cars.

The company officials included W. J. McWain, M. G. Delaney, E. D. Snowden and J. J. Marks. The "Steely" had a four cylinder engine with 35 hp and was a tourabout that could be converted into a 1000 lb delivery car. The venture was short lived.

WHEELS

The first automobiles used wagon wheels with wooden spokes. Some builders preferred wheels which resembled spoked bicycle wheels, but were much stronger. As cars became heavier and more powerful, the wheel showed signs of stress sideways, or at right angles to its plane. The so called "Artillery" wheel, a heavier spoked wheel which resembled a mobile gun carriage wheel, came into use.

By the 1920's, the "Disteel" wheel, which was shaped like a plate came into use. It was concave, tapered pressed steel. It was very good for side load such as going around corners because it is difficult to thrust a cone inside out. Pressed steel spoked wheels also were in use, which were much more dimensionally true and rugged than the wooden artillery wheel.

Steel spoked wheels made a comeback on the basis they had spring from the ground to the axle giving a smoother ride.

Because of frequent flat tires, solid tires were used with novel approaches of spring. The Detroit Wheel Company made a version of the "Moore" spring wheel with shock absorber spring used with the spokes. The "Airhart" wheel had leaf springs inside the tire rim, connected to each spoke. The Standard Auto Company made springs all along the body and the frame to isolate the body from the wheels with solid rubber tires.

Wheelwrights made wheels for automobiles. Oak was used for spokes, which were between 1.125 inch to 1.75 inch in diameter. A good grain was important since the wood would shrink, and make loose hubs and joints. Elm was used for naves, and ash for the felloes.
    courtesy Tinder

VAN DYKE MOTOR CAR COMPANY
DETROIT, MICHIGAN
1910-1912

The Van Dyke Motor Car Company was organized in 1910 with $1,000,000 capitalization. Frank G. Van Dyke was president, George A. Troutt was general manager, and the car was designed by M. Davis. The **Palms** estate was a major shareholder among several prominent Detroit capitalists. Seven acres of land was secured at West Fort and 36th streets to build a factory. Three 60 x 600 foot buildings were planned for an annual production volume of 1000 units.

The "Van Dyke" was a delivery wagon with a 1000 lb capacity. It used a water cooled 12 horsepower, two cylinder opposed motor, and was capable of any speed, under load, up to 16 mph. One of the chief features of the Van Dyke was its friction drive. It not only simplified the construction of the chassis, but it eliminated the risk of the inexperienced driver stripping the gears.

The company believed it was the first to offer a vehicle specifically designed for light delivery work, just as there were trucks specifically designed for heavy loads. The company was in receivership in 1912 and went out of production.

The Van Dyke delivery wagon courtesy AAMA

The Van Dyke factory courtesy NAHC

HUPP-YEATS ELECTRIC CAR COMPANY
DETROIT, MICHIGAN
1910-1919

HUPP CORPORATION
DETROIT, MICHIGAN
1911-1912

R.C.H. CORPORATION
DETROIT, MICHIGAN
1912-1915

While he was vice president and general manager of the Hupp Motor Car Company, **R. C. Hupp** teamed up with **R. T. Yeats** and formed the Hupp-Yeats Electric Car Company in June of 1910. The "Hupp-Yeats" sold for $1750 and was produced in a factory located at 285 Monroe, Detroit. The car was powered by a series wound motor suspended by axle and ball joint. The battery was an Exide hycap with 27 cells producing 48 volts and 27 amperes. Maximum speed was 16 to 18 mph with five forward and two reverse gears. Mileage was given at 73 miles per charge. It had an 86 inch wb and sat four people.

In May 1911 the Hupp-Yeats Electric Car Company, Hupp-Turner Machine Company, Hupp-James-Guyman Foundry Company, Hupp-Johnson Forge Company, R.C. Hupp Sales Company, Rotary Valve Motor Company, Hupp-Ellis-Rutley Construction Company, and the Hupp-Detloff Pattern Company were consolidated into the Hupp Corporation. It was capitalized at $700,000, with **Robert Hupp** as president, Charles Hastings was VP, and **Louis Hupp** was secretary and treasurer. The Hupp Corporation supplied castings, forgings, and machined parts for the Hupp Motor Car Company.

In 1911 production increased to 6079 units and a larger factory was needed. **Robert Hupp** wanted to expand into a multi-product line like GM and US Motors. Both Drake and Hastings were against expansion at the time because the fledgling Hupp Motor Car Company had neither funds nor managerial skills to support additional operations. **Robert C. Hupp** sold his Hupp Motor Car Company stock and quit in August 1911.

A new gasoline car was introduced by the Hupp Corporation in August 1911 in a plant on 115-185 Lycaste Street. It was named "R.C.H.," to avoid any confusion with the "Hupmobile" manufactured by the Hupp Motor Car Company. The first R.C.H., the Model F, looked very much like the Model 20 Hupmobile. The Model F was a runabout with extensive use of drop forgings and nickel steel construction. The motor was the long stroke type with 22 hp. With lamps, top, windshield, and other accessories, it sold for $700. It was claimed to be the first car under $1000 to have a self starter (initially gas, then electric).

In September 1911 Hastings and Drake of the Hupp Motor Car Company filed suit in the Wayne County Circuit Court to prevent **Robert** and **Louis Hupp** and the Hupp Corporation from using the Hupp name in conjunction with the manufacture of a gasoline automobile. In February 1912 a decree was handed down from the court requiring the Hupp Corporation to change its name to the R.C.H. Corporation. **Robert Hupp** remained president and general manager while B. Q. Hazelwood became VP, and **Louis Hupp** was secretary and treasurer.

The R.C.H. was a phenomenal success with 7000 cars sold the first year, with representation in 35 countries and orders for over 15,000 units for 1913. **Hupp** lacked the capital to handle the increased demand. J. F. Hartz was appointed general manager and treasurer in January 1913. Hartz was well known in Detroit for his business acumen. Charles P. Seider moved to VP. The company was still in trouble in July 1913 when it fell into receivership. Seider took over as president and the **Hupp Brothers** left the company. By November 1913 three creditors were pressing to place the company into involuntary bankruptcy. The assets for the R.C.H. Corporation brought in $295,000.

Charles P. Seider led a group and purchased the bulk of the R.C.H. Corporation assets for $100,000 in February 1914, including stock parts and a portion of the factory. Seider was elected president of a newly formed R.C.H. Corporation that included the Hupp-Yeats Electric Car Company. Albert H. Collins, who was an Indiana R.C.H. distributor, was elected VP. Albert F. Edwards, purchasing VP at the Detroit United Railway, was elected secretary and treasurer. By May, there were 100 men turning out 60 cars a week.

The new R.C.H. sold for $900 fully equipped. It had a full streamline body on a wheelbase of 110 inches. The engine was a four cylinder cast in a block with a 3.25 x 5 inch bore and stroke. For domestic use, the steering was on the left side and for export it had right hand steering unless otherwise specified. Two body types were available, one affording accommodations for five people and a two passenger roadster. Equipment included electric headlamps fed from a storage battery, Sears-Cross speedometer, demountable rims, windshield, top with jiffy curtains, and the usual complement of tools.

The first quota of 300 cars was given a complete test out before shipment. The motors were given a block test varying from 4 to 48 hours, and if they were not right, they were torn down and given another test. Rear axles were tested for noise before being mounted in the cars, and each car was road tested. The strenuous testing was relaxed as production demands increased. Quality was sacrificed and cars were falling apart until the reputation was severely damaged. The R.C.H. Corporation went out of production in 1915. By 1917, spare parts were furnished by the Puritan Machine company, which had purchased the manufacturing records as well as a large stock of parts for the R.C.H. This brought the number of orphan cars for which Puritan supplied parts to 108.

The assets of the Hupp-Yeats Electric Car Company were purchased from the Security Trust Co., trustee-in-bankruptcy, by a number of Detroit automobile and electrical industries men. J. A. Mathews, who was with the R.C.H. Corporation when it began, was in charge of business and manufacturing. The Hupp-Yeats was manufactured under the same name and along the same design except for a worm driven rear axle instead of the bevel gear drive. The new company made an exchange proposition to convert the 1000 Hupp-Yeats previously produced to the new worm gear axle. An electric charger was made part of the standard equipment for the new Hupp-Yeats.

The Ford Model T and the electric self starter were in demand. The gasoline engine had superior range and afforded lower pricing over electric's. These factors brought about the end for the Hupp-Yeats in 1919.

R-C-H Service Station located at Lycaste St. and Jefferson Ave., Detroit, Michigan

*1912- Hupp-Yeats
Note that the paint scheme of the electric and the gasoline motor car below is the same
courtesy AAMA*

*1912- R-C-H
It came with electric lights, 12 inch bullet headlights and 6 inch bullet sidelights. The battery was a 100 ampere hour made by Exide.*

CROWE MOTOR CAR COMPANY
DETROIT, MICHIGAN
1911

The "Crowe Thirty" was placed on the market in the fall of 1911 by W. A. Crowe. It was designed by W. W. McIntyre as a four cylinder with thirty horsepower. Later in September, Crowe was reported to be negotiating with the Industrial Association in Grand Rapids to move production there. He did not move and production was stopped all together.

PAINT

Early automobiles were painted in the same manner as horse carriages with up to 40 coats of sealer, primer, undercoat, body color, finishing color, and varnish applied with a brush. After each application, the surface was re-sanded. The paint was air dried. It was hampered by pits and running. It typically took a month to complete the process. After the automobile was in service, it was often re-painted each year, at a cost of $25.

Metal panels for the external surfaces soon replaced wood and required still further painstaking effort. The metal was dipped or sprayed with acid to clear the surfaces of all foreign matter, which also ate away some of the surface to allow the paint to adhere. Usually, body filler and sanding were also required to obtain a smooth surface. The end result looked like an off-white primed surface and was reffered to as a "body-in-white."

The paint department always determined the manufacturing capacity of the car factory. By 1911, development with hot air ovens was used for drying paint on metal automobile bodies. At a woodworking plant in Ohio, 15 ovens, each the size of a railroad boxcar, were used to bake automobile bodies like "Hot Cross Buns."

The Ford Motor Company began using gravity painting for the Model T which reduced the drying time to days instead of weeks. Paint was poured on and dripped into troughs, then the body was placed into a drying oven then sanded.

The oven drying process was called japanning, which was a process of using enamels or varnishes colored with mineral pigments and baked to 307 degrees. By 1918, the Young Brothers Company of Detroit furnished over 97 percent of the japanning and drying ovens used in the auto industry. Using gas, the large ovens provided up to three heats per day.

On July 4, 1920, nitrocellulose lacquer was first observed in the laboratory, which led to a lacquer base paint developed by Dupont that could dry in one hour. To help sell the new product for production at GM, Alfred Sloan's car was painted a different color while he was at lunch, without telling him. Needless to say, he was amazed and soon approved the new paint. The 1924 Oakland used a laquer for production called "Duco," for the first time. Duco could carry more pigment in suspension than either enamel or varnish and gave a vivid color. The Oakland only came in blue and was called the "true blue."

B.O.S.S. COMPANY
DETROIT, MICHIGAN
1911

The B.O.S.S. Company filed articles of incorporation on March 4, 1911, with capital stock of $250,000, of which $135,000 was paid in. The name was an acronym based on the names of Frank A. **B**owen, John A. **O**lsen, Frank A. **S**mith, and Frank **S**trattton. It was Strattton who held patents on the car.

The upholstery could be removed at the rear seat and it could be turned into a delivery car. A second change gave it a further extension and furnished a five foot box. The B.O.S.S. could also be turned into a runabout. The B.O.S.S. came in three grades at $800, $1100, and $1400. There was a small volume production, but the venture was limited.

DeSCHAUM AUTOMOBILE AND MOTOR MANUFACTURING COMPANY
BALTIMORE, MARYLAND
1900-1903

DeSCHAUM AUTOMOBILE COMPANY
BUFFALO, NEW YORK
1907-1908

DeSCHAUM-HORNELL AUTOMOBILE COMPANY
HORNELL, NEW YORK
1908-1909

DeSCHAUM MOTOR CAR COMPANY
DETROIT, MICHIGAN
1910-1911

SUBURBAN MOTOR COMPANY
ECORSE, MICHIGAN
1911-1912

PALMER MOTOR CAR COMPANY
ECORSE, MICHIGAN
1912-1913

PARTIN-PALMER MOTOR CAR COMPANY
CHICAGO, ILLINOIS
1913-1914

COMMONWEALTH MOTORS CORPORATION
CHICAGO, ILLINOIS
1915-1922

CHECKER CAB MANUFACTURING COMPANY
CHICAGO, ILLINOIS
1923

CHECKER CAB MANUFACTURING COMPANY
KALAMAZOO, MICHIGAN
1923-1975

The Schaum Automobile and Motor Manufacturing Company was organized in March 1900 in Baltimore, Md. **William A. Schaum** was president of a company that produced automobile accessories, including spark plugs with three contacts to generate a double spark. The firm also produced a gasoline runabout with a one cylinder motor that produced four to seven hp. Side chain drive was used, but it did not come with brakes. The "Schaum" could be purchased for seating two, four, or six people, and could attain speeds of 20 mph. The Schaum Company also produced 10 vehicles for the Autocarrette Company in Washington, DC. There was a level of dissatisfaction between the two companies that involved a balance due on a $40,000 purchase price. **Schaum** left Baltimore and headed for Buffalo, NY. He became interested in producing automobiles, and formed the DeSchaum Automobile Company in 1907. The first automobile was a highwheeler called "Seven Little Buffaloes." The name was changed to the "DeSchaum Motor Buggy" for 1908. Sales were good enough to secure more financial backing to expand operations. The DeSchaum-Hornell Motor Car Company was formed in 1909, and a factory was to be erected in Hornell, NY. Cars were never produced, and **William** moved to Detroit in the fall of 1910 where the DeSchaum Motor Car Company was established. A derivation of the "Hornell" was a prototype of a two seat roadster with 20 hp. By September 1911 the company was re-organized as the Suburban Motor Car Company. A 350 acre site located in a suburb of Detroit, called Ecorse, was being considered for a factory along with a village for housing employees. When it was discovered that **DeSchaum** rendered a bogus check for $150,000 he was replaced. In November 1912 Randel A. Palmer resigned as VP of Cartercar and took over the defunct Suburban Motor Company. The company was re-organized as the Palmer Motor Car Company that completed the last of a total 25 "Suburban" automobiles. In June 1913 R. A. Palmer joined the Partin Manufacturing Company in Chicago to produce the "Partin-Palmer," as an updated version of the Suburban.

In 1915 Partin-Palmer was bankrupt and was re-organized as the Commonwealth Motors Corporation. For 1922, the Mogul Taxicab was added to the product lineup. The Checker Taxi Company used the Mogul Taxi chassis and bodies from the Markin Auto Body Corporation of Joliet, Ill. In 1921 the Markin Auto Body Corporation and the Commonwealth Motors Corporation merged, and moved to Chicago. In May 1922 a receiver was appointed for the Commonwealth Motors Corporation and it was taken over by the newly formed Checker Cab Manufacturing Company. Morris Markin was in control of Checker, but he was in the middle of a taxi war. His house was bombed and anything could have happened, so the company was moved to Kalamazoo, Mich.

1910-Suburban courtesy AAMA

DAY AUTOMOBILE COMPANY
DETROIT, MICHIGAN
1911-1913

Thomas Whitfield Day was born in Mt. Clemens, Mich., in 1865. He was adopted by Thomas and Martha Patterson and assumed the name of Day. He went to common school, and then graduated from the University of Michigan Law School in1894. He was reared on the farm until he was 16 years old, then learned the trade of wheelwright, and bought out his employer. He engaged in real estate and insurance, and became the attorney for Butler County, Neb., from 1897 to 1900. He went to Kansas City, Mo. in 1901 and sold Locomobile cars. He located to Detroit in 1907.

Hugh Jennings was the VP, and was also the manager of the **Detroit Tigers** ballclub. He was born in Moosic, Penn., in 1873 in humble circumstances. He went to Baltimore in 1893 and became the king of the shortstops in the country for the Oriole baseball team. His manager, Edward Hanlon, taught him how to bat safely. His fault was that he pulled away from the plate. From a batting average of .192 in 1893, he was batting .397 in 1896. But the story of his batting career would not be complete without mentioning his habit of getting hit by the ball. This was a trick for which he was far more famous for than hitting the ball. From pulling away from the plate, he became adept at stepping into the plate--and getting hit. At that time, when a batter was hit by a pitched ball he automatically went to first base. No one had *dreamt* that a batter would voluntarily allow himself to get hit if he could escape the punishment that the blow was bound to inflict. It can be said with assurance that Jennings was soley responsible for the rule that compels the avoidance of a pitched ball. Jennings was a right handed batter and always wore heavy pads on his left hip, but his body was always covered with big black and blue bruises.

He attended Bonaventure College during the winter months while he was with Baltimore, and when he switched to Brooklyn, he entered Cornell University Law School, coaching the baseball team there as well as studying. He graduated in 1903, and spent that winter pursuing the practical side of law offices. He began working for a law firm in 1905. He started managing the **Detroit Tigers** in 1910.

The Day Automobile Company was organized in 1910 with **Thomas Day** as president, **Hugh Jennings** as VP, Cameron Roosevelt as secretary, and Wallace E. Brown as treasurer. A factory on Trumbull Avenue was used initially, but sales were good and the company moved into larger quarters on 25 East Milwaukee in the spring of 1912. Day designed the car for farmers. It had a box body with 34 x 24 inch carrying space with a tail board, and a rear seat that came out easily. The wheelbase was 100 inches and it had extra-heavy 30 inch wheels. The engine was a four cylinder, four cycle type, with splash lubrication and thermo-syphon cooling, producing 21 hp.

The Day Company began to have cash-flow problems, and could not get the support it needed in Detroit. In August 1913 Day reported that he found financial support in Spokane, Wash., and planned on moving, but the deal fell through, and the Day Company never resumed production.

1912- Day Utility touring

Hugh Jennings: He was known for standing on his right foot with his left leg pulled up and bent at the knee, with both arms upraised, fists clenched and head thrown back, emitting a shrill cry to battle "Eh-Yahhhhhh!"

FARMERS ATTENTION!

THIS CAR
IS DESIGNED AND BUILT ESPECIALLY
FOR YOU

PRICE, $1000.00, FULLY EQUIPPED
Including Magneto, Five Lamps, Windshield and Horn

The First 1000 Purchasers will receive a Special Stock BONUS

It will Pay You to Investigate this Proposition *first*

See Exhibit at State Fair Grounds, ~~Second~~ Floor, Automobile Building or Write for Particulars to

DAY AUTOMOBILE COMPANY
54 HOME BANK BUILDING   DETROIT, MICH.
— AGENTS WANTED —

MILLER CAR COMPANY
DETROIT, MICHIGAN
1911-1914

The Miller Car Company was organized in September 1911, with a capital stock of $50,000. Theodore Miller was president, Enos L. McMillan of the National Can Company was VP, and J. C. Hallock was secretary and treasurer. The factory of the Detroit Excelsior Works on the south west corner of Custer and Richmond streets in Detroit, of which Hallock was proprietor, was used to produce the Miller. Former factory manager for the Van Dyke Motor Car Company, **Guy Sintz**, was appointed general factory manager.

The Miller was introduced at the 1912 Detroit Automobile Show, available as a roadster with a 110 inch wb for $1250, and a touring car with a 116 inch wb for $1350. Both models had metal bodies in which *the straight line effect prevailed.* An L-type Wisconsin long stroke motor was used with a bore of 3.75 inches and a stroke of five inches, rated at 30 hp. A three bearing crank with the well known Wisconsin force feed system for oil lubrication was used. Cooling was by thermo-syphon with large diameter water connections. A flat tube cooler was supported on a cross member of a frame, but did not touch the side rails of the main frame, and was protected by a leather pad. Regular equipment included a silk mohair top, windshield, full set of lamps, combination oil and electric side and tail lamps with a 100 ampere hour battery, or gas headlights with a gas tank or generator. For 1913 a delivery van was added with a 1000 lb capacity.

In January 1914 the Miller Car Company was in receivership, with about $10,000 involved which included a trust mortgage of $4000. The Kosmath Company bought what was left of the Miller Company in February and continued production of the delivery wagon. **Guy Sintz** moved to Pittsburgh and worked for the Pennsy Motor Car Company which produced a refined version of the Miller called the Pennsy. In 1917 he moved to York, Penn., and worked for the Pullman Motor Car Company which later became the Bell Motor Car Company. In 1923 he returned to Detroit and worked as a factory superintendent for **Claude Sintz Incorporated** until 1929.

Miller roadster

The Miller Car

Touring model, completely equipped, $1350
Roadster - - completely equipped, $1250

Backed by a solid year of owners' satisfaction

We have purposely delayed, for a year, a widespread public announcement of the Miller Car.

We wanted to go before the dealers of the country with a certainty instead of a doubtful proposition.

So this tells you about a certainty.

Before the first Miller was sold, more than a year ago, we were sure of it in our own minds.

But we wanted the endorsement and support of Miller owners.

This we now have.

Understand that we were not experimenting at the owners' expense.

The car a year ago was a finished product; specifications and materials are the same today as then.

The car has had a good year's tryout.

SPECIFICATIONS
Motor .... Four-cylinder; full 30 H. P. Stroke 5 in.; bore 3¾ in.
Cooling ...... Thermo-Siphon.
Ignition ...... Mea or Bosch Magneto.
Carburetor ... Special design.
Oil Tank ..... Gravity feed direct to motor.
Clutch .... Cone, leather-faced.
Transmission . Three speeds forward and reverse; F. & S. Annular ball bearings; Chrome-vanadium steel gears, 1-inch to 1¾-inch face.
Springs ...... Semi-elliptic front; full-elliptic rear.
Wheel Base .. Touring car 116 inches, Roadster 110 inches.
Tires .... 34 x 3½ and 34 x 4.
Brakes ...... Internal and external; 12-inch drums.
Equipment. ..STANDARD ALL MODELS; Top, windshield, Horn, 100 ampere lighting battery or Gas Tank, Enameled combination electric and oil lamps.

It has demonstrated itself to be a first-class sales proposition.

Its long-stroke motor, equipment with ball and roller bearings at important points, the use of standardized gears, etc., and its price make it particularly attractive to the car buyer; and therefore attractive to the dealer.

We are assigning territory now; and have a limited number of cars yet to be contracted for.

Get in touch at once, if you are an established dealer.

MILLER CAR COMPANY, 1638 Russell St., Detroit, Mich.

PHIPPS-GRINNEL AUTOMOBILE COMPANY
DETROIT, MICHIGAN
1911

PHIPPS ELECTRIC COMPANY
DETROIT, MICHIGAN
1912

GRINNEL ELECTRIC AUTOMOBILE COMPANY
DETROIT, MICHIGAN
1912-1915

Ira L. Grinnel was born in Niagara N. Y., in 1848, the son of Ira and Betsy Grinnel. He was educated at Academy, N. Y., and Manchester, Mich. He began his business career with the sewing machine business in Ann Arbor, Mich., from 1866 to 1882. Ira and his brother Clayton started in the music business in 1880 under the name of Grinnel Brothers. The firm moved to Detroit in 1882, with Ira as senior partner, and began to manufacture pianos in 1901.

Clayton A. Grinnel was born in Albion, N. Y., in 1859 and moved to Ann Arbor, Mich., in 1866. Ira graduated from the University of Michigan in 1879, then joined his brother in the music business.

In 1910 the brothers were joined by their cousin, Albert A. Grinnel, from Elba, N. Y. In 1911 the trio joined with an electric car designer named Joel Phipps, and formed the Phipps-Grinnel Automobile Company. The company only produced two cars and one delivery wagon all year. The Grinnel Brothers decided to buy out Phipps's share.

Phipps joined up with C. W. Whitson and formed the Phipps Electric Company. Phipps stretched the wheelbase of the Grinnel Electric and produced the Phipps Electric in 1912, but insufficient funding forced the company out of business by the end of the year.

The Grinnel trio formed the Grinnel Electric Automobile Company, and produced the Grinnel from 1912 to 1915. Ira was president of the Grinnel Brothers music business, realty business, as well as the car company. Clayton was first VP and Albert was second VP. The Grinnel was essentially a carryover from the 1911 Phipps-Electric. By 1915 electric car popularity was declining. The electric self starter was an important factor for the decline. Production cessation was announced in January 1916.

Grinnel Model K courtesy Tinder

BRIGGS-DETROITER COMPANY
DETROIT, MICHIGAN
1911-1915

DETROITER MOTOR CAR COMPANY
DETROIT, MICHIGAN
1915-1917

DETROITER MOTORS COMPANY
DETROIT, MICHIGAN
1917

Claude Strait Briggs was born in 1872. In 1901 he founded the Briggs Dental Company, and married Virginia Ann Hupp. In 1909 he founded the K-R-I-T Automobile Company but sold his holdings and withdrew to join U. S. Motor Company. In 1910 he became general manager for the Brush Runabout Automobile Company.

In November of 1911 **Claude Briggs** and **John A. Boyle** organized the Briggs-Detroiter Company capitalized with $200,000. W. S. Lee was chief engineer for a car that was to compete with the Brush Runabout. The factory was at 461 Holbrook Avenue by the Grand Trunk railroad and had 75,000 square feet of floor space.

The "Detroiter" was introduced at the Detroit Automobile Show in January 1912. It had a number of distinguishing features, such as a long stroke motor, platform rear springs, double universal joints, three point suspension of the motor, a pressed rear axle housing instead of the usual built-up tubular construction, and extremely large internal brakes. The motor had four cylinders cast in a single block. The stroke was 4.75 and bore was 3.4, giving a stroke-bore ratio of 1.4. This was deemed to be within the long stroke class without cavil. The camshaft ran in three long bearings with the end ones 2.25 inches long. The cams were forged integral with the camshaft and were ground to a master cam, and cooling was by thermo-siphon circulation.

The Detroiter sold well initially (1100 cars in 1912, 2750 cars in 1913, 1600 cars in 1914), but then declined in 1915. That same year **A. O. Dunk** of the Puritan Machine Company purchased the company and re-organized it as the Detroiter Motor Car Company. **Claude Briggs** departed and went to the C. R. Wilson Body Company.

In 1917 **Dunk** made himself chairman of the executive board and a New York banker named J. S. Kuhn, who had participated in the re-organization, was made president. In March of 1917 Kuhn again re-organized as the Detroiter Motors Company. By October the company was in receivership again, and by the end of the year all of the stock was sold to **Sam Winternitz** of Chicago.

1915- Detroiter "Eight"

Briggs-Detroiter factory

Claude S. Briggs

KING MOTOR CAR COMPANY
DETROIT, MICHIGAN
1911-1923

KING MOTOR CAR COMPANY
BUFFALO, NEW YORK
1923-1924

"THE CAR OF NO REGRETS"

The King Motor Car Company was organized in February 1911, with a capital stock of $500,000, and a rented factory at 1559 West Jefferson Avenue, in Detroit. H. Kirke White was president, **Charles B. King** was VP, E. C. Hough was secretary, and H. Nelson Dunbar was general manager.

The "King 35" was the first product of the new company and was hailed as a car of European descent and Northern Automobile lineage. (**King** had returned from studying European automobiles for two years after he left the Northern Company.) It had a cast *en bloc* four cylinder engine built into a unit power plant with a three speed selective sliding gear and shaft drive. The engine was of the long stroke type with 35 hp. The exhaust pipe was unusually large and was cast integrally with the motor. The valves were 1 1/16 inches in diameter and their action was silenced by the complete enclosure of the operating mechanism by an ingenious cover plate. The plate also carried a pipe connection for air drawn into the carburetor. With this design, the air was warmed slightly, while the hissing of the carburetor was silenced. One of its strong points was left hand drive with right hand gear control. The brakes were pedal operated so there was only one lever placed in the center of the foot board. The car was equipped with a closed-front body of the flush-sided type, with concealed door latches and a torpedo style of dash. Equipment included quick detachable rims, mohair top, Bosch dual ignition, windshield, gas lamps, and a complete tool outfit. It was priced to sell for $1565.

By the end of 1911, the King Company re-located to the former Hupp Motor Company factory at 1300-1324 Jefferson and Concord. Shortly thereafter, the King Motor Car Company fell into receivership. An advertising man and chewing gum manufacturer, Artemus Ward Sr., from New York City purchased the King Company outright, and installed **J. G. Bayerline** as president. **Bayerline** was soon replaced by Artemus Ward Jr. In December, 1914, the King Motor Company brought out the King Model D with a V-8 engine. **Charles King** had studied European cars with V-8 engines. (The Clement Adler racer used a V-8 in 1903 for the Paris to Madrid race. The 1910 De Dion was the first production car to use one, and Cadillac introduced a V-8 in September 1914.) The King was advertised as the "World's First Popular-Priced V-8." The King V-8 sold for $1350, which was $625 less than the Cadillac Type 51 with a V-8 engine.

Charles King left the King Motor Company to become the head of the US Army's Division of Engine Design in 1916. The King Motor Car Company survived the war period as one of the few individually controlled car companies in the country. In 1920, the King-Eight came as a Foursome, Road-King, Touring, and a model that was part limousine and part sedan, called a "Limoudan." A redesigned instrument board with a mahogany finish contained a hinged center cluster, which made the electrical connections accessible. The vacuum tank was moved to the front of the dash to prevent any drippings from falling on the electrical wiring. A horn and an eight day clock were standard on all models.

In the fall of 1920, Artemas Ward led a group of five King company directors and requested dissolution. A lack of working capital and failure to secure loans were cited as the reasons. The King company assets, and debts totaling $1,000,000, were sold in early 1921 for $500,000 to Charles A. Finnegan of Buffalo. The claims were paid off in late 1922. In 1923, the King Company was moved to a smaller factory in Buffalo, with only 240 cars built that year. The company entered bankruptcy in 1924, and the last Kings were sold in England in 1925.

courtesy AAMA

courtesy AAMA

King Motor Car Company.

AUTOMOBILE RACING

For centuries, horse drawn wagons and carriages were slowly hand-crafted by artisans using time proven methods and materials. Occasionally, something new would come along and artisans would adopt it carefully. The quality of the wagons and carriages was easily measured by the consumer. The size and type of wood, the number of spokes in the wheels, the tire material, springs, seats and paint finish were understood. On the other hand, the automobile was so complex that even the designers and manufacturers didn't understand everything about them. The entire front suspension and steering had to be thought out because the motor pushed the front wheels instead of pulling and pneumatic tires became necessary to steer at higher speeds. During the first decade of the automobile, many designs, materials and methods of construction were developed by trial and error. Many of these "experiments" resulted in poor performance, poor quality and perhaps an automobile that was unreliable. Many companies had only one car to test before it was manufactured in volume and the only driving tests were made on public roads.

Considering that horse racing was popular for centuries, automobile racing was inevitable. It gave the automobile manufacturers an opportunity to test durability and develop new ideas. The first formal automobile competition took place in the 1894 reliability trial in France between Paris and Rouen. The first motorcar race in the world took place in June, 1895 between Paris and Bordeaux and back to Paris, covering 732 miles. The first motorcar race in the United States took place in Chicago on November 18, 1895. It was held on the exact sight as the Great Chicago Exposition of 1893. **Henry Ford** could not afford to attend the race but **Charles King** participated as a judge. **King** even drove the "Mueller" and came in second place, after the driver collapsed from fatigue. It soon became apparent that speed and endurance racing had mass appeal.

For all-out speed, Ormond Beach, near Daytona, Fla., was the choice of many early automobile manufacturers, starting in 1902, with a souped-up Olds runabout called the Pirate. There simply were no roads smooth enough to support high speed, so Ford, Maxwell, Packard and many other companies used the level surface of the beach in order to extol their car's performance. It was expensive to go there, but it was very fascinating and eventually special railroad passenger cars would be used to take the company representatives to Ormond. Most of the early races were dominated by European cars and one day a French Renault blew an engine as **Henry Ford** was watching. He picked up broken pieces of the valve stems and took them back to Detroit. He asked a metallurgist working for him, named **Harold Wills,** to try and determine the metal composition. He determined that it was Vanadium steel, which was much stronger than stock steel in use at the time by **Ford**. This alloy became the foundation for the Ford Model T which had to be light and strong to survive the rough country roads that farmers used.

During race week in 1906, **Ford's** car could develop 100 hp and he took **James Couzens** and **Horace Dodge** along with his driver, Frank Kulick. The Ford team missed the first race and started in the mile run event, hitting 40 mph. Later in the week it spun out in the 30 mile race. The world's record was boosted to 127 mph by a Stanley Steamer called the "Turtle" that week. The internal combustion people cried, "freak." The races at Ormond never attracted the gatherings of this proportion afterwards. In the following year, Fred Marriot tried to beat his own record. His run was not officially timed but he was up to 150 mph when the car flipped longitudinally, overturned and disintegrated. Marriot was not killed and no more steam cars attempted to beat the land speed record after that.

Oval tracks with dirt pavement were becoming very popular, mainly because they were all over the country and accessible by most people. There were a number of exceptional drivers, such as Earl Cooper, Gil Anderson, Eddie Pullen, Harry Grant, Spencer Wishart and Billy Carlson. But for years police officers would ask a driver being ticketed for speeding, "who do you think you are, **Barney Oldfield**?" He was probably the most colorful driver of them all and he even staged accidents at the race track. He once staged a crash using a fake guard rail made of cardboard, then was picked up by an ambulance just to stir up the crowd. **Oldfield** began losing races to **Louis Chevrolet,** who was driving a big Fiat owned by a Standard Oil millionaire, Major C. J. Miller. **Oldfield** said: "That man gets the real speed out of that car and he won't be beat until he goes through the fence once or twice or is killed." **Oldfield** was rather put out, in fact so much that at Chicago he was successful in keeping **Chevrolet** from entering a $1000 race on a protest.

After **Chevrolet** quit racing, one of **Oldfield's** strongest competitors was **Ralph De Palma**, who was one of the best drivers of all time. The two had a feud going both on and off the track. **De Palma** wouldn't even use Firestone tires because **Oldfield** endorsed them. **De Palma** was also an engineer who built special order cars in New York and Detroit. He worked as an engineer for Packard Motor Company and helped perfect the Liberty aircraft engine.

1905-
A Renault driven
by Maurice Bernin
at Ormund Beach
courtesy AAMA

1906-
Vencenzo Lancia
checks his F.I.A.T.
at Ormund Beach
courtesy AAMA

*1910-
On the far right,
James Couzens
and Henry Ford
(extreme right)
watch the racing
at Ormund Beach*

*(L) Spencer Wishart
and (R) Ralph DePalma.
Wishart was killed in
the Elgin National
road Race on August 22,
1914. His mechanic,
John C. Carter later
died from his injuries.
The Mercer Automobile
Company discontinued
racing for the remainder
of the year.
             courtesy AAMA*

*In mid-1917, Barney
Oldfield unveiled a
$15,000 racing car
called "The Golden Sub."
It was designed by
Harry A. Miller and
had an enclosed cockpit
that was supposed to be
safer for the driver.
It was capable of 125 mph
and Oldfield vowed to
settle the score once and
for all with Ralph DePalma.
             courtesy AAMA*

LOUIS JOSEPH CHEVROLET

Louis Chevrolet was born in La Chaux de Fonds, Switzerland in 1878. While **Louis** was still an infant, the Chevrolets moved to Burgundy, France. He and his brothers, **Gaston** and **Arthur**, learned the value of precision engineering from their father, who was a watch and clockmaker. **Louis** had instinctive mechanical know-how without a formal education. In his teens, he worked for a blind winemaker in Beaune, France and invented a wine barrel pump that became a modest success. While the bicycle craze was sweeping over Europe, he began designing and building racing bicycles, called **Frontenacs**. He rode in bicycle races himself, which eventually led him to the automobile industry. He learned about the internal combustion engine while working for Darracq, de Dion Bouton and Hotchkiss, and Mors automobile companies in France.

In 1900 he immigrated to Montreal, where he worked as a chauffeur for six months. He went to New York and worked in the experimental and repair department of the de Dion Bouton Motorette Co. in Brooklyn, N.Y. Following the death of their father in 1901, **Gaston** and **Arthur** joined **Louis** in New York. In 1902 **Louis** went to work as a mechanic for a F.I.A.T. car dealer named E. Rand Hollander, who later introduced **Louis** to race promoter, Alfred Reeves. With the help of Reeves and Hollander, Fiat agreed to sponsor **Louis** as a F.I.A.T. race car driver. He won the first event at the Hippodrome oval in Morris Park, N.Y. **Louis** quickly established himself as a premier driver in a series of F.I.A.T. sponsored races, including the Vanderbilt Cup race, when he made the F.I.A.T. team. He smashed into a telegraph pole and wrecked a 110 hp F.I.A.T. during the warm-up laps. He was given another car for the race itself and after 150 miles hit a second telegraph pole. **Louis** was a heavy footed driver (it was said that all of the Chevrolet brothers were rough on machinery).

In 1905 **Louis** held the world's land speed record of 111 mph, set at Sheepshead Bay, N.Y. in a F.I.A.T. racer. He also beat **Barney Oldfield** three times in 1905, once in a 10 mile race at Brunot's Island, Pittsburg, once at Hartford, Conn., and at Morris Park, N.Y., where Walter Christie, the inventor of the front wheel-drive, also raced that day.

In 1906 **Louis** left F.I.A.T. and organized the Chevrolet-Kenen Auto Co. in New York. **Louis** also joined Walter Christie, and together they built a land speed record machine powered by a 200 hp V-8 Darracq, which **Louis** drove to a new record of 119 mph at Ormond Beach, Fla. and later broke the "two miles a minute" barrier, driven by Victor Demogeot. This firmly established **Louis** as a daring race car driver, as well as an engineer. **Gaston** and **Arthur** were both above average drivers and the **Chevrolet** name was well known throughout the racing scene.

The Buick Motor Company was earning a reputation with their valve-in-head engine racers. **Billy Durant** recognized the opportunity to have the **Chevrolet** brothers drive Buick racing cars and hired **Louis** and **Arthur** in 1907. Along with Bob Burman, Buick had numerous racing successes with racers 16A and 16B, nicknamed the Buick "Bugs," which they helped design. In the 1910 running of the Vanderbilt Cup, **Louis** narrowly escaped death when his Marquette Buick left the road. **Louis** and **Arthur** completely rebuilt the totaled Marquette Buick and took it to Indianapolis for the first running of the Indy "500" in 1911. **Louis** qualified at 93 mph, which was 20 mph faster than the winning average set by the eventual winner **Ray Harroun**, in the Marmon Wasp. **Louis** was disqualified from entering the race after the Marmon team took note of his practice time and protested that Louis' car had been registered late. **Arthur** also made an excellent showing until mechanical failures forced him out in the 36th lap.

Louis Chevrolet in a F.I.A.T. courtesy NAHC

Louis Chevrolet in a F.I.A.T.

A heavy mist hung over the road with Chevrolet going at high speed. He failed to keep his car on the road where it curved to the right and he crashed into a telegraph pole. He was thrown out uninjured, but his "mechanicien," Henry Schutting sustained three broken ribs. The 110 hp car was completely demolished.

CHEVROLET MOTOR COMPANY
DETROIT, MICHIGAN
1911-1913

CHEVROLET MOTOR COMPANY
FLINT, MICHIGAN
1913 -

In December 1910 **Billy Durant**, at the age of 49, was ready to start another car company. He lost control of General Motors in September 1910, but was still the vice president. He had a reputation among the Flint auto men, as having a magnetic personality and as being wealthy. He financially backed **Louis Chevrolet**. **Chevrolet** let it be known that he was retiring from Buick racing, and wanted to design and build his own car. **Durant** recognized that **Chevrolet** had one of the best known names on the racing circuit, the name was pleasing to the ear and had a French connotation. France was a leader in the automobile industry and had been the largest automobile producer in the world until it was surpassed by the U.S.A. in 1906. **Chevrolet**, in turn, enlisted the support of **Etienne Planche,** whom he had met when both worked for William Walter's Brooklyn Motor Shops in 1900-01, and who had designed the Roebling-Planche automobile. **Louis** immediately set up shop on the second floor of a building at 3939 Grand River Avenue in Detroit.

Development work continued secretly, although the press started to pick-up on the venture by May 1911. By June the press announced that **Chevrolet** was designing a car, and that he was preparing to leave for Paris to drive in the Grand Prix race. He also planned to participate in the innaugural Indianapolis "500" mile race.

By now **Billy** was ready to let the public know about his plans. First he bought the Flint Wagon Works factory from James H. Whiting, who was anxious to retire. Then he asked Arthur C. Mason, a Buick engineer from the early days, to build engines for his cars in the Wagon Works factory. On July 31, 1911, he incorporated the Mason Motor Co. in Flint. He then asked William H. Little, a former general manager at Buick, to share space with Mason and manufacture an inexpensive car named the "Little." The Little Motor Company was incorporated on October 30, 1911. Finally the Chevrolet Motor Company was incorporated on November 8, 1911. Durant secured a factory that had been used by the Corcoran Detroit Lamp Company, manufacturer of automotive lamps at 1145 West Grand Boulevard in Detroit for initial production. He also purchased about 40 acres of land on Woodward Avenue, directly across the street from the Ford Motor Co. factory in Highland Park, Mich. He even erected a large billboard proclaiming it as the future home of the Chevrolet Motor Company. Work started on the Woodward site, but was interrupted by a strike. **Durant** ultimately decided to concentrate Chevrolet manufacturing in Flint (the 40 acres were later sold to Ford who purportedly intended to use it for manufacturing electric vehicles).

Louis was still busy at the drawing board, so Billy sent Bill Little to Detroit to help out. One of **Durant's** associates from the wagon manufacturing days, Alexander Brownell Cullen Hardy (who liked to be referred to as A.B.C. Hardy), took over the factory in Flint for Little. By the summer of 1912 there still was no car from Chevrolet, but the Little had arrived. The Little sold for $650, had a 90 inch wheelbase, a 20 hp vertical four cylinder engine with the cylinders cast in two pairs, an integral motor-transmission and thermo-syphon cooling. Durant ordered a 25,000 mile road test which it barely made.

Meanwhile **Louis** told **Billy** that he had road tested one of his latest six cylinder prototypes at four in the morning and had it up to about 110 mph (60 mph is probably more accurate). The noise from his testing alerted a constable, who placed a log on the road to detain the speeder in case he returned. **Louis** did return and was arrested. At his summary trial he pleaded guilty, figuring that his reputation would spare him. Instead he was fined $30. "What for?" protested **Louis**. Five dollars for speeding and twenty-five for impersonating a famous race driver!

He told the story to **Durant,** but he was not amused, and thought that **Louis** was taking too long and the car was too big and expensive. But at least he had a car, and Durant ordered it into production. The first model was designated as the "Chevrolet Six Type C Classic." It had a T-head six-cylinder engine with a 5.6 inch bore and a 5 inch stroke, totaling 299 cubic inches. The wheelbase was 120 inches with the motor supported at three points on the chassis. The connecting rods and the camshafts were drop forged with integral lobes. The motor was cooled by a centrifugal pump and fan, lubrication was by a combination of force feed and splash. The motor had a dual magneto with an English compressed air starter mounted on the rear axle. The transmission was also mounted on the full floating rear axle, and a worm and gear was used for steering. The car was extremely well constructed, and sold for $2150. Once the car was in production, **Louis** went to France for an extended vacation.

In September 1912 **Billy** organized the Sterling Motor Company to supply six cylinder engines for the Chevrolet and the Little. The Classic Six and the Little Six hit the market at the same time. The price of the two cars was about $1000 apart, and accordingly, the Little outsold the Chevrolet. Although there were orders for 3000 Littles, there were far fewer for the Chevrolet being produced in Detroit. There was some concern that the name "Little" would eventually pose problems for its salesmen. Before the Little was on the market long enough to test its endurance, **Durant** decided to combine the two cars by using the best points of each. He decided to call it a Chevrolet and thus be able to challenge Ford. At the same time he also moved Chevrolet production to the Flint Wagon Works factory.

When **Louis** returned from vacation, he did not like the Chevrolet marketed as a competitor to Ford, nor did he want to move to Flint. He told **Durant** that he didn't want his name on any cheap economy car. **Durant** pointed out to **Louis** that he no longer had ownership of his own name.

Durant had a deep disdain for **Louis** constantly walking around with a cigarette hanging from his lower lip. The final straw came one day when **Durant** suggested that **Louis** switch to cigars, which were much more befitting for a businessman of stature in **Durant's** organization. **Louis** hit the ceiling and screamed, " I sold you my car and I sold you my name, but I'm not going to sell myself to you!" **Louis** left in a rage and soon sold his Chevrolet stock at a reported loss.

In 1914 **Louis** went to Indianapolis and organized the Frontenac Motor Company. One of his principal backers was French-born **Albert Champion,** who had made a fortune developing a ceramic process for spark plugs. **Champion** was a friend who shared interest in bicycle racing. In fact, he assumed his last name after he won a bicycle race in France and was called "Le Champion" by the press. In 1908 **Durant** tried to set up his own Champion spark plug company and hire **Albert Champion**. The Champion Company refused and went to court to retain the rights to the trade name "Champion." **Billy** then invited **Albert** to join GM and set up a subsidiary company using **Albert's** initials-AC.

Because **Champion** was a financial backer for **Louis,** and at the same time working for **Billy,** it placed a strain on the relationship between **Champion** and **Chevrolet.** It caught up to **Albert** sometime in 1915 when **Louis** stormed into his office and beat him to a pulp, then left with the promise that if their paths ever crossed again, he'd finish the job!

"Frontenacs" were very succesful racers. In 1919 they finished seventh and tenth in the Indianapolis "500." **Gaston** took the checkered flag in 1920 (although tradegy struck during practice and **Arthur** was seriously injured when he collided with Rene Thomas' Ballot racer). Then in November **Gaston** was killed in a crash at the Beverly Hills speedway in California, and **Louis** never got over it. In 1921 Tommy Milton won the Indy "500" in a Frontenac, giving **Louis** a victory two years in a row. In his later years he would say these two victories gave him his greatest sense of pride.

From there **Louis** made modified cylinder heads for Ford Model T engines. A "Fonty-Ford" surprised the big money entries in the 1923 Indy "500" when it came in fifth beating both the Mercedes and the Bugatti teams.

Louis and **Arthur** then turned their attention to the budding aircraft industry, and teamed up with ex-Ford dealer, **Glenn L. Martin,** who had air frame know-how, to form the Chevrolet Aircraft Corp. In 1926 the brothers developed a high output engine, dubbed the Chevrolair engine. Both claimed credit for the design which ultimately led to a break up between the two brothers in 1927. **Louis** left the company in 1929, when no chance of success appeared in the foreseable future. The Martin Company became a dominant player in the aircraft industry.

In 1933, without a job and having suffered many setbacks, **Louis** went to work as a mechanic in one of the Chevrolet plants in Detroit. In 1934 his son, Charles, died. Later he had a cerebral hemorrhage. He and his wife Suzanne moved to Florida. In 1941 he returned to Detroit for surgery because of circulatory problems resulting in the amputation of his leg. **Louis** died on June 6, 1941 in virtual obscurity.

Arthur resided in Slidell, La. for several years, where he worked for the Higgins Ship Building Company. He took his own life on April 17, 1946, at age 61.

The Shulte Garage on Grand River Ave. in Detroit where the first Chevrolet was made. It was a six cylinder, although a four cylinder was started but never completed.

Louis Chevrolet seen on the Grand Boulevard in Detroit. This is the first Chevrolet, completed in mid-1911. Note the lack of a windshield and gas side lamps.
courtesy NAHC

The Corcoron lamp factory was purchased by Durant
courtesy NAHC

The former Corcoron plant being converted into the first Chevrolet factory
   courtesy AAMA

The first production Chevrolet. Behind the car (left to right), Alanson Brush, Louis Cheverolet (with white duster), William Little (in derby hat), unidentified, Etienne Planche (no hat), A.B.C. Hardy (derby hat), and W.C.Durant (derby hat). Cliff Durant, son of W.C.Durant, and his wife sat in the front seat. The car is parked next to the Chevrolet factory on the Grand Boulevard.
   courtesy NAHC

February 24th, 1923.

Mussy-Lyon Company
Detroit, Michigan.

Attention Mr. David W. Rodger.

Gentlemen:

Answering your inquiry about the connecting rods which you lined for us with Mogul Metal for our racing cars, would say that they behaved splendidly and as far as we know they will keep on doing so, as after quite a bit of running they do not show any wear and we certainly expect to see them last a long time.

The conditions under which those bearings have to work is rather unusual except on racing motors, where they are subject to a very high bearing pressure and very high heat, as we turn those racing motors above 4000 R. P. M.

Yours very truly,

Louis Chevrolet

The origin of the "bow-tie"- Durant told the story of seeing it on wallpaper in a hotel in France, but the explanation given by his wife after his death was that they were vacationing in Hot Springs, VA... While reading the Sunday newspaper, Billy saw the design and said I think this would be a very good emblem for Chevrolet.

SOCIETY OF AUTOMOTIVE ENGINEERS

In the early days of the automotive industry, parts' suppliers produced steel tubing's, iron castings, nuts and bolts, radiators, tires, wheels, bodies, and hundreds of other parts required for the construction of an automobile. Some suppliers made parts to their own design, but most had to make special parts that were unique in size, shape, and weight for the manufacturer they supplied. More often than not, the automobile engineer drew up specifications for his car that dictated the quality and dimensions of the parts that were to be made to order.

In 1902 an article in one of the trade magazines suggested the formation of an organization that could help standardize parts. Fred L. Smith, of the Olds Motor Works, was its first president.

Because the A.L.A.M. served only license holders of the Selden patent (steamers and electric vehicles were also allowed), there was a need for an organization that would serve everybody. This need was filled by the Society Of Automobile Engineers (S.A.E.), which was formed on January 18, 1904, with A. L. Riker becoming its first president in 1905. The term Automobile was not in the dictionary so the founders drafted their own definition as, "Anything that moves on the earth, under the earth, in the water, under the water, or in the air." (*Automobile* was later replaced by *Automotive,* coined by Elmer Sperry.) The two associations co-existed with duplication of effort for over six years, (while the management of the S.A.E. was on a part time basis).

The fourth S.A.E. president was, Howard Coffin, the chief engineer for the Hudson Motor Car Company. He believed in standardization and took on the presidency of the S.A.E. on a full time basis starting in 1910, when the A.L.A.M. terminated, commensurate with the Selden patent loss in the courts. Coker F. Clarkson, former assistant general manager of the A.L.A.M., became secretary of the S.A.E. One example parts maker made 800 different sizes of lock washers, 1600 different sizes of steel tubing, and 135 steel alloys. From the work of Coffin and Clarkson, 16 standard sizes of lock washers were developed, 210 types of steel tubing, and steel alloys were reduced below 50.

Coffin promoted an expedition to England in 1911 to visit the Institute of Automobile Engineers in Great Britain. In turn, the I. A. E. visited the S.A.E. in the United States in June of 1913. There were over 600 engineers involved in that event. The best known among the British party was the venerable J. B. Dunlop (pictured at left) who was 74 years old.

The I. A.E. and S. A. E. Visit the Packard Plant in Detroit.   courtesy AAMA

CENTURY ELECTRIC MOTOR CAR COMPANY
DETROIT, MICHIGAN
1912-1913

CENTURY MANUFACTURING COMPANY
DETROIT, MICHGIAN
1914-1915

John Gillespie was born in 1877 in Chicopee, Mass., and attended public schools. He began his career as an office boy at the Ameo Sword Company in 1890. He worked for the Armstrong Regalia Company in 1900, then began his own business in 1903--the Detroit Regalia Company. Gillespie, Paul Gray, F. M. Aiken, William Merry, and **Phillip Breitmeyer** formed the Century Electric Motor Car Company in April 1912 with a capitalization of $80,000. The factory was on the south west corner of Woodward and Lothrop, 100 feet by 200 feet, made of brick and steel.

The company officers were: president, John L. Wynn, Jr.; VP, Charles L. Weeks; secretary and treasurer, Edward Atkins; general manager, John B. Gillespie; additional directors, Wm. A. Jackson, Howard Strieter and **Philip Breitmeyer**, who was also the senior member of Breitmeyer & Sons, florists, and was visited often by **Mrs. Henry Ford** who loved flowers. He was also a VP of the German American Bank. In 1907 **Breitmeyer** turned down an offer to become a stockholder in the Ford Motor Company, when he was associated with **John S. Gray**. He asked Gray's opinion when he was offered stock and **Gray** replied, "Don't do it. I believe the Ford deal is the worst investment I ever made." **Gray** ended up cashing in $20,000,000 worth of stock for an initial $10,000 investment. **Breitmeyer** was mayor of Detroit in 1909 and 1910, and during this time became a convert to motoring.

The first appearance for the "Century" was at the Detroit Auto Show in January 1913. The Brougham sold for $1,950. The car had an underslung frame which was unusual for an electric vehicle. It was advertised as having Magnetic control using Cutler Hammer controllers, and operated by hand or foot from front or rear seats. It had electric brakes operated through a foot pedal. Slight pressure on the pedal slowed the car down, and continued pressure applied the mechanical brakes. The car could be changed from front to rear drive by inserting a master key in the front control, or rear control.

Sales expectations were never met and in May 1914 Edwin Denby bought control of the company and re-organized as the Century Manufacturing Company with a capital stock of $40,000.

New company officers were elected as follows: John Wynne Jr. was president and director of the German-American Loan Trust Company. Hon. Edwin Denby was VP and a member of the law firm of Chamberlain, May, Denby, and Webster. He was also treasurer and director of the Hupp Motor Car Company,

and also interested in the Federal Motor Truck Company. **Philip Breitmeyer** was secretary and treasurer, John Gillespie, general manager, and police commissioner of Detroit, Wm. M. Pagel was director and a member of the Gordon-Pagel Company. Garvin Denby was director and VP, secretary and treasurer of the Federal Motor Truck Company. Howard Streeter was a director, and a member of the law firm of Mills, Griffin, Seely and Streeter.

The selling price of the Century was increased to $3250 fully equipped F.O.B. Detroit. Even with company officials of the highest standing in the industrial and financial world, the Century Manufacturing Company declared bankruptcy in June of 1915 with liabilities of $48,000 and assets of $24,000.

COLONIAL ELECTRIC CAR COMPANY
DETROIT, MICHIGAN
1911-1913

The Colonial Electric Car Company was organized in September of 1911 by A. D. Stansell, Albert Webb, Jr., F. C. Willis, Dr. L. C. Moore, Mark W. Allen, and **W. E. Storms**, who was the mechanical expert. The factory was located at 66-70 Brainard Street, Detroit.

The "Colonial" Brougham sold for $2700, and was large and roomy with seating capacity for five. It was built on a 93 inch chassis and had shaft drive, and was capable of 25 mph with a 75 mile range per charge.

The Colonial was out of production in April 1913 after being petitioned into bankruptcy. Parts and materials were sold at auction and the creditors received a dividend of 25 cents to the dollar.

SELF STARTER

"THE CADILLAC CAR WILL KILL NO MORE MEN IF WE CAN HELP IT. LAY ALL OTHER PROJECTS ASIDE. WE ARE GOING TO DEVELOP A FOOL PROOF DEVICE FOR STARTING CADILLAC MOTORS."
- Henry Leland

In 1896 the first electric self-starter, weighing 100 lbs, was used on an English version of the German "Benz," in Kent, England. By 1905 there were a number of European car manufacturers with starters, but none proved very practical or dependable. A French starting device consisted of a cylinder of carbonic acid, which applied force to a track and pinion that engaged an engine shaft gear. This "soda fountain reservoir" did not last long either.

On a December day in 1910 a lady stalled her car going over one of the small arched bridges on Belle Isle. A man at the scene offered to crank the car for her. He did not ask for verification that the spark was retarded, and it wasn't. It appears that he must have used his right hand, and when it backfired, the crank broke his arm and jaw. As a result, he died of gangrene a few weeks later. Two Cadillac engineers had come upon the scene of the accident and reported it to **Henry Leland.** He asked his engineers to develop a self starter that was fool-proof.

The big break came when they reasoned that an electric device offered the best opportunity. The Cadillac engineers developed a design strategy to incorporate gear teeth on the flywheel for use in turning the motor over. The electric motor had to be around four horsepower when engines at that time were 10 to 20 horsepower. The physical size of a four horsepower electric motor was impractical. One of the engineers recalled that **Charles Kettering** at the National Cash Register Company had developed a small economical motor for cash registers. The concept was based on an overloaded motor. It had high torque, would heat up fast, but the cash register required only intermittent duty. Based on this application, **Kettering,** who was by then working for the Dayton Electric Company, was enlisted to develop an electric starter for the 1912 Cadillac.

Formal announcement was made in Detroit on August 20, 1911, that the 1912 Cadillac would have an electric self-starter. The starter was operated by retarding the spark lever and pushing down gently on the clutch. A fully charged battery could turn the motor over for 20 minutes. The announcement set off a burst of activity among the automobile manufacturers. Velie, Everitt, Pullman, and Chalmers announced the introduction of compressed air starters. The Hupp Motor Car Company came out with a splash ad proclaiming it would be the first to offer a self-starter with a touring car under $1000. Everybody suddenly decided that the public was weary of standing in the mud and turning over a heavy motor.

The astute Howard Coffin was in England on November 15, 1911, when he guessed wrong and stated that six months ago, the electric generating outfit had looked like a walk-away for lighting. But the threat of competition had brought about gas-lamp improvements. He predicted the industry would use the acetylene self-starter and gas lamp combination.

By March of 1912 one half of the cars in the United States had some type of self starter as regular or optional equipment. There were 13 different types of acetylene, six types of compressed air, seven types of electric, 14 types of mechanical and one exhaust gas (Winton). There were over 50 manufacturers of self starters, but one stood out that was to become the industry standard--the "Delco" model found on the 1912 Cadillac.

An arched bridge on Belle Isle courtesy MSHC

One of the first Cadillac electric starters. It was a combination starter and generator.

1912- A Cadillac with an electric starter on the left side at the rear of the engine. (a flag with numeral 1 is on top of the starter.) picture taken in 1994

STANDARD MOTOR COMPANY
NEW YORK, NEW YORK
1912

MAXWELL MOTOR COMPANY
DETROIT, MICHIGAN
1913-1921

MAXWELL MOTOR CORPORATION
DETROIT, MICHIGAN
1921-1925

"THE GOOD MAXWELL"

In late October 1912, **Walter Flanders** went to Florida for an extended vacation. He and several business associates were developing a strategy for reconciling the collapse of the United States Motors Company. **Flanders** had just formed the Flanders Motor Car Company with **Bill Metzger** and **Barney Everitt**. For **Walter's** services, U. S. Motors purchased the Flanders Company for $2,750,000 stock in the new company and $1,000,000 in cash. The new company was named the Standard Motor Company, made up of nine car lines: Stodard-Dayton, Maxwell, Columbia, Brush, Courier, Sampson, Alden-Sampson truck, Courier-Clairmont and Everitt. **Flanders** eliminated every carline except Maxwell since it was always the division that made money and had a good reputation. The company's name was changed to Maxwell Motor Company. Everything on the East Coast was sold off and the Maxwell production was transferred to Detroit.

The Newcastle, Ind., factory was geographically in the center of the United States population, and was converted to a spare parts distribution center for the former U. S. Motors cars. With nine car lines, there were a total of 164 chassis that required 146,000 different spare parts. At the time, there were 122,000 of these vehicles on the road. With a total of 165,000 square feet, $1,750,000 was spent for modifications in order to produce spare parts. The drawings, jigs, tools, and dies were scattered across the country and they were brought to Indiana. Some drawings were missing or could not be identified. People who were familiar with the cars were brought from the old plants to help sort things out.

On December 20, 1913, **Ray Harroun** entered a contract with **Walt Flanders** to build three racing cars to participate in the 1914 Indianapolis 500 race. The contract carried a stipulation that the cars would be capable of 93 mph. The engine had four cylinders with a bore of 4.2 inches and a stroke of eight inches, producing 140 hp at 2400 rpm. The most unusual part of the engine was that it didn't have a flywheel, instead the crankshaft was counterbalanced along the entire length by counterbalance weights opposite each connecting-rod bearing. Although they never placed first, they came close and **Harroun** was made chief engineer, and **William Kelly** his assistant, of the Maxwell Motor Company.

Flanders inaugurated a one-chassis philosophy to keep cost and complexity low, starting with the 1914 season. Under his leadership, sales were bolstered by the racing victories, and reached over 50,000 per year. The Maxwell Company needed more production space and the entire Chalmers factory on Oakland Avenue in Highland park was leased in 1917. **Flanders** negotiated a complicated lease with various clauses that allowed Maxwell to share profits, property and assets of the Chalmers Company for five years. Maxwells and Chalmers were produced on the same assembly line. At Maxwell, **Flanders** was elected chairman of the board of directors in the fall of 1917, and **W. Ledyard Mitchell**, of Cincinnati was elected president. **Flanders** began disposing of his Michigan properties, and by 1918 left the company.

In 1918, The Maxwell "25" chassis was continued and was still the company's sole offering since 1914. The wheels were demountable instead of just rims, and the car came with an extra wheel. The frame was six inches deep instead of three inches and the body rested directly on it instead of projecting side pieces. The electrical system was changed to twelve volts with a single-wire system. The wb was lengthened 109 inches. The chassis was equiped with five different body styles: the five passenger touring, staggard-seated three passenger roadster, six passenger Berline, three passenger coupe, and the sedan. The roadster and touring cars sold for $745 and the closed models were $1095 f.o.b . Detroit.

Under **Mitchell**, the Maxwell Company began to decline. The Maxwells had severe quality problems with the rear axles, resulting in gear teeth falling onto the road. Sales plummeted and the Maxwell Motor Company went into receivership. **Walter P. Chrysler** was hired to revive the Maxwell Company in 1921 and it was re-organized as the Maxwell Motor Corporation. The banks had invested $18,000,000 in the Maxwell Company, and they owed creditors $35,000,000. **Chrysler** advised that it would take $15,000,000 to get things turned around. There were 26,000 Maxwells sitting in warehouses waiting to be sold. He ordered every Maxwell into the factory and trussed the rear axles. He slashed the price to the bone, and sold every one at a profit of $5.00.

For 1922 the Maxwell got a new advertising slogan " The Good Maxwell." People believed the ads and sales shot up to 48,850, and the company earned $2,000,000. The Maxwell and Chalmers organizations split into two factions that filed lawsuits against each other over bookkeeping practices.

By mid-1923 **Chrysler** assumed the presidency of Maxwell and began excercising stock options for Chalmers. By August 1923 Chalmers was a 90 percent owned subsidiary, and production was terminated. In its stead, the first Chrysler was announced at the National Automobile Show in New York in January 1924. By the beginning of June 1924, production was running at 150 cars per day, and in September the Chrysler line outsold Maxwell. A new record for the first year production was set with 32,000. On April 9, 1925, a special meeting of Maxwell Motors Corporation approved a takeover, through a change of securities, by **Chrysler.**

A Maxwell in front of the factory office building on Oakland Avenue courtesy AAMA

A Maxwell with a Martin-Parry high-side express body

A Maxwell Sedan

*1925-
The last year for the Maxwell*

The Maxwell factory on Oakland Ave. photographed in 1994

KEETON MOTOR CAR COMPANY
DETROIT, MICHIGAN
1912-1913

AMERICAN VOITURETTE COMPANY
DETROIT, MICHIGAN
1913-1915

Forest M Keeton owned a bicycle business in Chicago. In 1903 he moved to Toledo, Ohio, and became manager for the Pope Motor Car Company. From there he went to Detroit and worked for Deluxe Motor Car Company. In 1908 he designed a taxicab and organized the Keeton Towncar Works. The Forest City Motor Car Company in Massillon, Ohio purchased the taxicab and re-organized as the Jewel Motor Car Company which proceeded to manufacture the "yellow-bonnet taxicab." Some were sold to John Hertz in Chicago, and became the forerunner of the Yellow Cab. One of the Yellow Bonnet cabs was designed with a left-hand drive. The reason was that cabs were permitted to drive on the left side of street cars, which gave the driver better vision of oncoming traffic.

The man behind the Jewel Motor Car Company was Herbert A. Croxton. In 1909 he joined with **Forest Keeton** and re-organized into the Croxton-Keeton Motor Company, of Massillon, Ohio. The partnership was dissolved by 1910, and the company was renamed Croxton Motors Company. In 1914 the company was relocated in Washington, Penn. as the Universal Motor Car Company, maker of the Universal cyclecar until 1915.

Forest Keeton moved back to Detroit and in 1912 formed the Keeton Motor Company. At first the company used the former plant of the Seitz Auto and Transmission Company in Wyandotte. The Keeton had a "French-like" appearance and was the first American car to have a sloping hood like the Renault. It began selling well, and in January 1913 the Oliver Motor Truck Company was purchased. There were four one-story buildings, with a fifth being planned. They were located at 464 Lawton, at the corner of Brechenridge by the M.C.R.R.

The car was such a big hit that the czar of Russia bought one as well as the president of Panama. **Bob Burman** drove a special Keeton that Memorial Day at the "Indy" and broke several records, but the car caught on fire during the 55th lap.

The radiator was mounted in back of the engine directly in front of the dash, using the flywheel for a fan. The air was forced out through the sides of the radiator, like the Renault, Darraccq, and Charron cars. The Keeton came with a four cylinder, but it also was one of the first cars to offer a six cylinder engine. It was cast *en bloc* using a special grade of gray iron that was properly aged after casting. The engine was an L-head with 364.35 cubic inches rated at 33.75 SAE hp. It used a Zenith carburetor with a float feed and employed a "strangler" valve for making it easier to start in cold weather. The ignition system was a true high-tension magneto, with electric lighting and starting, wire wheels and left-hand drive with center control. The capacity of the oil reservoir was three gallons, using a pump forcing the oil directly to each main crank and connecting rod bearing. It used a thermo-syphon cooling system. The steering gear was a worm and full gear semi-irreversible type. The body was made of sheet steel over a wooden frame. The upholstery was hand buffed, pebble grained, black leather stuffed with curled hair over special springs. Pockets were provided in the doors and shroud for storing maps and gloves. The Riverside model was a four door, seven passenger. The Meadowbrook was a four door roadster, and the Tuxedo was a fully enclosed coupe.

In June of 1913 Detroit's first completed cyclecar made its appearance on Woodward Avenue and was seen daily. The new car was called the CAR-NATION, manufactured by the American Voiturette Co. ("little car"). Charles B. Shaffer was president . He was known as the greatest oil producer in the world, with oil wells in every state.

The CAR-NATION had a 44 inch tread and a 100 inch wheelbase. It had a four cylinder air-cooled engine with a 3.5 inch bore and a 3.75 inch stroke, rated at 18 to 24 hp. The transmission had three forward and one reverse gears with center control, and had left hand steering. It came as a four passenger touring car, a two passenger roadster and a tandem, all under $500.

In late 1913 Charles Shaffer bought the Keeton Motor Company and an addition was made to the Keeton plant. Both companies were re-organized in January as the American Voiturette Co.

Narrow tread cars were not selling well and financial strain resulted in receivership by September 1914 when all creditors were notified that there were liabilities of $509,000 and assets of $477,000. The creditors had agreed to finish making 350 combined cars with material on hand and selling them. But in November 1914, the Samual L. Winternitz Co. of Chicago was allowed to purchase all of the assets except the real estate and cash for $100,000 with the stated purpose of continuing production. About three months later **Forest Keeton** bought the assets from Sam Winternitz with the intention of continuing low volume production and repair. Production was never started and he only got into the repair business. **Forest Keeton** did not manufacture any other cars, although in 1923 he tried to form a company to build taxicabs.

*1913-
A Keeton in front of the Hotel Ponchartrain
courtesy AAMA*

The Keeton factory

*Front view-
Carnation showing
the streamline effect.*

*three-quarter view-
Carnation showing
side doors and
tandem seats*

*1914-
Carnation being
photographed
for an advertisment
courtesy AAMA*

CHURCH-FIELD MOTOR COMPANY
SIBLEY, MICHIGAN
1912-1913

The Church-Field Motor Company was formed by Austin Church and H. George Field in 1911, in Sibley, Mich. They introduced a new electric car at the Detroit Automobile Show in February 1912. It came as a coupe or a roadster with an aluminum body, a Renault-type hood and an underslung frame which gave it a lower center of gravity. The company claimed the "Church-Field" was the first electric automobile to use a two-speed planetary transmission, enabling it to travel up to 25 mph and still climb hills in low speed.

The Church-Field Company was in financial trouble by September 1913, and production was stopped. Assets were sold for $600 in May 1915.

1912- The Church-Field at the Detroit Automobile Show

The Wayne Hotel where the Detroit Automobile show was held in 1912. The Hotel erected an annex for its pavilion to show 75 makes of cars that year.

S & M MOTOR COMPANY
DETROIT, MICHIGAN
1912-1913

BENHAM MANUFACTURING COMPANY
DETROIT, MICHIGAN
1914

The S & M Motor Company was organized in September 1912 by two New Yorkers, Edward E. **Stroebel**, and Walter C. **Martin**. Stroebel had previously been a furniture maker while Martin used to sell Cadillacs in Manhattan. A third man, Owen R. Skelton, worked for S & M Motors as the only engineer. The S & M factory was located at 1890 Mt. Elliott Avenue, Detroit, Mich.

Stroebel and Martin applied a European flair to their vehicles by employing the talents of Goodwin. Mr. Goodwin designed many streamline bodies on display at the Importer's Salon of 1913. The vehicles all had 130 inch wb and featured large doors, mahogany side panels, cowl boxes, Brewster fenders, left hand steering and center gear shift. Instead of being manufactured, S & M assembled their cars.

The "S & M" used a 6-48 Continental Light six. The unit power plant, combining the motor, clutch, gearset, and electric starter, was suspended at three points, the single point being at the front of the motor, and the other two at the rear formed by lugs which were part of the crankcase casting. This allowed the frame to move independently from the motor, thereby eliminating crankcase failure. The single point of suspension was lubricated by a grease cup mounted on the bronze-lined bearing support of the motor.

The cylinders were cast in two blocks of three, of close grained gray iron. The motor developed 38 hp at 1500 rpm on block test. The engine was water cooled, and oiled with a combination of force feed and splash. The gearbox was a selective 4 speed by Warner, and the steering gear by Gemmer. The electrical system was divided into two distinct units: the ignition, and the starting and lighting. The Jesco system was used for starting and was really a combined starter and generator.

Production didn't begin until late 1913, and after only 40 cars, the S & M Company was petitioned into bankruptcy in January 1914. George Benham, of the Benahm Manufacturing Company, acquired the assets later in January and renamed the car "Benham." Priced at $2484 for the touring and roadster bodies, and $2535 for the limousine, it was the same car as the S & M.

After building 60 cars, the Benham Manufacturing Company signed a voluntary petition for dissolution, and by October was dissolved. The parts and materials constituting the main assets of the Benham Manufacturing Company were purchased by the Puritan Manufacturing Company, of which **A. O. Dunk** was the head.

*S &M -
6 cylinder touring car*

*Benham-
2 passenger*

BENHAM SIX

ROADS

IT APPEARS THAT EARLY MOTORISTS GOT BOGGED DOWN IN MUD DUE TO THEIR OWN ENTHUSIASM--APPARENTLY, THE OWNERS OF HORSE CARRIAGES DID NOT VENTURE OUT IN CONDITIONS THAT DOOMED THEIR TRIP

In the early United States, roads started nowhere and ended nowhere. They were merely paths worn in the ground from many years of horse travel over the same terrain. The horse was allowed to pick its own way over hills and around wooded stretches. Eventually, farmers began to clear paths by chopping down trees. If a wagon was too low, it could become caught on a tree stump, which is the origin of the expression "I'm stumped." The dirt road then came into being, and could be hard to negotiate when muddy. After it rained, ruts were created as the ground dried out. Wagons fell into these ruts, which led to the expression "I'm in a rut." Roads were not graded and reached inclines of over 30 degrees.

The first road in Michigan stretched along the coast from Toledo to Detroit. The first "interior" road in Michigan was Moravian, which was laid out in the winter of 1785-86 to connect a village on the Clinton River with Fort Detroit, 23 miles away.

Detroit's first street plans were created by Judge **Augustus Woodward** in 1805. Before coming to Detroit, **Judge Woodward** lived in Virginia and was a friend of Thomas Jefferson, and he arrived just after Detroit was completely burned down. He was familiar with the street plan of Washington D.C. and intended to use it for rebuilding Detroit. Due to prior property holdings of Detroit's residents, compromises were made, and he ended up with Grand Circus Park and Campus Martius as two hubs with streets emanating from them, instead of four. He figured the most important street in the city would be at the crest of the east-west river embankment, so he named it Jefferson. He later named streets for Madison and Monroe. He also named the north-south center street in Detroit **"Woodward."** He became upset when two judicial associates pointed out he had named a street after himself. He rebutted "I did not, I named it **Wood-Ward**--to the woods."

The first paving in Detroit was cobblestone, laid in 1825 on Jefferson Avenue. Cobblestone was used for horses to gain a foothold, but metal carriage tires made a lot of noise. As more streets were added, brick became the pavement of choice. Most of the streets, however, were paved with round cedar blocks, which were quiet and aromatic, but they had a tendency to float away during a hard rainstorm. By 1892 asphalt came into use to bind around the cedar blocks. Sidewalks were made of 10 inch boards laying on a cinder base. They were used at cross sections of some of the non-paved major intersections.

In 1903 Detroit had almost 300 miles of paved roads made of the following materials: cedar on concrete-156 miles (worn out cedar was about the most agitating surface for a car), cedar on planks-33 miles, cedar on sand-33 miles, brick on concrete-35 miles, asphalt on concrete-31 miles, granite on concrete-2 miles, cobble on concrete-1 mile, medina on concrete-1 mile and macadam-.8 miles

The first white dividing line in the automotive era was applied on River Road (Jefferson was originally called River Road west of Woodward) near Trenton, Mich., in 1911. **Edward Hines**, who was the road commissioner, was aware of accidents involving vehicles going around curves. He got the idea from seeing milk spilling from a truck (the oldest known centerline marking was on a road between Mexico City and Cuernavaca dating back to 1600). The first dividing line on a rural highway in North America was applied between Marquette and Ishpeming, Mich., in 1917.

In Europe, natural cement and lime, which was used by the Romans, gave way to concrete, which used a synthetic plastic material called Portland cement (it resembled natural rock quarried in the island of Portland). The first mile of concrete "highway" in the United States was poured in September 1913 on Woodward Avenue between Six and Seven mile roads, which was originally a toll road. It was just north of the Ford Highland Park Plant, the home of the Model T. Henry Ford and James Couzens were involved with the road commission at the time. People from all over the world came to see this road, and in five years, 3.5 million vehicles passed over it. It wore less than .25 inches from its original six inch thickness.

*1875-
What Detroit
cedar block streets
looked like
    courtesy AAMA*

*circa 1960-
The last cobblestone
street in Detroit,
with stones the size
of two fists,
100 ft x 30 ft wide
Pierce and Riopelle
    courtesy Burton*

*road sled-
Farmers paid taxes
by putting a few days
of work on the road
every spring
    courtesy AAMA*

*1903-
Fort Street
courtesy NEWS*

*1906-
The Grand Boulevard
courtesy NEWS*

*First concrete
highway in
the USA*

THE LINCOLN HIGHWAY

"LET'S BUILD IT BEFORE WE'RE TOO OLD TO ENJOY IT"
-Carl Graham Fisher speaking to his wife. The phrase was later used on Lincoln Highway stationery.

Henry B. Joy

In September 1912 Carl Graham Fisher and his partner, James Allison, presented a plan to 300 Indianapolis industry leaders to build a highway from the East Coast to the West Coast. It was to be completed in time for the 1915 Pan-American Exposition in San Francisco. Within two weeks they received pledges for over a million dollars. In December the Packard Motor Company telegraphed Fisher and Allison that they were joining the movement.

The president of Packard, **Henry Joy** tried to persuade Congress to re-allocate $1,700,000 for this road instead of a Lincoln Memorial in Washington D. C. **Henry Joy's** father was a personal friend of Abraham Lincoln from the Michigan Central Railroad days. The family was an ardent collector of Lincoln memorabilia and what better name for the new road. The Lincoln Highway Association was formed in July 1913, and **Joy** gradually reduced his involvement with the Packard Company.

Detroit industrialists and banking personnel led the movement, and soon cities and states all wanted the highway to pass through their area. President Wilson petitioned for having the highway pass by the Lincoln Memorial in Washington, D. C., and several cities offered to change their name to Lincoln.

The highway was dedicated with parades and festivities in October 1913. Markers with a large letter "L" on a white background with a red stripe above, and a blue stripe below, were placed along the route.

Early markers were painted on telephone poles, and later, steel enamel markers were placed from Omaha to California except Ohio, which used brick markers. All of the markers were financed by Willy-Overland and Autocar. By 1917, 8000 markers were completed from New York City to North Platte, Nebraska. The most dangerous and formidable obstacles were found west of the Great Salt Lake, "The Great American Dessert." It would eventually reach 3385 miles long.

The first attempt to use the Lincoln Highway for commercial application was in June 1914 when **H. F. Vortkamp**, a representative of Marvel Carburetor, established the **National United Service Company** office in the Ford Building in Detroit. He sold auto parts and accessories through a chain of garages along the route of the Lincoln Highway.

Before long, the Dixie Highway was started from Sault Ste. Marie, Mich., through Detroit down to Miami, Fla. By 1926 there were 17 named interstate routes and in 1927, the L.H.A. ceased to exist, as the Federal government took over and began assigning numbers instead of names to the main highways.

Henry Joy (L.) on test trip out west to establish the route for the Lincoln Highway. Later, it would become part of Interstate 80.
courtesy Bentley

Return back to the Packard plant-Henry Joy behind the wheel.
courtesy Bentley

advertisement

READ MOTOR CAR COMPANY
DETROIT, MICHIGAN
1913

The Read Motor Car Company was incorporated in the summer of 1913 with $50,000. The principal stockholders were Joseph Beatty, Ray **Read** and Roy Herald. A factory was secured to build the "Read" at 68 Champlain Street and an office and salesroom were located at 541 Woodward Ave. The Read model "30" had a four cylinder 20 hp motor, a 115 inch wb, and sold for $850.

In December 1913 Albert Schneider sued the Read Motor Car Co., alleging that the company was insolvent and that he had been swindled out of $1000. Although the Read was sturdily constructed and had a conservative price, the company went out of business.

*The body was
painted Read grey,
striped in black: wheels were Read grey striped
in black: fenders, hood
and running board were
black enamel.*

WAHL MOTOR COMPANY
DETROIT, MICHIGAN
1913-1914

WAHL AUTOMOBILE COMPANY
DETROIT, MICHIGAN
1914

The Wahl Motor Company was formed in the late spring of 1913 by a group of capitalists including George Wahl, Alvan M. Dodge, and Joseph Hofweber. The Hofweber Motor Car Co., in LaCrosse, Wisc., was trying to produce a car during this same time period, and considered producing it in the Wahl factory at 3089 E. Grand Blvd., but never got beyond the development stage.

The "Wahl" was in production by late summer 1913, with a two passenger roadster, and a five passenger touring car, using one standard chassis. The expression "standard" chassis could be interpreted in two ways. First in the usual manufacturing sense, and second in regard to the entire absence of any radical or experimental features about the car. Right hand steering control and a "cane handle" lever in the center of the floorboard was used for shifting the three forward speed transmission. This arrangement gave easy access to the driver's seat from either side of the car.

The Wahl had a 108 inch wb and a four cylinder Hazard engine with 24 hp at 1,400 rpm. The cylinders were cast in pairs with a bore and stroke of 3.5 x 5.5, giving it an honest claim as a long stroke engine. The engine was lubricated with the splash system with a pump to maintain the oil level. A 108 inch wb allowed for a body of sufficient length to give plenty of leg room in both compartments, and further permitted making all doors full size without the corners cut off in order to clear the fenders. Standard equipment included a mohair top, Prest-O-Lite tank, gas headlights, and oil side lamps and tail lamps.

The Wahl Company had been under capitalized with $85,000 which resulted in early cash flow problems. In an attempt to spur sales, the Wahl Company advertised that automobile dealers could sell the car with their own trademark on the radiator. By the fall of 1913 creditors were complaining, and on October 6, George Wahl drank a bottle of carbolic acid, ending his life at 35 years of age. The Wahl Company was in receivership by December 1913, and taken over by St. Louis automobile men for $100,000. James Root became president, Joseph Nulsen VP, and A.G. Nulsen, Jr., secretary and treasurer. The price of the car increased from $750 for the roadster and $790 for the touring car, to $850 and $890. The M.W. Bond Automobile Company of St. Louis became the chief distributor of the Wahl car.

Production resumed in January 1914, and in February Alvan Dodge left to form his own company- the Dodge Motor Company. In the same year, the Dodge Bros. Motor Company was formed and Alvan sued for the rights to the name, but lost (to the Dodge brothers.)

New features for the 1915 model year were reported in July 1914 as chiefly minor refinements. A full streamline body with flush doors, concealed side door hinges, wider seats, deeper upholstery, two-wire electrical system, double-bulb headlights, electric horn and an optional electric starter. The wheelbase was extended to 112 inches and the car weighed 2250 lbs. The engine had a higher displacement with 28.5 hp, and had timing gears that were spirally cut and of proper pitch to eliminate noise when running at speed.

The Wahl Automobile Company was declared bankrupt in October 1914. The assets were purchased by Albert C. Barley, and the factory was taken over by the Massnick Phipps Manufacturing Company in 1915.

(Barley brought out a car called the Halladay, in Streator, Ill. in 1907, moved production to Ohio in 1918, and went out of business in 1920. He also brought out the Roamer, in Streator Ill., in 1916, then moved to Michigan in 1917 and went out of business in 1930).

MONARCH MOTOR COMPANY
DETROIT, MICHIGAN
1913-1916

In February 1913 a new company called the "Monarch Motor Car Company" was introduced, with **Louis G. Hupp** as president (the brother of R. C. Hupp). In April 1913 Joseph Bloom (a brother-in-law of R. C. Hupp) made application for incorporation papers also under the name "Monarch Motor Car Company," on behalf of **Robert C. Hupp**. The rights to the Monarch name were granted to **Robert C. Hupp**. **Louis G. Hupp** was then forced to switch the name of his company to "Tribune Motor Car Company."

By August 1913 the Monarch Company moved into the vacated Carhartt factory on Jefferson Ave., and by the end of the year a small 16 hp four cylinder engine on a 110 inch wb selling for $1000 was in production. By May 1914 approximately 150 Monarchs were built.

In March 1914 the cheapest and lightest six cylinder car on the market selling for $1250 was announced by Monarch, but gravitated to $1400 by June. It retained the same general lines as the four cylinder, but with a longer chassis and a wheelbase of 118 inches. Standardization was the word from **R. C. Hupp**, using a Continental motor with 40 hp at 1200 rpm, a Salisbury rear axle, Warner steering gear, Mott wheels and a Detroit Gear and Machine Company gearset. The driveshaft was unenclosed with universal joints at front and rear. Equipment was complete with jiffy curtains, rain vision ventilating windshield, tire carriers, speedometer and headlights.

In September 1914 a "light" four cylinder was introduced at a cost of $675. It had a streamline body and had *creditable unbroken smoothness* from front to rear. Although it was a four passenger vehicle, there was no rear door. Instead, the right front seat could fold up and allow room for passage to the rear compartment. The wheelbase was 103 inches with a 56 inch tread. The rear axle was the gearless differential type. Each rear wheel was fitted with a form of roller clutches which had the same action as the usual bevel and spur gear arrangement. Electric cranking of the Ward-Leanord make was available for $25 extra.

In May 1915 the Monarch Company was re-organized in order to expand into the eight cylinder field. Eastern capitalists backed the new company under Delaware law with $400,000. **Hupp** remained president, Arthur Frost Spaulding was VP and George B. Tanner of the Bankers Trust Company was treasurer. They planned to make 3000 to 5000 eight-cylinder cars for the 1916 season, listing at $1500, followed by a 12 cylinder motor.

In April 1916 the Monarch Company went into bankruptcy and by October was acquired and made a subsidiary of the Carter Bros. Company of Hyattsville, Md., with **Robert C. Hupp** as president. A small car, known as the Monarch Midget, was offered that sold for under $600. The whole line of Monarch automobiles was out of production by the end of 1917. (The "Cartermobile" was produced by the Carter Motor Car Company of Hyattsville Md. from 1921-22.)

In the meantime, **Louis'** Tribune never went into production. The company was capitalized at $10,000 with a factory located on Scotten Ave. The Tribune was a small touring car with a 116 inch wb and a four cylinder Buda engine and was to sell for $1200. But the factory was leased to the Mercury Cyclecar Company in November 1913, and **Louis** joined the Briggs-Detroiter Company.

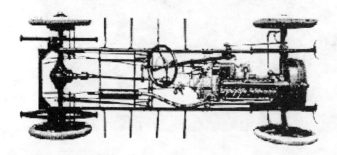

1913- Monarch

1914- Monarch showing folding front seat

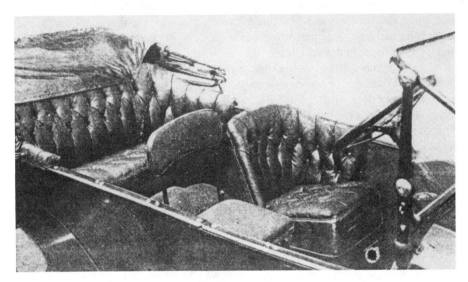

Tribune- Never made it to production

CYCLE CARS

The ancestor of the cycle car dates back to 1902, when in Europe various light weight vehicles with three or four wheels made their appearance. In 1909 the first true cycle car, called the Bedalia, was constructed in Paris by Henri Barbeau. It had tandem seats and was driven in the rear. An in-line front axle revolved around a central point in front of the body, similar to a fifth-wheel.

No absolute definition was ever possible for describing a cycle car, but example specifications were a 37 inch tread, 67 inch displacement engines, total weight of less than 800 pounds, 30 mpg and sold for less than $400. The cycle car became popular in the United States in November 1913 when the Detroit Cycle Car and Motorette Club was organized in the parlors of the Hotel Pontchartrain. In January 1914 **William B. Stout,** father of the American cycle car, was at the New York Auto Show with America's first cycle car, the IMP, made in Auburn, Ind.

Newspapers reported that the "poor man's" car was here at last. There were very creative ideas to help reduce weight and parts. One vehicle powered the left front wheel, obviating the need for a differential. There were inherent flaws with many of these machines. Many had hammock seats and no floor pan. If the hammock broke, the occupant could fall directly to the ground.

Steering was generally either "rack and pinion" or "cotton reel," which was a drum connected to the tie rods and rope that wound around the steering shaft. Seating was side by side, but mostly tandem in order to maintain a narrow tread. Some had steering wheels in the back seat. Most were powered by motorcycle engines, with drive belts that stretched and slipped. Frames were often made of wood that did not hold up. Eventually, there were over 400 manufacturers of cycle cars in the United States and the same in Europe. The simplicity of this class of vehicle allowed too many substandard products on the market. The Ford Model T was approaching the cycle car in cost, and by 1915 the United States market collapsed. It held on much longer in Europe for several reasons including: licensing taxation, cost of gasoline, and perhaps better fleet quality. The French Bedalia was manufactured until 1925.

William B. Stout

MERCURY CYCLECAR COMPANY
DETROIT, MICHIGAN
1913-1914

The Mercury Cyclecar Company claimed it was the first cyclecar company in Detroit to start actual manufacturing, the first to ship cyclecars from Detroit, the first to record an actual retail sale, and on March 4, 1914 the first to ship a carload of cars.

The Mercury Cyclecar Company was organized in early November 1913 by W.J. Marshall, the head of the company, and R.C. Albertus, the director of sales. After they decided to enter into the cyclecar business, and to start manufacturing immediately, the former factory of **Louis Hupp's Tribune Motor Company**, at 807 Scotten, was secured. The first car was built in just one week, before Detroit *was awake* to the fact.

The "Mercury" sold for $375. It had a torpedo body style with 100 inch wb, a 36 inch tread, 28 x 2.5 wire wheels, and weighed 450 lbs. The engine was an air cooled Spacke twin with a bore of 3.5 and a stroke of 3.67, with a displacement of 70.6 cubic inches. A friction drive was used with the final transmission of power by a V- belt on each side of the car. Transverse springs were used in front, and cantelever springs in the rear.

A Chicago agent named Steele, and Mercury's chief engineer, H.J. Woodall, pulled out of Detroit at 7:30 pm on Friday, January 22, 1914 and arrived in Chicago at 9:30 am Sunday on January 25, 1914. No attempt was made to establish a speed record, with the aim of the crew to show the reliability *and to work up a little advertising on the way.* It was pouring rain when they left the Wolverine City, and the roads soon became gummy. Still the little car plugged along without a hitch. The 36 inch tread enabled the driver to pick his own course so successfully that many of the worst mudholes were avoided. When the car pulled up to the coliseum on Sunday morning it was covered with mud, but was otherwise O.K.

The Mercury Cyclecar Company was petitioned into bankruptcy in September 1914 by the Grafton and Knight Manufacturing Company, creditor in the sum of $1244. In November 1914, the Michigan State Auto School purchased the Mercury Company. They had been using Mercurys for their driver training course. The Michigan State Auto Company intended to assemble Mercurys in its school and sell them for $200, but never did.

Mercury cyclecars in front of the Hotel Ponchartrain courtesy AAMA

The Mercury factory

The Mercury which drove from Detroit to Chicago over frozen roads

publicity picture of a Mercury

CRICKET CYCLECAR COMPANY
DETROIT, MICHIGAN
1913-1914

MOTOR PRODUCTS COMPANY
DETROIT, MICHIGAN
1914

The Cricket Cyclecar Company was incorporated with $400,000 of capital stock by George B. Burton, William L. Davis, and William R. Demory. Tracy Lyon was the factory manager, and O. C. Hutchinson was named sales manager. The original "Cricket" had a metal body and side by side seating, per a design by an automobile engineer in England named Anthony New. There were claims that the Cricket had only about half the number of parts as other small automobiles. A number of test miles were driven around London in August of 1912 before the car and drawings were brought to Detroit.

In February 1913 work commenced to redesign the car to meet American conditions. A longer wheelbase and heavier springs gave the Cricket better perfomance on the rough roads in the United States. The first model was ready in November 1913. *This was given a test to develop any possible weakness or improvement that could be made to advantage.* O. C. Hutchinson claimed the painful distinction of being the first man hit by the back-kick of a cycle car motor. The starting crank hit him between the eyes when he was cranking his car.

The American version hit the streets in November 1913, and a factory, with 12,000 feet of floor space, was secured at 80 Walker Street. The Cricket was produced starting in the spring of 1914 with a selling price of $385.

The Cricket had a 78 inch wb, 47 inch tread, and weighed 475 lbs. It was powered by an air cooled Spacke two cylinder V motor with a displacement of 70 cubic inches, that developed nine horsepower. Independent side frames of tubular construction, connected by flexible cross members helped make the car light. The engine, disc clutch, and sliding gear transmission were mounted on the driver's side frame and powered only one wheel. As a self-contained power plant, it could not get out of line, and was very accessible. Both pleasure and commercial models were available of the Cricket.

In October 1914 the Cricket Cyclecar Company was absorbed by the Motor Products Company for the purpose of increasing the manufacture of the Cricket Cars. The Motor Products Company had been building and supplying engines for the Cricket. Plans called for the continuance of both models, and a reduced selling price, but production ended in 1914 with the collapse of the cycle car market.

The Cricket cyclecar

The Cricket factory

SAXON MOTOR COMPANY
DETROIT, MICHIGAN
1913-1915

SAXON MOTOR CAR CORPORATION
DETROIT, MICHIGAN
1915-1922

SAXON MOTOR CAR CORPORATION
YPSILANTI, MICHIGAN
1922

Harry W. Ford was a graduate of the University of Chicago, and a former newspaperman who had been with **Hugh Chalmers** at the National Cash Register Company and had gone with him to be advertising manager for the Chalmers Motor Car Company. Ford had been convinced for some time of the possibilities for a good two passenger car that would sell for $400. The first model was out August 10, 1913. He succeeded in interesting **Chalmers**, who organized a syndicate of 10 men, each of whom put up $10,000 to form the Saxon Motor Company on November 1, 1913. The company was capitalized at $350,000, with $250,000 in common stock. The officers were: H. W. Ford, president and general manager; G. W. Dunham, VP; L. R. Scafe, secretary and treasurer; R. E. Cole, chief engineer. The factory of the former **Demotcar**, with three stories and 50,000 square feet was leased for initial production.

The Saxon venture got off to a spectacular start in December 1913. The "Saxon" had a cyclecar price of $395, but was a standard design with a 96 inch wb with a standard tread. It was a well designed small automobile, that was light weight, had low upkeep, oval fenders, and a streamline body. The Saxon was a left hand drive car with control levers in the center, a foot controlled accelerator, and a spark control located on the dash. The steering mechanism was of the double gearing type with dropped forged steering connections, which resulted in a four cylinder (Ferro) engine of the L head type, with all four cylinders cast integral. The bore and stroke was 2 5/8 x 4 inches. The engine was lubricated by a vacuum feed oiling system with splash distribution. The cooling was of the thermo-syphon type with a tubular radiator and fan. The transmission was carried on the rear axle and was of the sliding gear progressive type, with two speeds foward and one reverse. The drive was by shaft through a torque tube with one universal joint. The clutch consisted of three dry plates, steel on raybestos. There were two sets of brakes on the rear wheels. The service brake was eight inches in diameter, lined with heat proof material. The emergency brake was of the internal expanding type, steel on steel.

The Saxon Motor Company stood out under the direction of Harry Ford, the boy president, as he was known. The first completed chassis came out in February 16, 1914. There were 13 cars produced in February, 470 in March, and 762 in April. The factory was running at 40 cars per day in May with 200 employees, and plans were being made to add a night shift.

With a factory hopelessly too small, the former **Abbot factory** was leased for five years in October 1914. (Abbot production continued in a plant owned by A. C. Knapp Company on Beaufait Avenue.) In 1915 production exceeded 21,000. Harry Ford bought out Hugh Chalmers' interest in the company, and re-organized as the Saxon Motor Car Corporation. In 1916 a six cylinder engine was added and Saxon ranked eighth in production, and industry observers commented that Harry was coming on big like **Henry**. Although production reached 28,000 in 1917, the Saxon Company could never get its cost down to the point that it could be competitive with the Model T. A capital investment of $100,000 and a leased factory were no match for the millions that **Henry Ford** poured into the Highland Park plant to create mass production.

During the First World War, the Saxon factory burned down and a new and larger plant was under construction. Harry Ford was struck by the influenza epidemic in Detroit and resigned in December 1917. He was replaced by Benjamen Gotfredson. Harry died the following year.

Production was discontinued in 1917. With wartime material shortages, only 3426 Saxons were built in 1919, and Gotfredson resigned. C. A. Pfeffer, a former Chalmers man, took over as president and began re-organizing the company. The Saxon stock was held by a large number of people with small amounts of shares, who seemed indifferent as to what happened to the company. By 1921 the Saxon parts business for former models was sold to pay off loans. The new factory was sold to the Industrial Terminal Corporation which completed the plant and sold it to GM. In 1922 the former **ACE** factory in Ypsilanti was leased in order to continue production. Pfeffer resigned in the summer and the company was bankrupt by the end of the year.

*1916-
A woman editor drove this Saxon on a west coast tour
    courtesy NAHC*

*M. A. Croker drove 135 miles per day for one month straight.
    courtesy AAMA*

Harry Ford

Named after the "Saxon" people. Saxon men were sturdy, great lovers of the out doors, nimble, thrifty and imbued with integrity.

DETROIT CYCLECAR COMPANY
DETROIT, MICHIGAN
1913-1914

SAGINAW MOTOR CAR COMPANY
SAGINAW, MICHIGAN
1914

In August 1913, the Detroit Cylecar Company was organized by Aurthur R. Thomas, who was to become president of the company. Benjamin C. Bradford became VP. The "Detroit Speedster" was designed by Ernest Weigold, former engineer for the E. R. Thomas company, and chief engineer for the Herreshoff Company. Offices were located at 504-508 Elks Temple Building in Detroit.

The Detroit Speedster was a *two-seated affair*, similar to the bevy of cycle cars in Detroit. It had a four cylinder engine, water cooled with a 2.75 x 4.0 inch bore and stroke, and shaft drive. The radiator was pointed to direct the air currents around the engine to a greater degree than ordinary radiators. The wheelbase was 92 inches and the tread was 44 inches. The drive was left-hand with center control. The body was of the open speedster type, straight dash, wherein the gas tank was installed. The mud-guarding was considered very efficient, and wooden wheels set it apart from other cyclecars.

In 1914, Phillip M. Cale, attorney and promotor for the Detroit Cyclecar Company, was involved with fraud in stock sales that violated the Michigan "blue sky law." In May 1914, Thomas moved to Saginaw, Michigan and organized the Saginaw Motor Car Company. In July 1914, Thomas secured the factory of the Brooks Manufacturing Company which produced a light delivery car under the auspices of Charles Durea. The "Saginaw Speedster" was a rebirth of the Detroit Speedster, but was not successful and the company failed before years end.

The Detroit Speedster

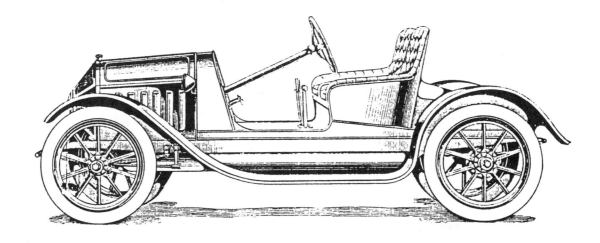

PRINCESS CYCLE CAR COMPANY
DETROIT, MICHIGAN
1913

PRINCESS MOTOR CAR CORPORATION
DETROIT, MICHIGAN
1914-1919

In December 1913 the Princess Cycle Car Company announced that its first deliveries would be made on February 1, 1914 at a list price of $375. L.N. White headed up the company with a suite of offices located at 1311 Dime Savings Bank Building. The "Little Princess," as the car was called, would be available as a two passenger model at first, followed by a four passenger model and then a light delivery van. The whole design had a decided foreign effect, mainly because C. J. Thornewill was one of the engineers who had prior experience at Wolseley-Sideley and the Thornewill Brothers companies in England. The body was streamlined using metal panels with tangent fenders (although future plans called for round fenders), and weighed 620 pounds.

The engine was an air cooled four cylinder block-cast with 92.8 cubic inches. The cooling fan was incorporated in the flywheel. Details of the differential were guarded but it was known that it was incorporated in the wheel hub instead of the usual bevel and spur gears. A planetary transmission with two forward and one reverse and shaft drive was used. It had rack and pinion steering and wire wheels with a 44 inch tread and an 86 inch wheelbase.

Production problems beset the company and production was halted. In August 1914 it was re-organized as the Princess Motor Car Corporation. A new group of investors including some from Tennessee took over and named C. D. Smith as president and W. H. Brennan as vice-president. They leased a building at 348 Clay (with 10,000 square feet) in order to have production capacity of 100 cars per month.

By September 1915 the company moved to a larger plant at 1305 Bellevue. It was formerly used by Demot from 1909 to 1912, and Saxon when they first started out in 1913. For the 1916 model year, a touring body was available with fenders that followed the curve of the wheel for added streamline effect. From this plant they also furnished a version of its Model 30, with different trim, to the Times Square Automobile Company in New York who sold it as the Mecca. It only survived for a year. By early 1919 the Princess Motor Car Corporation could not survive either, and went out of business.

1915-Princess

10,000 square feet factory at 348 Clay

MALCOM-JONES CYCLECAR COMPANY
DETROIT, MICHIGAN
1913-1914

MALCOM MOTOR COMPANY
DETROIT, MICHIGAN
1914-1915

In November 1913 the Malcom-Jones Cyclecar Company was founded with offices in the Hammond Building. Stockholders included E. Malcom Jones, C. H. Lawrence, and Charles H. Bennett who was vice president and general manager of the Daisy Manufacturing Company. The "Malcom" cyclecar was to be a two-cylinder tandem two seater with provision for a third person. It had a single headlight mounted atop the radiator. The price was projected to be $395, but only experimental models were built.

In April 1914 production still had not started. E. Malcom Jones attributed the delay to the foresight of his engineers who were designing for the future as well as the present (the cycle car market was beginning to collapse). They eliminated many features of the first design which had belt drive and friction transmission, which was changed to a shaft drive. A friction differential was added, as well as a planetary transmission. All of the additional experimental work and testing played havoc with the production plans.

In August 1914 the Malcom-Jones Company considered moving to a brick and concrete factory 18 miles from Detroit in Plymouth, Mich. A decision was made to stay in Detroit and re-organize as the Malcom Motor Company.

In February, 1915 a 100 inch wb Malcom described as a "light roadster" was exhibited at the New York Automobile Show. It had a four cylinder with 22 hp, sliding gear transmission with three foward speeds and one reverse, and a bevel gear rear axle. It had a single seat with a capacity for three using a special construction. It sold for $425 but was not manufactured beyond 1915.

*left-
A Malcom roadster arriving in New York in 1915*

DOWNING CYCLECAR COMPANY
DETROIT, MICHIGAN
1913

DOWNING MOTOR CAR COMPANY
DETROIT, MICHIGAN
1914-1915

The "Downing" was a tandem two seat cyclecar that weighed 670 lbs and sold for $450. It had 103 inch wb with a 56 inch tread, powered by an air cooled, twin cyclinder, V-motor with 12 hp. The transmission was a three speed selective and the frame was pressed steel. Wooden or wire wheels were available.

In 1914 H. O. Carter, general manager of the Downing Company in Detroit, secured the factory of the Penn Motor Car Company, in New Castle, Penn. The acquisition gave the Downing Company the largest cycle car plant in the United States, with a floor space of 84,000 square feet. The company was renamed as the Downing Motor Car Company, and the cars produced in the Detroit plant were renamed "Downing-Detroit."

In 1915 a 103 inch wb side-by-side two seater was added that weighed 870 lbs and sold for $480, along with a four passenger model that sold for $530. They came with a top, curtains, windshield, and a Prest-O-Lite tank. The motor was water cooled with four cylinders cast *en bloc*. It had a 2.75 x 4 inch bore and stroke, with 18 hp at 2,000 rpm. The inlet and exhaust valves were on opposite sides, operated by direct lift from the crankshaft. It was lubricated with a constant-level splash system, with pockets under each connecting rod. The oil level was maintained by a cam-operated plunger pump. It used a Bosch magneto and a Schebler carburetor. Semi-elliptic springs were used in the front and rear with 28 x 3 inch clincher tires.

The Downing Company did not make it through the collapse of the cyclecar market and quit production in late 1915.

1914- After the first model was on the market, an experimental chassis for 1915 was on the road.

La VIGNE CYCLECAR COMPANY
DETROIT, MICHIGAN
1913-1914

La VIGNE MOTOR COMPANY
DETROIT, MICHIGAN
1914

J. P. La Vigne returned to the automobile industry in 1913 with a prototype cyclecar housed in a small shop behind his residence. The press reported it as a light-car with an underslung construction with a 50 inch tread and a 96 inch wb. A factory was leased in January 1914, at the corner of Commonwealth Avenue and the Grand Trunk Railroad, with space to accomodate other companies that would manufacture parts for the "La Vigne."

For 1913 the La Vigne had a four cylinder separately cast L-head motor with a 2.625 x 3.75 bore and stroke. It was air-cooled, and used a disc clutch with a two speed progressive gearset driving to a semi-floating rear axle with a worm drive. The car was marketed in three forms; a roadster, a coupe selling for $425, and a touring selling for $500.

In February 1914, the La Vigne was shown at the New York Automobile Show and contracts were received for over 5000 units. For example, The Cutting Larson Company of New York signed up for 1000 units to distribute to New York, New Jersey, and New England. There were three models; a roadster, a cabriolet, and a parcel delivery car, selling for $425, $650, and $600 respectively. The stroke was increased to 4 inches, yielding 15 hp. The steering wheel was on the left, the control lever was in the center, and the toe board accelerator pedal was also fitted.

The frame was made of pressed steel section, and the hood was made of sheet steel with *wire cloth sides*. Cooling was by direct air circulation through these panels, assisted by a 15.5 inch fan on the exhaust side of the motor. The hubs were fitted with Hyatt roller bearings. The starter was mechanically operated from the driver's seat, and steering was the worm and split-nut type, irreversible, with a 14 inch steering wheel. The electric lighting was unique: the headlights were built into the front fenders which turned with the front axle.

The La Vigne Company suffered as a result of the collapse of the cyclecar craze, and began marketing the La Vigne as a "light" car. The company was then renamed the La Vigne Motor Company. In June 1914, the Pullman Motor Car Company of York, Pa., considered manufacturing the La Vigne, but decided against it.

The market collapsed for cyclecars as well as light cars, and the La Vigne company went out of production before year's end.

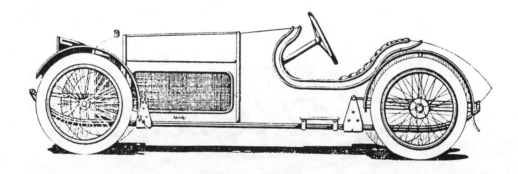

J. P. L. Cyclecar

J. P. La Vigne courtesy NAHC

1930- Olive La Vigne is seen on the left with her original "La Petite," and J. P. La Vigne is seen on the right with his La Vigne Cyclecar. They are in front of their home on Commonwealth Ave. courtesy NEWS

TRAVELER MOTOR CAR COMPANY
DETROIT, MICHIGAN
1913-1914

The "Traveler" was developed by **J. P. La Vigne**, and backed by W. K. McIntyre and F. W. Barstow. Initially, Travlers were produced in a factory at the corner of Commonwealth and the Grand Trunk Railroad, along side the La Vigne cyclecar. The two companies were distinctly separate, although La Vigne was involved with both. In March of 1913, the Traveler Motor Company purchased a two-story red brick factory at 1146-1148 Grand River, in Detroit, with an 80 foot frontage and 100 foot depth, and an adjoining lot, for $15,000.

The model 36 was a five passenger touring car that listed for $1180. The wb was 118 inches. It had a 22.5 hp four cylinder engine, cast in pairs, with a 3.75 x 5 bore and stroke. It had a three point suspension, and was water cooled with a honeycomb radiator. The ignition was a Deaco magneto run on dry cells, with one set of spark plugs. Lubrication was force fed with a splash system. It had a selective sliding transmission located in the middle, with shaft drive. There were two sets of rear brakes, expanding and contracting, with Raybestos linings. The frame was pressed steel and the gasoline tank was located under the seat. Equipment included top, windshield, speedometer, electric lamps and a horn.

The model 48 was a five passenger touring car with a 130 inch wheelbase that listed for $2000. The steering was on the left with center controls. There were also a two passenger roadster for $2000, and a three passenger coupe for $2500. The engine had six cylinders cast 3 *en bloc*, 3.75 x 5.25 bore and stroke. Water cooling was by centrifugal pump and a honeycomb radiator. A Rayfield carburetor was used with a single system Bosch magneto, Disco dynamo and Willard battery. The transmission was the selective type with three forward speeds and one reverse, with direct drive on the high speed. Final drive was with a shaft and a bevel gear. The Traveler Company failed late in 1914.

REX MOTOR COMPANY
DETROIT, MICHIGAN
1913-1914

The Rex Motor Company was incorporated in late 1913. The principal stockholders were **C. H. Blomstrom**, W. H. Frazier, Albert Robinson, and Frank Lemories. The "Rex" was a light car designed by **Blomstrom**. They first used a factory on Junction Avenue in Detroit, until they constructed a plant on West Jefferson near the River Rouge, in Ford, Mich.

The Rex was a side-by-side standard tread vehicle with a 100 inch wb. It had an unusual transmission arrangement, in that it drove the left front wheel through a medium of spur reduction gearing and a friction change speed, arranged to operate on the face of the flywheel, located at the front of the engine. It was claimed that this arrangement had the advantage of securing ample traction without the expensive construction of the differential. Freedom from skidding was also claimed due to the fact that the propelling power was pulling instead of pushing. The four cylinder block motor was 2.75 x 4.5, with a Stevens ignition system and a mechanical starter. The Rex was capable of 45 mph and cost $425.

The Rex Motor Company did not last through 1914. **C. H. Blomstrom's** next venture was with the Batemen Mfg. Co., in New Jersey, maker of the Frontmobile; a front wheel drive car that was manufactured in 1917 and 1918. **Blomstrom** died in February 1923.

Rex car demonstration of front wheel drive

EXCEL DISTRIBUTING COMPANY
DETROIT, MICHIGAN
1914

The "Excel" weighed 1000 lbs, which was heavy for a cyclecar, and had a 96 inch wb. The engine was a water-cooled 91.5 cubic-inch four-cylinder that developed 12 hp. The rear wheels were powered by a friction transmission and belt drive. A two passenger roadster sold for $450, and was the only model offered. Production ceased along with the whole cyclecar market.

NATIONAL UNITED SERVICE COMPANY
DETROIT, MICHIGAN
1914

The National United Service Company produced many small automotive parts for other car producers. In June of 1914 the production of a cycle car called the "Arrow" was announced, with **H. F. Vortkamp** as president and manager.

The Arrow was a side by side, fully equipped and sold for $395. It was considered to have a streamline body. The Arrow was powered by a water cooled four cylinder engine with 18 hp. It had friction change speed and double belt drive. Lubrication was by splash, with a gravity feed from a two quart tank under the cowl. An electric lighting outfit consisted of two headlights and one tail light supplied by a 40 ampere hour storage battery. Ignition was furnished by an Atwater-Kent Unisparker which would give up to 3000 miles per set of dry cells, *so it was stated*.

The Arrow had left hand steering through a 15 inch steering wheel. The wheelbase was 100 inches, the tread was 40 inches, and it weighed 850 lbs. It came with wire wheels equipped with cup and cone ball bearings with 28 x 3 inch tires. The frame was pressed steel mounted on quarter elliptic springs in front and full cantilever in the rear.

1914-Arrow

Advertisement- supplier of motors, axles, mufflers, batteries, transmissions, carburetors, piston rings, spark Plugs, Radiators, and tire pumps.

NATIONAL-UNITED-SERVICE

Parts and Fittings for the Small Car Manufacturers and Dealers

AUTOMOBILE CYCLE CAR COMPANY
DETROIT, MICHIGAN
1914-1915

A new single passenger roadster called the "Auto-Cyclecar" was announced in December 1913 by **William A. De Schaum**. When the car arrived on the market in June 1914, it was a light car and came in three models in the $300 price range, called the "Tiger," alluding to the Detroit baseball team.

The Tiger entered the market as Model W, a two passenger sociable; Model A was a four passenger deluxe, and Model D was a parcel post light delivery car. The car had a 90 inch wb and a 48 inch tread. The frame was three inch deep pressed steel channel section. Semi-elliptic springs were used in the front and rear. Wire wheels with 28 x 3 inch tires were used. The engine was a four cylinder with a 2.75 x 4 inch bore and stroke, giving a displacement of 98 cubic inches. It used thermo-syphon cooling and a carburetor of the company's own design. The transmission was a three foward and one reverse speed selective shift, with a leather coned clutch. In October 1914 the Tiger was entered in several races, driven by **De Schaum**. At the light car meet at Combination Park, Medford, Mass., over two thousand people watched a Dudley, a French Sigma and a Detroit Tiger battle it out on an old half-mile trotting track.

A Tiger was entered in the five mile race at Englewood, N.J., with **De Schaum** up against a French Sigma, a Zip, and a Coey. The Tiger went up front from the start and the other three let it set the pace. In the seventh lap, at 3.5 miles, the Sigma came up front *with a rush*, with the throttle wide opened and flashed by with a speed *that couldn't be denied.* **De Schaum** came in second place. An interesting event that day was a handicap between a <u>man</u>, a <u>horse</u>, and a <u>car</u>. A Morgan light car was made to go one mile, while the horse, N'Importe, did a half a mile and the man, Festus J. Madden, winner of the recent Brockton marathon, ran a quarter. The race was extremely close, with the man winning, and the horse and Morgan neck to neck at the tape. The runner's time was 51 and 3/5 seconds, the horse one minute and three seconds, and the Morgan one minute and 30 seconds flat. The Tiger won the 50 kilometer free for all that day in 48 minutes and 20 3/5 seconds.

William De Schaum died in March 1915, and E. M. White took over charge of affairs. Production of the Tiger was soon discontinued.

October 1914- The light car meet at Combination Park. Seen lined up for the 50-Kilometer Race, from left to right is an IMP, Dudly, Trumbull, Tiger (with De Shaum gripping the steering wheel), Zip, and a Coey.

GRANT BROTHERS AUTOMOBILE COMPANY
DETROIT, MICHIGAN
1913

GRANT MOTOR COMPANY
DETROIT, MICHIGAN
1914-1916

GRANT MOTOR CAR CORPORATION
DETROIT, MICHIGAN
1916-1922

The Grant Motor Company was organized in Detroit, Mich. in 1914 with $10,000 capital. The car was manufactured in Findlay, Ohio, starting in late 1913 with two cars for testing. The president was George D. Grant, and the vice president was Charles A. Grant. George D. Grant was also president of a successful auto dealership, the Grant Brothers Auto Company, on 1000 Woodward Avenue in Detroit, and he was also president of the Grant Brothers Foundry.

Former treasurer for the Simplex Motor Company, David Shaw was treasurer for the Grant Motor Company. James Howe, a Cornell graduate, was chief engineer with prior engineering positions with Thomas, Cunningham, Selden, and Studebaker. Factory manager George Salzman built an experimental car in 1897. The entire management staff had excellent automotive and engineering background. When word got out in 1913 that a new fully equipped car would be produced for under $500, it caused a real sensation.

The Grant was an assembled car and was one of the first compact cars made in the United States. The Model M roadster for 1914 had a 96 inch wb, a 56 inch tread, and weighed 1050 lbs. It had a four cylinder engine, and an unusual curved radiator. **Corcoran** single filament electric lights were available with Dietz kerosene cowl lamps.

Model M production ended in 1915. A six cylinder Model T replaced the model M, and sold for $795, and was advertised as the first six cylinder under $1000. In 1916 Grant added a plant in Cleveland, Ohio, and changed its name to Grant Motor Car Corporation. Business started falling off after the First World War until an economic depression resulted in the Grant Motor Car Company going into receivership in October 1922.

Front view of the Grant showing the radiator

(L) George Grant
(R) Charles Grant
courtesy NEWS

below-
A Grant six-cylinder is put through a severe test for power and flexibility. A steel frame with a 50% grade was erected behind a sales office in Chicago. The car started on a runway 24 feet back and then went up the 36 foot incline. The brakes were set and the wheels were locked, but the car rolled backward down the wooden ramp. This exhibition of the hill climbing powers of the Grant Six attracted wide attention.

SHARP ENGINEERING & MANUFACTURING COMPANY
DETROIT, MICHIGAN
1914

The Sharp Engineering and Manufacturing Company produced a cyclecar for $295, making it one of lowest priced cars on the market. Initially, it was going to be named the S.E.M., but was changed to "Sharp" by the time it was on the market. It only came as a roadster with a 90 inch wb, and staggered seating for two passengers. It had a two cylinder engine with a 2.87 x 3.5 bore and stroke, producing 7 hp. Power was coupled through a two speed transmission and shaft drive.

In October 1914, the company announced its plans to add a standard sized car with a 120 inch wb and a six cylinder engine, to sell for $1000. The company stopped all production before the end of the year.

Sharp 2-pasenger roadster

GADABOUT MOTOR CORPORATION
NEWARK, NEW JERSEY
1914-1915

GADABOUT MOTOR CORPORATION
DETROIT, MICHIGAN
1915-1916

HEZELTINE MOTOR CORPORATION
BUFFALO, NEW YORK

The Gadabout Motor Corporation was formed in the middle of 1914. The company officers were as follows: president, Robins A. Law; first VP Charles Vail; second VP, A Vernon Clark; third VP, J.B. Brandon; fourth VP, John J. Bush; secretary, Edwin M. Simpson; treasurer and general manager, Philip Hezeltine; designer and consulting engineer, Walter A. Gruenberg.

The company executive offices were set up in New York City with a temporary office in Detroit so that Hezeltine, formerly of Detroit, and Gruenberg could arrange for supplies to be shipped to the Gadabout factory in Newark, N.J.

The Gadabout Sales Corporation was formed in Detroit to handle sales in Michigan and Ohio. Charles F. Gazley, general manager of the Detroit Axle Company was president.

The Gadabout was termed the "aristocrat of cyclecars," and used a reed (wickerwork) body to give the car light weight, instead of the usual wood or metal construction. The body was easy to clean by first removing the spring seats. Waterproof pads behind the back of the seats and similar waterproof material at each side of the footboards were attached similar to glove fasteners. A hose was turned on and water went through the body to thoroughly clean it.

It had an 86 inch wb and a 46 inch tread. The engine was a Sterling water cooled four cylinder with splash lubrication, producing around 12 hp. It had cone clutch and a two speed sliding gear with reverse, with shaft drive to a bevel gear solid rear axle, supplied by the Detroit Axle Company. The wheels were the artillery type, and seating was side-by-side.

The Gadabout Company re-organized in April 1915, and moved to a factory on Lafayette Street in Detroit by the summer. The capital was increased to $10,000,000, and the president was Charles F. Gazley, with Hezeltine staying on as treasurer and general manager. A sales plan was developed to have 27 zones, each having a sales manager. In turn, each manager would have 150 selling agents. The Gadabout was shown in the Chicago autoshow in January 1916 with a 104 inch wb and lost its distinctive wickerwork body. Sales went into a decline and the company was sold off before the year's end.

In the late summer of 1916, the Hezeltine Motor Corporation was formed in Buffalo, N.Y. Philip Hezeltine was president, and Walter Gruenberg was the chief engineer. The Hezeltine was intended to move away from the cyclecar and into the light car classification. It had four cylinder with 27 hp, and was a two seater with a choice of two wheelbases. The Hezeltine Company was out of business in 1917.

1915- Gadabout

Gadabout factory on the corner of Badger Ave. and Runyon Sts. in Newark, N.J.

LIGHT MOTOR CAR COMPANY
DETROIT, MICHIGAN
1914

The Light Six roadster weighed 1800 lbs. and sold for $1050. The demi-tonneau weighed 2000 lbs and sold for $1150. There was also a touring car available that weighed 2705 lbs and sold for $1250. The tread was 56 inches and the wb was 115 inches. Left hand drive was used with a center control selective shift transmission.

All models came with an L-head engine having cylinders cast in pairs, with a 3.5 x 4 bore and stroke. Four crankshaft bearings and splash and plunger pump lubrication were used. A lubricated wet clutch with a three speed roller bearing transmission were on the engine. One universal joint was used with a 3:1 standard gear reduction to a semi-floating axle, with radius rods. All springs were 1.5 inches wide with 3/4 elliptic rear springs. Tires were 32 x 4 inches (32 x 3.5 on the runabout) with q.d. rims on wood wheels.

The car was introduced in October 1913 for the 1914 season, but was out of production before the season ended.

HIGHWHEELERS

Highwheelers were essentially a resurgence of the original "Horseless Carriage" with technological improvements. There were at least 75 manufacturers in the United States, mostly in the mid-west. (**They were not popular in Detroit.**) They used a carriage type body and had large narrow wheels with solid tires. Wheel diameters of 36 to 40 inches made it hard to steer in ruts, especially since they were very light. The narrow tire was said to cut down on road dust, which often included horse droppings. The engine was usually under the seat, fairly simple and of low power. Fifteen miles per hour was typical. Many had a leather dash even though there was no horse to dash and splash mud on the occupants. Steering was conventional with knuckles on each side at the front axle. At least one manufacturer had the entire front axle turning like a fifth wheel. Novel approaches to transmit power to the wheels were used, including manila rope, but friction drive or a chain were more common.

They had created a market for an inexpensive car that was easy to maintain which could traverse poor road conditions without the fear of getting a flat tire. It is somewhat strange they found the popularity they did, using a retrogressive technology. The Ford Model T, more than anything else, gave them a run for their money which they could not beat.

C. J. FISCHER COMPANY
DETROIT, MICHIGAN
1914

FISCHER MOTOR VEHICLE COMPANY
YORK, PENNSYLVANIA
1914

The C. J. Fischer Company was organized in December 1913 in Detroit, Mich. The car was designed by C. J. Fischer and was originally called the "Fischer-Detroit Cyclecar." It was soon changed to "Fischer-Detroit" and sold as a light car. The chassis had a wheel base of 104 inches with a tread of 56 inches. It was powered by a Perkins four cylinder water-cooled engine with a selective transmission and shaft drive.

The Fischer-Detroit was first shown at the Grand Central Palace Automobile Show in New York City in late December 1913. It came in five streamline type bodies. Prices ranged from $595 up to $845 for the sedan.

In October 1914, the Fischer Company announced that it would relocate to York, Penn. A group of "Yorkers" joined in and raised $50,000 capital to launch the company. Operations were to start at the Glen Rock Stamping Company at Hill Street and the Maryland and Pennsylvania Railroad. Sample cars from Detroit were on-hand, and production was to start in two weeks. The car was to feature battery lighting, electric horn, one-man top and a power tire pump. An electric starter was available at extra cost. Few if any cars were produced in York, and the company went into receivership and assets were sold for $8000.

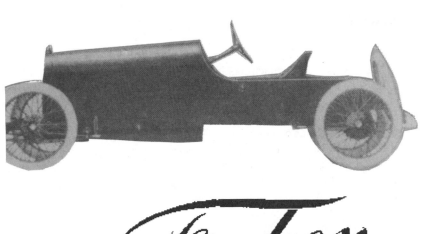

"The BIGGEST Light CAR"
Detroit, U. S. A.

AMERICAN MOTORETTE COMPANY
DETROIT, MICHIGAN
1913

LINCOLN MOTOR COMPANY
DETROIT, MICHIGAN
1914

After the Keeton Motor Car Company was dissolved, a failed attempt was made to form the American Motorette company. A second try resulted in the Lincoln Motor Company, formed in late 1913.

The Lincoln was a light car designed by H. D. W. Mackaye. It was powered by a four cylinder engine with a two-speed progressive transmission and a cone clutch. A Renault-type hood gave it a distinctive look. The Lincoln had left-hand steering, but from the back seat. There was a folding seat in front.

The first car was tested in January 1914. Production ensued shortly after, but was very low until the company was dissolved.

DODGE BROTHERS MOTOR CAR COMPANY
DETROIT, MICHIGAN
1914-1928

CHRYSLER MOTOR CORPORATION
DETROIT, MICHIGAN
1928-

"DEPENDABLE"

The **Dodge** brothers were born in a little wooden cottage on North Fifth Street in Niles, Michigan. **John Francis** was born in 1864 and **Horace Elgin** in 1868. The brothers did everything together and were inseparable. **John** graduated from Niles High School, and **Horace** quit when he was 15. They worked in a foundry and machine shop on the banks of the St. Joseph River. It was established by their grandfather and operated by their father, Daniel, and his brother. They mastered the skills of the forge and machine shop, repairing marine engines. **Horace** developed into one of the most mechanical minded people in the US. **John** was mechanically gifted also, but became adept at management and organization.

In 1886 the brothers worked at the Murphy Engine Company in Detroit, manufacturing boilers. **John** became foreman in six months. They worked there for four years, learning machinery and manufacturing operations. **Horace** spent two years working part time for **Leland-Faulkner**.

In 1890 they worked in Windsor, Ontario at the Canadian Typothetae Company. They soon established themselves as machinists, with **John** becoming superintendent. **John** married Ivy Hawkins in 1892 and **Horace** married Anna Thompson in 1896. In 1897 **Horace** patented an adjustable ball-bearing bicycle hub. The brothers leased shops from the Canadian Typothetae shops to manufacture the Evans and Dodge bicycle, with financing from Detroit manufacturer, Fred S. Evans.

John and **Horace** sold the business and returned to Detroit in 1901, building engines for the Olds Motor Works in the Boydell building with 13 men. Before the end of 1902 they risked everything they had and threw their faith in **Henry Ford**. In February 1903, the Ford Motor Company contracted the **Dodge** brothers to supply two-cylinder engines, transmissions and axles. In 1906 they opened a larger shop at Monroe Avenue and Hastings Street in Detroit, eventually employing 200 men.

As the Ford Motor Company expanded, the Dodge brothers kept pace. In 1913, they were together in a downtown department store in Detroit and they met Howard B. Bloomer, a friend and legal counselor. Bloomer suggested "Why don't you brothers build your own car?" "Why should we?" returned **John Dodge**, "We've got a heavy stake in the Ford Motor Company and, besides, we have a contract for building parts." Bloomer explained that **Ford** could stop buying from them anytime he wanted to, and they would be left idle. The bickering went on without an outcome. The following day, Bloomer visited the brothers to continue the argument. "I've got you, Howard," **John** admitted, "You're right." **Horace** had taken no part in the discussion, but had been listening. Grinning at his brother, he drawled: "My heavens, **John**, it has taken you a long time to see that."

The Ford contract stipulated that a year's notice had to be given by either party wishing to end the contract and in July 1913, the **Dodges** notified the Ford Motor Company they would not continue after July 1, 1914. In August 1913, the **brothers** announced to the press that they would double the size of their huge Hamtramck plant and manufacture their own car. **Ford** decided to make his own transmissions and axles at the end of the Dodge contract and had to purchase larger generators for his power plant to handle the added electrical load for manufacturing equipment.

John and **Horace** began getting together an organization. They hired Arthur I. Philp and gave him authority to gather the finest executive talent he could find. After screening and rejecting, he occasionally hired. A dozen advertising agencies were called in and were all dismissed. Philp again summoned one of those he had rejected. He said the young man did a great job advertising the two-cycle Elmore and if he could do as good of a job with the "Dodge" he would not talk to another advertising man in America. The two men worked..... together.

By the time the **Dodge brothers** entered automobile manufacturing, the larger companies owned entire forest ranges and were committed to wooden frame bodies with sheet metal. The **Dodge brothers** were mechanics with little expertise in the fabrication of wooden car bodies. At the same time, Edward Garwin Budd was only beginning to recover from a financial low period after the failure of the Garford company, in Indiana. The **Dodge brothers** became good friends with Edward Budd, who could offer bodies made from pressed steel.

(Just by coincidence, America's first all steel body was used for the Eastman Electric Automobile made in Cleveland, Ohio. In 1898 a silent three-wheeled Eastman stopped at the alley door of the Edison Electric Company in Detroit. At the sound of a bell, a young man stuck his head out. The driver asked, "Can I buy some electricity?" The young man withdrew into the building and returned with two long black lines and hooked them up in the back of the vehicle. The young man at Edison was **Henry Ford**.)

Even though the Hupmobile had a head start, the Dodge became popularly known as the "first all-steel bodied automobile." Budd and his brilliant engineer, **Joseph Ledwinka** began to work long hours to fulfill the first order for 5000 bodies. The four cylinder, 2200 lb Dodge selling for $785 debuted in November 1914. Publicity touted its strength, sturdiness, and dependability. There were over 22,000 applications for dealer franchises. Production could in no way meet demand. The **Dodge brothers** had clearly earned a reputation for high standard and machining expertise throughout the automotive industry. (Despite their faith in Edward Budd, the **Dodge brothers** ordered rivets for reinforcing all the joints.)

In 1917, the Budd company and the **Dodge brothers** discussed an all-enclosed steel sedan. In 1919 the **Dodge brothers** introduced a steel four door sedan with a wooden top and door frames. It weighed 2800 lbs and cost $1900. The new sedan was shown in New York at the Grand Central Palace. Two days later, **Horace Dodge** was stricken with "grippe," as influenza was known as at the time. **John** fell to grippe two days later, developed pneumonia, and died at age 55 in his New York hotel room that January. **Horace** survived but remained in a weakened condition, and succumbed by the end of the year.

The Budd company continued supplying bodies, fenders and wheels for Dodge. Frederick J. Haynes, a trusted friend of the Dodge brothers, was appointed president upon the death of **Horace**. Shortly after, he contacted the **Graham brothers** of Indiana and arranged to ship Dodge chassis to the Graham plant to be fitted with heavier suspension to be sold through Dodge dealerships. In 1922 Dodge was in third place with a production record of 152,673. The Dodge Company made automotive history in 1923 when it introduced all steel construction, starting with a three-window business coupe. A business sedan followed, and sales were very good even though 90% of the market was open touring bodies.

The Dodge family was not able to duplicate the synergy of running and operating the company like **John** and **Horace**, and financial advisors urged them to sell their interest. The widows yielded to the advice and the company was up for sale. The banking house of J. P. Morgan, representing GM, bid $124,000,000. The company was sold to the banking house of Dillon, Read & Co. in April 1925. A 42 year-old Clarence Dillon casually wrote a check for $146,000,000, the largest financial transaction in business up to that time. Business writers marveled that $50,000,000 of the selling price was for good-will the **Dodge brothers'** name still had.

At first, Dillon had visions of merging Dodge, Hudson, Packard and Briggs body to rival GM. After finding the proposal unwelcome, Dillon decided to consolidate the Dodge position in the truck market, and bought 51% interest in the Graham Brothers in November 1925. The **Graham brothers** moved into the Dodge organization as vice-presidents with **Ray** as general manager, **Joseph** in charge of manufacturing, and **Robert** in sales. In the spring of 1926 **Walter P. Chrysler** offered to buy the Dodge Company, and after being turned down began planning the Desoto as a Dodge competitor. In April 1926 the Dodge Company purchased the remaining 49% of the Graham Truck company and the brothers resigned.

A close associate of Dillon's, E. G. Wilmer, switched from chairman of the board to president, and Haynes was removed from active direction of Dodge and made chairman of the board. One of Wilmer's first steps was to end Dodge's dependence on the four cylinder it started out with in 1914. A six cylinder car called the Senior Six was brought out in 1927, but Dodge had slipped to sixth place in vehicle registrations. Dillon realized that running a truck and car company was considerably more than he could handle.

In the spring of 1928 Dillon approached **Walter Chrysler** with the intention of selling Dodge. For five days in New York's Ritz Hotel, **Chrysler** feigned indifference and Dillon concealed his eagerness to sell. Finally an agreement was struck and the Chrysler Corporation purchased Dodge for $170,000,000 in new Chrysler stock and the assumption by Chrysler of all Dodge liabilities.

1914- There was excitement and anticipation before the arrival of the first Dodge. After they built the best car they knew how, they placed a moderate price on it.
courtesy AAMA

Dodge factory on Jos. Campau called "Dodge Main" built in 1910 by Kahn
courtesy MSHC

Final inspection courtesy MSHC

*1914-
Dodge with
one-man top
courtesy AAMA*

*1915-
North of the
foundry
courtesy MSHC*

*1915-
Test track
courtesy MHSC*

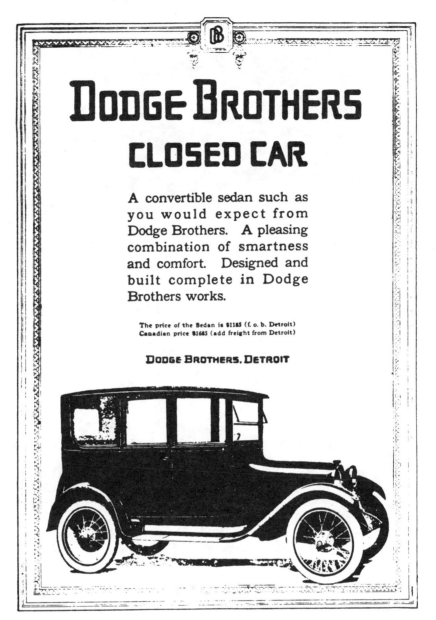

John and Horace Dodge

The interlocking triangles of the Dodge Marque represented the close working relationship between the two brothers. (People said that the Dodge did not have to be equipped with a horn because the emblem already carried the warning "Dodge Brothers."

COLUMBIA ELECTRIC VEHICLE COMPANY
DETROIT, MICHIGAN
1914

COLUMBIAN ELECTRIC VEHICLE COMPANY
DETROIT, MICHIGAN
1915-1917

The **Columbia** Electric Vehicle Company was formed in February 1914 by C.F. Krueger, who had considerable experience in the electric vehicle field. The vehicle was specifically designed by E.T. Birdsall to have a low price with two models from the one chassis originally offered. Offices were 1705 Dime Bank Building, Detroit.

When the **Columbia** Motor Car Company in Hartford went out of business in late 1914, Krueger changed the name of his company and automobile to **Columbian** to avoid confusion.

In 1915 "Columbians" were built in the former Grabowsky Power Wagon plant on Mt. Elliott Ave. in Detroit. Three models with aluminum bodies were available. The body width was 44.5 inches, the doors were 26 inches and battery space under the hood was 30 inches by 29 inches. A two passenger roadster sold for $785, a three passenger couplette sold for $985, and a four passenger brougham sold for $1450. The roadster and couplette had a wb of 88 inches and the brougham had a 98 inch wb, and all three had wire wheels.

The selling prices had to be raised to maintain a profit. In 1916 the roadster was raised to $1175, the couplette was raised to $1375, and the brougham raised to $1485. Sales began to lag, and by 1917 the company went out of production.

Left-
Columbian runabout

Right-
Columbian brougham

STORMS ELECTRIC CAR COMPANY
DETROIT, MICHIGAN
1915

William E. Storms had been associated with the Colonial Electric Car Company and the Anderson Electric Car Company before deciding to build an inexpensive electric automobile. Along with **Storms**, Ferdinand H. Zilisch, a Milwaukee businessman, and F. T. King of Detroit formed the Storms Electric Car Company in late 1914, using the former Mercury cyclecar factory at 807-815 Scotten.

The "Storms" was an electric cyclecar with a 90 inch wb and a 44 inch tread. It had elliptical springs on all four wheels, and had a pressed steel frame. It was offered as a coupe for $950, a roadster for $750, and a light delivery for $650. It could get 50 miles on a battery charge, and had a top speed of 18 mph.

Like many other cyclecar companies, the Storms Company didn't make it through 1915. William Storms tried again the following year in Jonesville, Mich., with a new electric vehicle called the American Beauty, but it failed in 1916.

PILGRIM MOTOR CAR COMPANY
DETROIT, MICHIGAN
1915-1918

The Pilgrim Motor Car Company was incorporated for $250,000 in January 1915. The principal stockholders were Clarence H. Leete, organizer, **W. H. Radford**, engineer, and Leonard C. Welford, of London, England. Eastern capitalists were behind the company financially. **Radford** was formerly chief engineer of the Warren Motor Car Company, and was previously with the Olds Motor Works and the Hudson Motor Car Company. A temporary office was located in the Moffat Building in Detroit.

The "Pilgrim" was a light car weighing 1300 lbs and priced at $650 to $835. The wheelbase was 100 inches and the tread was 56 inches. The wheels were of wire, demountable and were fitted with interchangeable clincher rims. The body was streamlined and made of sheet steel, while the fenders and running boards were of pressed steel. The fenders were domed and followed the wheels closely. The Corsair Motor Company designed the water cooled engine. It was a block-cast four-cylinder of the L-head type with a 3.125 x 4 bore and stroke, with a rating of 15.6 hp according to the S. A. E. formula. Power was transmitted through an enclosed propeller shaft and a three speed gearset on the rear axle, which was the semi-floating type. It came with an electric starter and lights using an Auto-Lite generator.

The first car had been on the road and logged 3000 miles by September 1914. Production was originaly to start in December 1914, then changed to June 1915. In May 1915 Clarence Leete was arrested in his office in the Moffat Building, on complaint of other officers of the company. The arrest was based on peculiar methods Leetes used in obtaining cash from "unsuspecting workingmen." It was alleged that he used his position as president to sell jobs to unsophisticated persons as foremen at from $100 to $200 each. Although there were only 15 foremen in the company, he accepted money from 80 persons for the positions. Leete was released from custody a week later under $500 bail. He issued a statement and said there is "*no dissension whatever in the firm and that the entire matter is a piece of spite work and an apparent endeavor on the part of interested parties to disrupt the concern.*"

Radford left the company and **R. C. Aland** was hired to redesign the Pilgrim. The first Pilgrim designed by **Aland** was completed in May 1916. It was priced at $735 and produced in a plant in Highland Park, Mich. The Aland design had a 112 inch wb and a motor with 38 hp at 2800 rpm. Production followed with several models, and ended in 1918.

*Pilgrim 37
5 passenger
touring*

THE DETROIT ATHLETIC CLUB

Detroit was always a sports town, going back to foot racing between the Indians and French fur traders. As Detroit grew, horse racing became popular at Highland Park and Grosse Pointe. Several athletic institutions sprang up, including the Detroit Boat Club, and on land, the Detroit Athletic Club. Highwheel bicycle racing became very popular, with **Tom Cooper** one of the most successful racers at the club. In 1889 the DAC had one of the best track teams in the country. Among the stars on the team was **Harry M. Jewett**, soon to join the automobile game. **John C. Lodge** was a star baseball player on the semi-professional championship teams of 1889, 1890 and 1892. **George P. Codd** was a star pitcher for the team and became mayor of Detroit.

The automobile pioneers used to gather in the bar at the **Ponchartrain Hotel**. Everyone brought automobile parts, such as fenders, headlamps, brakes and goggles, into the bar to look at and to discuss. **This was the heart of the automobile industry until 1915** when a new athletic club was built on Madison Avenue in downtown Detroit. This was largely due to the efforts of **Henry B. Joy**, to improve the business environment. The tenor and climate of the DAC was predicated on strict rules for attire and membership, presumably restricting patronage to the industry leaders and the affluent.

When the DAC opened its new building on April 17, 1915, it immediately became the meeting place of the automobile industry.

The DAC located on Madison Ave., near Woodward. courtesy MSHC

ECLIPSE MACHINE COMPANY
DETROIT, MICHIGAN
1907

In December 1906 the Eclipse Machine Company, producer of automotive parts, announced it would manufacture a light four cylinder car. The company was organized with a capital stock of $30,000 by A. S. Keller, A. P. Schulte and P. J. Murphy. The "Eclipse" came in four models including a roadster, and three touring cars. The roadster had a 96 inch wb and a four cylinder motor with 28 hp. A five-passenger touring version had a 110 inch wb and a 34 hp four cylinder motor. The other two were seven passenger models with a 115 inch and a 40 hp four cylinder motor. The seven passenger models could also be ordered with a V-8 motor. A truck, motorcycle and a motorboat were also produced by the Eclipse Company.

The Eclipse Company was affiliated with the Detroit Correspondence Institute of Motoring, and offered a free car to its dealers, (provided they attended the Eclipse driving course). The Eclipse did not last past its first year.

ALAND MOTOR CAR COMPANY
DETROIT, MICHIGAN
1916-1917

Cyril R. Aland was the engineer and the designer of the "Aland." He had experience as the designer of the S & M in 1913 and in 1916, before the Aland went into production, **Aland** was asked to re-design the Pilgrim for the Pilgrim Motor Car Company, which was having troubles.

The Aland was not scheduled for production until December 1916, but the word was out in August that it would have a 16-valve 4-cylinder engine. The Aland Motor Car Company was incorporated with a total capital of $500,000. There were several body styles and one chassis, with a five passenger touring car and a two passenger roadster produced first. The high speed 16-valve aluminum engine attracted a lot of attention. High tensile steel was also used to keep weight as low as possible.

Although the engine was rated at 15 hp with the SAE formula, **Aland** stated that it would produce 65 hp at 3200 rpm. The brakes were expanding types on all four wheels, connected diagonally. Two activated with a foot pedal, and two with the hand brake. This concept was first used by the Argyll Company in Scotland. The theory was that when one or both wheels locked up, the opposite front wheel would still have control over the steering. The body was sheet aluminum over a sub-structure of wood. The steering wheel was an 18 inch walnut knobby grip.

The Aland sold for $1500, which for an overhead cam with hollow push-rods, roller tappets, Raybestos-faced multiple disc clutch, and three speed selective-type gear box was well thought out. Sales were low and the company stopped production by 1918.

1917- Aland four-cylinder, five passenger touring car

BOUR DAVIS MOTOR COMPANY
DETROIT, MICHIGAN
1916-1918

FRANKFORT MOTOR CAR COMPANY
FRANKFORT, INDIANA
1918

LOUISIANA MOTOR CAR COMPANY
SHREVEPORT, LOUISIANA
1919-1922

Charles J. Bour was born in Chicago in 1865 of French parentage. At 14 years of age, he worked for the Illinois Central Railroad in the car accountant's office. He transferred to the news service of the Illinois Central suburban passenger trains, and in the station with the news service he published a suburban time table in book form with advertising of Chicago concerns. This started him in the advertising field, and in 1888 he contracted with the Illinois Central Railroad for advertising privileges of its suburban trains. **He originated the first 16 inch by 24 inch advertising cards** in passenger coaches. Their success brought about a myriad of suburban, surface elevated and subway car advertising businesses throughout the world. The name "Chas. J. Bour, Advertising" soon became known to Chicago businessmen. This business continued until 1910.

Bour got into the automobile business with a car designed by A. A. Gloetzner. The idea for the car came from Robert G. Davis, an engineer for the Chicago, Duluth, and Lake Superior Steamship Company, who along with E. C. Noe, the general manager of the Chicago Elevated Railroad, moved to Detroit to produce a car. The initial public acceptance was enthusiastic. In July 1916 the first vehicles were produced in temporary plants of 30,000 square feet on East Fort Street and East Towbridge. Meanwhile, **Bour** tried to build a large plant at Kercheval and the Outer Beltline Railroad, a location within the territory occupied by Hudson, Chalmers and Continental Motor Company, but they could not secure the necessary steel work. An alternate plant was selected on the N. W. corner of W. Fort Street and 23rd. This building originally started as a two story structure 200 feet by 130 feet. It was increased to a four story concrete factory.

The "Bour-Davis" was an assembled car using a Continental engine. It had a distinctive look with the radiator and windshield both starting at the same angle, otherwise it was conservative. In 1918 "Bour-Davis" was absorbed by the Frankfort Motor Car Company in Frankfort, Ind., owned by the Shadburne Brothers of Chicago. In 1919 "Bour-Davis" was re-organized as the Louisiana Motor Car Company of Shreveport, with a factory in Cedar Grove, La. Sales were good initially, with advertising focusing on the southern manufacturing location. But the recession of the early 1920s caught up with them and they went into receivership in June 1921. In 1923 they re-organized as the Ponder Motor Manufacturing Company, but never went into production.

1916- Bour Davis

Bour Davis Detroit factory picture taken in 1994

Charles Bour

DOBLE MOTOR VEHICLE COMPANY
WALTHAM, MASSACHUSETTS
1914-1916

GENERAL ENGINEERING COMPANY
DETROIT, MICHIGAN
1916-1918

DOBLE-DETROIT STEAM MOTORS COMPANY
DETROIT, MICHIGAN
1917-1918

DOBLE MOTORS INCORPORATED
SAN FRANCISCO, CALIFORNIA
1921

DOBLE STEAM MOTOR CAR CORPORATION
SAN FRANCISCO, CALIFORNIA
1922-1930

Abner Doble tried making a steam car while he was in high school. He went to MIT and set up a machine shop in Waltham, Mass. Doble visited the Stanley plant in Newton, Mass., but he was not impressed with what he saw. He completed the Model A in 1912 using a water tube boiler that raised steam fast. The Stanleys used it three years later. He sold four model A's and formed the Doble Motor Vehicle Company on October 30, 1914, to manufacture the Model B. He needed more capital and toured through industrial towns until he came across C. L. Lewis in Detroit. With Lewis, Abner organized the General Engineering Company to manufacture the "Doble." Lewis was president and the model C was scheduled for production in 1917. The ignition was electric instead of a Bunsen burner, which allowed it to be started in less than three minutes.

The new Doble was announced at the New York Auto Show for 1917 models. The volume of mail was so great that the post office refused to deliver and Doble had to send cars and pick up 50,000 pieces of mail. More than 11,000 orders were placed in three months. Production began but the National War Emergency Board stopped allocating steel to Doble. This ended the G.E.C.-Doble and Abner was forced to reorganize. After the government re-allocated steel, the Doble-Detroit Steam Motors Company was organized with a new plant and new vehicles. The "Doble-Detroit" sold for $3750 and orders were received for over 5000 cars, but less than 100 were built due to wartime conditions and financial maneuvering.

Abner Doble moved to California in 1919 and organized the Doble Steam Motors incorporating it in 1921. The company was re-organized in 1922 as the Doble Steam Motor Car Corporation. In 1923 the model E was shown at the California Auto Show and production began. Compared to Packards, Cadillacs, Minervas, Rolls-Royces, and Mercedes, Howard Hughes said "I know them all, but I prefer the Doble."

As business grew, stock traders began trading Doble at a 9:1 profit margin with a stock worth less than 1/8 its stated value. The Doble company was unaware of the trading, but it became the target of several lawsuits involving 800 plaintiffs in 1924. The company was forced to reduce the workforce and settle at 30 cents on a dollar. Financial support was no longer available and the company began to sell land, machines, and mortgaged the factory. Doble continued to sell cars at a greatly reduced volume, and by 1930, the depression forced them into liquidation.

The Doble factory at the corner of Fourth and Porter streets, in Detroit.

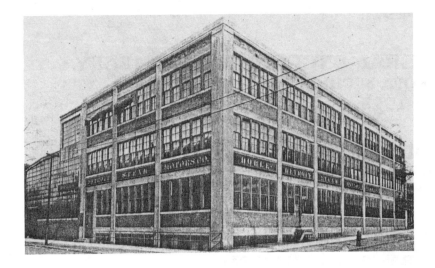

The series E Doble

Abner Doble

LIBERTY MOTOR CAR COMPANY
DETROIT, MICHIGAN
1916-1923

"ALL THE WORLD LOVES A WINNER"

Percy Owen was born in Oswego, N.Y. in 1875. He worked in the real estate and insurance business until 1898. In 1899 he became an agent for the Winton Motor Carriage Company in N.Y. From 1900 to 1906, he was the Eastern sales manager and opened up the first gasoline automobile salesroom on Broadway, N.Y. He traveled crossed the Atlantic twice and studied European design and methods of manufacturing. From 1907 to 1908, he imported Bianchi (1898-1937) automobiles from Italy. From 1908 to 1910, he was general manager for Carl H. Page & Company, Chalmers' distributor, and became the Eastern sales manager in 1910. He moved to Detroit as sales manager for Chalmers, then for Saxon as a vice-president and sales manager.

Percy Owen had his own ideas on what a motor car should be while he was at the Chicago Automobile Show in February, 1916. He had convictions regarding what an ideal car should embody. With his long years of experience in the automobile business, he heard many comments and criticisms from automobile drivers and owners. He looked around the automobile show for the ideal car and could not find it. From the depths of his ensuing disappointment, he resolved to produce an ideal car himself.

The Liberty Motor Car Company was organized in February, 1916, with a capital stock of $400,000, all of which was subscribed. R. E. Cole, builder of the first notably successful six-cylinder car, that sold for a popular price, was the chief engineer. Before joining the Liberty company, Cole was with the Saxon Motor Car Company, and had previously contributed to the Peerless, Olds, and Chalmers. James F. Bourquin, a manufacturing engineer from the University of Michigan and formerly of Paige-Detroit and Chalmers, was named VP. H. M. Wirth, formerly of Saxon, was named purchasing agent. The former R.C.H. factory at 101 Lycaste in Detroit, was taken over and completely remodled. Cement floors were laid throughout and equipment for progressive assembly was installed.

The "Liberty" was introduced in the lobby of the Hotel Ponchartrain in Detroit, during the summer of 1916. It had a continental six cylinder engine and was hailed as having a *colonial tendency* with beautiful flowing lines. The lines and curves were carefully proportioned, producing a very harmonious effect. The straight lines of the hood blended into a graceful cowl. The usual bulge at the sides of the cowl were eliminated. The wheelbase was 115 inches and the body was particularly roomy for five passengers.

As a result of suggestions made by car owners during **Owen's** 16 years experience, every detail to provide comfort and convenience to the car owner was exercised. The windshield upright supports were perfectly straight and the windshield fitted very closely with the help of a weather strip. The steering wheel was carefully placed in an easy driving position. The position of the control levers, buttons and pedals was thoughtfully studied so they would be in easy reach. The dash instruments were conveniently grouped to be easily read from the driving position. The upholstery was genuine leather with a generous amount of hair, making it soft and comfortable. Cushions were well slanted to support the passenger's comfort.

The company produced 733 automobiles in the first year. In 1919, 6000 Liberty's were produced, and by 1921 production reached 11,000. A new factory was built on a 12 acre site at Charlevoix Avenue and Connor's Lane. The factory was built in units, with a main assembly building 120 by 600 feet. A two-story colonial type structure nicknamed "Independence Hall" was included for general offices.

The company was in receivership in 1923, and according to **Percy Owen**, it was due to the "inability to take advantage of the facilities," and "by the failure of large part makers to make delivery in the quantity ordered." Re-organization failed, and in September 1923 the assets of the Liberty Motor Car Company were acquired by the Columbia Motor Car Company.

*1903-
Alexander Winton(L), and Percy Owen(R), on the day before the Bennet Cup race in Ireland. Winton was unable to finish the race due to wax build-up in the carburetor. A dinner was held in Dublin afterwards, where the toastmaster said of Winton, "he's plucky, but unlucky."*

*1916-
At the start of the assembly line in the Liberty factory. Note the engine being lifted on the right and the overhead runways delivering wheels from the stockroom. This plant was later sold to Kessline.*

*1916-
Liberty cars being prepared for shipment.*

1917- 733 cars were produced in its first year.

Offices for the new Liberty factory, dubbed "Independence Hall," can be seen in the background located at Charlevoix and Connor. The first floor held production offices and a showroom and the second level held administration and sales offices. The entire site consisted of 12 acres.
  picture taken in 1994

Percy Owen

The car's name stood for liberating its owner from the speed and power manias and brought him the freedom to really enjoy his car.

COLUMBIA MOTORS COMPANY
DETROIT, MICHIGAN
1916-1924

"A CAR WORTHY OF THE NAME"

The Columbia Motor Company was incorporated in January 1916, with a capital stock of $500,000. **J. G. Bayerline** was president, and was joined by A. T. O'Connor, W. L. Daly, and **William Metzger**. Ray Long was chief engineer. The company purchased the Argo Electric Vehicle Company of Saginaw, Mich. which had a factory well adapted for automobile manufacturing. The first "Columbia" was built with a degree of secrecy early in the summer of 1916, and driven 60,000 miles before the 1917 season. A factory was secured on Jefferson Avenue in Detroit, and two months later, a larger plant formerly occupied by the B. F. Everitt Company was acquired at Beaufait and Mack in Detroit.

The Columbia was an assembled car moderately priced at under $2000. The wheelbase was 115 inches, with 32 x 3.5 inch tires. Rear tires were non-skids. It was powered by a Continental six cylinder motor with a 3 x 4.75 bore and stroke, and had a Stromberg carburetor. It was water-cooled using a Harrison radiator, and had a thermostat placed above the fan which automatically opened the radiator shutters as the temperature increased. It also had cowl ventilators and a Boyce Moto-Meter.

The motor was built as a unit with a three speed Warner gearset, and Borg and Beck multiple disc dry clutch. The frame was made of chrome-nickel with Timken axles and roller bearings. The suspension utilized Detroit self lubricating cantilever rear springs and semi-elliptic front springs.

In 1920, the capital stock was increased to $4,000,000 by a stock dividend of 700 percent, and then to $5,000,000 by subscriptions to shareholders. In the middle of 1923, the capitalization was reduced by an exchange of stock, and then 83,000 shares of no par stock were offered to stockholders at $6 per share. In 1923, the company had its production peak of 6000 automobiles sold.

The defunct **Liberty Motor Company** was purchased with the intention of continuing its production. Less than 100 "Liberty's" were produced in 1923, all made using parts on-hand. The Columbia Motor Company had over-extended itself, and the equipment was sold to **Winternitz & Tauber**, of Chicago, for $112,500 in November, 1924.

J. G. Bayerline

E. A. NELSON MECHANICAL ENGINEER
DETROIT, MICHIGAN
1917-1918

E. A. NELSON MOTOR CAR COMPANY
DETROIT, MICHIGAN
1918-1920

E. A. NELSON AUTOMOBILE COMPANY
DETROIT, MICHIGAN
1920-1921

Emil A. Nelson was born in 1880, in Cleveland, Ohio. He worked in automotive related jobs as soon as he was out of school. He moved to Detroit and worked for the Olds Motor Works until 1906, then moved to the Packard Motor Car Company. **Nelson** was chief engineer for the Hupmobile Company in 1908 and designed the Hupmobile "20" and "32" models using pressed steel panels. In 1910 and 1912 he traveled to Europe to study automobiles. From this experience he developed a concept for a small, high efficiency automobile for the Hupmobile Company, but they were not interested. In 1913 he left Hupp and returned to Europe to study design practices and manufacturing methods.

In 1914 he built his first car, followed by two more later that year. For the next two years he tested and modified the design. In 1917 he went into production under the name of E. A. Nelson-Mechanical Engineer. He financed the entire enterprise, and built a factory at Bellevue and Kercheval Avenues with a production capacity of ten cars per day.

The "Nelson" weighed 2200 lbs with empty tanks, sold for $2200, and got an average of 25 mpg. The finished vehicle carried only five quarts of oil and three gallons of water. The vehicle was designed for economy, which meant lightness. It had an engine designed along aircraft engines with an overhead cam, four cylinders and a counterweighted crankshaft. Full pressure oiling was used with no splash to any of the bearings. The oil was circulated through the hollow crankshaft and camshaft, where it kept the bearings continually bathed in oil. In the body lines, the cowl was completely eliminated, with the hood in contact with the body proper. All of the windows could be lowered into the body sash with a leather strap, and a double windshield was used. On January 18, 1918, the business was incorporated as the E. A. Nelson Motor Company.

The First World War resulted in material shortages throughout the country and the post war depression severely impacted the Nelson company. In 1919 he tried unsuccessfully to consolidate with the Gray Company of Detroit. **Nelson** suffered his first bankruptcy in March 1920. In September 1920 he re-organized as the E. A. Nelson Automobile Company. A second and final bankruptcy was filed in September 1921. A total of less than 500 vehicles had been built. *Below-Note cross scroll rear spring.*

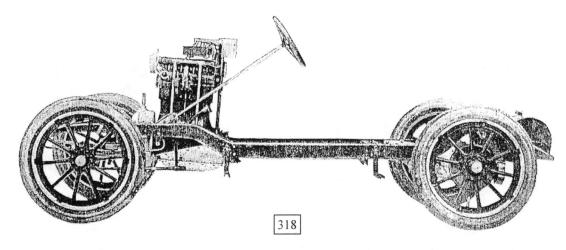

1917-
Nelson

1920-
Nelson
The experimental version of this model only had one door, located on the passenger's side.

E. A. Nelson

HARROUN MOTORS CORPORATION
DETROIT, MICHIGAN
1917-1922

After **Ray Harroun** won the first "Indy 500" in 1911, he went into the carburetor business. He did so well that a "Harroun" carburetor was stock equipment on the 1913 Marmon automobile. **Harroun** developed ideas including venturis heated with exhaust gasses. **Harroun** designed and built Maxwell racing cars that were very successful. In late 1914, he was made chief engineer of the Maxwell Motor Company, which was previously headed up jointly by R. E. Benner and **William Kelly**, the latter having much to do with the design of the motors. Benner left and **Kelly** stayed on as an assistant to **Harroun** of the Maxwell racers and the Harroun "Kerosene Carburetor."

The Harroun Motors Corporation was incorporated in September 1917, with $10,000,000. John Guy Moniham, formerly with Premier and Marion, was president. The former Prouty and Glass Carriage Company factory in Wayne, Michigan was used for initial production, while construction of a new factory with a capacity for 24,000 vehicles per year was begun immediately.

Roadsters for three passengers or for five passengers were available selling for $595, and a sedan was offered for $895. The wheelbase was 107 inches. The engine was an ohv four cylinder with 16.9 hp SAE, and 43.1 hp at 2400 rpm. The engine was combined with a three speed transmission and a cone clutch. The "Harroun" contained a greater proportion of pressed steel than any other car on the market. Some of the components included the frame, radiator shell, hood, fenders, running boards, hub caps, rim carrier, crank case, oil pan, clutch and clutch housing, and the instrument "board." The bodies were also made with pressed steel panels, along with many parts in the rear axle, engine, and various brackets and supports throughout the car.

Harroun declared the motor car of the future would be built almost entirely of pressed steel. He predicted that pressed steel would displace gray-iron, malleable castings, aluminum, wood, brass, bronze, and a large proportion of forging which were being used in automobile construction.

About 500 cars were produced by April 1918 before **Harroun** received a large government contract to make munitions for the war effort. (The Harroun Company funded $500,000 for material and machinery in order to start quickly.) After the war, it was difficult to settle war claims with the government, which caused postponement of automobile production. A new model was in limited production by 1922, but additional capital was needed. New investors could not be found during the post war depression and Harroun Motors went into receivership in June 1922.

1917-Harroun

APEX MOTOR CAR COMPANY
YPSILANTI, MICHIGAN
1919-1922

During the summer of 1919 F. E. Earnest was having difficulty trying to get enough new cars to sell at his dealership in Seattle, Wash. There was a shortage of cars following the First World War, so he traveled to Detroit, but could not get the cars he needed. It was during this effort that he met **Fred M. Guy** and **O. W. Heinz**. **Guy** was a former chief engineer for Hacket Motor Car Company in Jackson, Mich. The Hacket Company moved to Grand Rapids and re-organized in 1920 as the Lorraine Motor Corporation, where **David Buick** was employed.

While working at Hacket, Guy developed a unique design for the internal combustion engine using disc valves instead of poppet valves. These valves were in the cylinder heads and operated at one eighth crankshaft speed. The valves were shaped somewhat like a Maltese Cross and were keyed to the crankshaft by a gear. Thus, the camshaft and valve springs were eliminated.

Heinz became president, and **Guy** became chief engineer for the Apex Motor Car Company. Factory construction began in Ypsilanti, Mich., in October 1919. By April 1920 the first shipment of "Ace's" were headed for the West Coast, but by April 1921 both **H**einz and **Guy** left Apex to manufacture the "Guy" disc valve engine. H. T. Anover became president, and he was forced to use poppet valve engines such as the Hershell-Spillman six cylinder. Unfortunately, the company's fortunes began to diminish, and by May of 1922 the Saxon Motor Car Company took over part of the factory. Then in September 1923 the Commerce Motor Truck Corporation took over the entire factory.

1921-
Ace Coupe

Presenting—
THE Motor Car Achievement
The Ace

The Ace car represents the greatest advancement in motor car construction since the inception of the industry. It marks the first practical and proven elimination of the *unmechanical* from the gasoline engine.

The epoch-making "Guy Disk-Valve Motor," found only in the Ace, eliminates every objectionable feature of the poppet-valve motor and embodies the virtues of the slide-valve type with improvements which far overshadow it.

The almost unbelievable will be presented to you and proved to you when you visit the Ace exhibit at the Commodore. A motor with no valves to grind or adjust will surely interest you—but you will be more inspired over the development of a motor which finally attains that long-sought-for feature of high torque (pulling ability) at low speeds. In this respect the Ace motor compares favorably with the steam engine—tremendous pulling ability at low piston speed. And in the Ace there is a perfect fuel combustion with consequent great economy. There is a silence of operation which is marked even in this day of quiet operating motors. In fact, this truly remarkable motor achieves every factor which engineers have long been striving for.

But the Ace represents more than a motor achievement. It is a car of faultless chassis design, and possessing a singular beauty which sets it apart from present day body types.

1921- The Ypsilanti factory

THE FISHER BROTHERS

Andrew and Stephina Rimille Fisher immigrated from Germany in the early 19th century and settled in Peru Township in Ohio, where Andrew worked as a blacksmith. Lorenz, one of their four sons, married Margaret Theisen and the couple moved to Sandusky, Ohio, where Lorenz made carriages. A son, Frederick was born in 1878 and Charles was born in 1880. In 1885, Lorenz moved his family to Norwalk, Ohio where William-(1886), Lawrence-(1888), Edward-(1891), Alfred-(1892), and Howard-(1902) were born. There were also four sisters: Anna, Mayme, Loretta, and Clara. As Lorenz made carriages and wagons, including painting, upholstering and repairing, his sons served apprenticeships.

Fred was the first brother to venture to Detroit when Barney Everitt hired him to make bodies for the curved dash Oldsmobile. He left Everett to become a draftsman for the C. R. Wilson Body Company in Detroit, and became superintendent within a few years. In 1906, Charles moved to Detroit and worked in the carriage manufacturing trade.

As automobile production expanded in Detroit, there were a number of body builders supplying the automobile manufacturers. One in particular was the Fisher carriage manufacturing company in Detroit. It was owned by Albert Fisher, a brother of Lorenz, and an uncle of Fred and Charles. The three organized the Fisher Body Company of Detroit in July 1908 with $50,000 capitalization, with $30,000 provided by Albert. They advertised both carriages and automobile bodies, and supplied Ford, Olds, and E-M-F. By mid-1909, Albert Fisher sold his interest in the company to Louis Mendelssohn, who owned a major share in the Herreshoff Motor Company. Fred Fisher became president and Charles was vice president.

Closed bodies were still in the experimental stages, being tall and awkward, heavy with glass, and costly. In 1910 the Fisher Body Company accepted an offer from the Cadillac Motor Car Company for 150 closed bodies. They were to be clad with sheet metal instead of the usual veneer wood and there was a significant volume, making it a tough assignment for a young company. By December 1910 the Fisher Closed Body Company was formed as a subsidiary.

Fisher Body of Canada was organized in 1912, the same year Lawrence joined the firm. He was followed by Edward and Alfred in 1913, and William in 1915. Some of their major customers were Buick, Hudson, Oakland, Chalmers, Studebaker, Regal and Maxwell. As long as composite bodies with wooden frames and metal skins were in use, Fisher was very competitive. Workers had to cut dies, cut lumber and sheet metal, screw and glue frames together, and install upholstery. They also painted and varnished, which meant long periods of storage for drying time. In 1916, the Fisher Body Company, the Fisher Closed Body Company, and the Fisher Body Company of Canada were brought together in the Fisher Body Corporation.

In 1919 GM acquired 60 percent interest in the Fisher Body Corporation with arrangements to supply all GM passenger car requirements it could sell. In 1926, the remaining 40 percent was purchased by GM and it was absorbed as a division. One of the main reasons for the acquisition was that Chevrolet's capacity was limited by the ability of Fisher to provide closed bodies. There were operating economies to be gained by coordinating body and chassis assembly. It was also important to bring the Fisher brothers closer to the GM organization.

By 1925, Fisher Body had 44 plants. In 1928 the Fisher brothers moved their offices into the family owned Fisher Building in Detroit. Frederick died in 1941, Charles in 1963, William in 1969, Lawrence in 1961, Edward in 1972, and Alfred in 1963.

One of the first Fisher bodies courtesy MSHC

An early Fisher body courtesy MSHC

circa 1920's Stanley Plewuch(L) Max Fick (R) working on a square head shaper for a deck lid. courtesy E. SZUDAREK

KESSLER MOTOR COMPANY
DETROIT, MICHIGAN
1920-1921

KESS-LINE MOTORS COMPANY
DETROIT, MICHIGAN
1921

Martin C. Kessler designed an engine for the Chalmers Motor Company in 1907, and worked as an automotive consultant afterwards. In 1917, the Kessler Motor Company was formed to manufacture aircraft engines during the First World War. The Kessler Motor Company announced its plans for automobile production in January of 1920. The factory was located at 1297-1309 Terminal Avenue, at East Jefferson and the Terminal Railroad. George B. Emmerson was president and Martin C. Kessler was VP and consulting engineer.

The "Kessler" had a 117 in wb and was powered by a tandem four cylinder engine that Kessler designed. With a supercharger, it had 70 hp. The selling price was $1995.

In September 1921, the Kess-Line Motor Company was formed with $500,000 of preferred stock and $250,000 of common stock. Martin Kessler was president, **W. H. Radford,** prominent in car manufacturing since 1903, was VP, and H. H. Scott, a former executive of Fisher Body, was secretary and treasurer. The former factory of the Liberty Motor Car Company, on Lycaste Avenue and the Detroit Terminal Line, was leased for production. The factory had 70,000 square feet of floor space with a production capacity of 60 cars per day.

The "Kess-Line" had a 119 in wb and used a 167 cubic inch tandem-four cylinder engine with a supercharger that produced 90 to 100 hp. The first car built was given 8,000 miles of the roughest tests to prepare for production. The car was shown at the Detroit Auto Show in February 1922, but never made it into production.

The Kess-Line Motor was reworked and used in the "Balboa" produced in Fullerton, California in 1924 and 1925, with the help of **W. H. Radford**.

1921-
Kessler touring

Kessler engine

Kessler factory courtesy NAHC

Kess-Line factory (former RCH and Liberty) courtesy NAHC

GRAY MOTOR CORPORATION
DETROIT, MICHIGAN
1920-1926

"ARISTOCRAT OF SMALL CARS"

Ora Mulford was the founder of the Michigan Yacht and Power Company, a manufacturer of boats that was using Sintz engines made in Grand Rapids, Mich. The Sintz Gas Engine Company relocated to Detroit and helped create the Pungs-Finch Automobile Company. **Mulford** left the company to design his own marine engine with investment from **David Gray**, son of **John S. Gray**. The Gray Marine Motor Company was formed in 1905, with a plant comprising 18 acres on Mack and the Terminal Railroad.

In the spring of 1920 **David Gray** backed Frank F. Beal, a former VP of Packard Motor Company, William H. Blackburn, a former Cadillac superintendent and took over the plant of the Gray Marine Motor Company for the purpose of producing an automobile called the "Gray."

After a year without production, **Frank L. Klingensmith**, former VP and treasurer of Ford Motor Company, became the president of the Gray Motor Corporation. **Klingensmith** was born in 1879 in Pittsford, Mich. He prepared himself for a business career at the Ypsilanti Business College, graduating in 1900. In 1903 he was employed as a bookkeeper at Standardt Brothers, a wholesale hardware house in Detroit. In 1905 he joined the Ford Motor Company as cashier, and married Julia Elizabeth Myhrs that same year. This was early in the formative stage of the great industrial enterprise in which he was soon to advance. In 1915 he became vice president, treasurer, and one of three directors. The other two directors were **Henry** and **Edsel Ford**. It was said that his strongest characteristic was his inherent love for a *square deal*.

The Gray automobile was introduced in November 1921, and went head to head with the Ford Model T. It had 20 hp and a 100 inch wheelbase like the Model T, but had additional refinements, such as a three speed selective transmission and improved suspension.

The sales volume never came close to the Model T. The 1924 model year provided improvements such as an increase in the number of wooden ribs, which supported the metal body panels, and locks that were built into all closed car doors. The 1925 model increased the wheelbase as well as the price. In 1926 rubber blocks were used to support the engine, and a new type of *oil rectifier* was mounted on the intake manifold which was controlled automatically by the air suction. The chassis lubrication was by the Zerk system (a one way grease valve, instead of grease cups).

In January 1925 **Frank Klingensmith** resigned. By June of 1926 the Gray Motor Corporation was unable to sustain itself and went out of business.

*1922-
San Francisco
to New York,
averaged 33.8 mpg
in 4819 miles of
travel, setting a
new transcontinental
record*

*Unapproached Economy ~
Style Leadership*

*1926-
Gray de luxe sedan
The exterior finish
was torpedo-boat
gray with black
upper structure
above the black
bead.*

Frank L. Klingensmith

C. H. WILLS & COMPANY
MARYSVILLE, MICHIGAN
1920-1923

WILLS SAINTE CLAIRE MOTOR COMPANY
MARYSVILLE, MICHIGAN
1923-1927

Childe Harold Wills was born on June 1, 1878, in Fort Wayne, Indiana. He was the youngest of three children born to John Carnegie and Mary Engelina Swindell Wills. His first name originated from the poetry of George Gordon, but he didn't like the name Childe and preferred **Harold**. After Harold's birth, John moved the family to Detroit where he was a master railroad mechanic. **Harold** received training as a mechanic from his father while attending high school. After graduation he became an apprentice toolmaker at the Detroit Lubricating Company, and studied engineering, metallurgy, and chemistry at night. He completed his four year apprenticeship at age 21, and went to work for a machine shop. In 1901, he worked at the **Boyer Machine Company** as a chief engineer.

Wills met **Henry Ford** who was already well known in the Detroit community. It appears that he helped **Ford** at the Henry Ford Company, then in the spring of 1902, joined **Ford** to work on race cars. **Harold** worked on the **999** and **Arrow** racers in the morning before working at **Boyer** and at night. **Wills** apparently contributed some of his own money to help **Ford**, and in return was promised a share of the money **Ford** would eventually make. The shop was not heated, which was common at the time, and when it got cold at night **Harold** and **Henry** put on boxing gloves to spar and warm up.

Ford received financial backing to form a car company in the summer of 1903. By October they were hiring workers to help develop an automobile. By the spring of 1903, an acceptable car was developed, mainly due to Wills' efforts. The Ford Motor Company was formed in June with **Ford** serving as VP and **Wills** became his right hand man.

While **Ford** was away in 1912, **Wills** modified a Model T with up-to-date design changes and features as a surprise. When **Ford** returned he personally destroyed the car, demonstrating his authoritarian management style. **Ford's** refusal to accept advice from his aides led to the departure of **Wills** on March 15, 1919. **Ford** had always kept his promises. In August, Wills received $1,592,128 as final payment from **Ford's** stockholdings.

Although **Wills** had enough money to last a lifetime, he wanted to develop a car that was ahead of its time. **Wills** was frustrated by his years of service to an ideal which only meant finding ways of reducing cost. He wanted to build the finest car possible, without regard to initial cost. He partnered with another Ford official named John R. Lee and began designing a car and selecting a manufacturing sight. Detroit was his first choice but it was a booming town and land was becoming too expensive. Housing was in critical demand, there was a labor shortage, and factories were delayed because of so much construction. A 3000 acre site was located near Port Huron, Mich. **Wills** created a new village for the workers and called it Marysville on behalf of his second wife, Mary Coyne. The C. H. Wills Company was organized with an authorized capital of $10,000,000 and a cash advance from **Wills** himself. The cornerstone for the plant was laid in November 1919.

As it turned out, the "Wills Saint Claire" didn't go into production until March 1921, and it sold for $4000. It was an excellent car made with the auto industry's first usage of molybdenum steel. It had a 60 degree V-8 engine along the lines of the World War I Hispano-Suiza aircraft engine. It had a one-piece cylinder head with dual overhead cams. A post war depression set in and only 1532 cars were sold in 1921, far less than the anticipated 10,000 units.

By the end of November 1922 the C. H. Will Company was over $8,000,000 in short term debt, and was forced into receievership. All of the company officers summarily quit, except Lee who quit a month later.

In 1923 the company was re-organized as the Wills Sainte Clair Motor Company, with backing from Kidder, Peabody and Company of Boston. Wills continued to make improvements and in 1924, four wheel hydraulic brakes were added. In 1925, in addition to the V-8, Wills added a six cylinder to help cut costs. The company never sold enough cars to make a profit. **Wills** was partially at fault for stopping the assembly process too often to make minor improvements, and by having a car designed to be functional but not styled enough.

In 1929, **Wills** helped in the formation of the New Era Motor Company and helped design the Ruxton in New York, City. In 1933 he was working as a consulting metallurgist for the Chrysler Corporation. The Wills factory was purchased by Chrysler in 1935. **Wills** tried to return to the Ford Motor Company only to be snubbed by Charlie Sorensen, and **Wills** continued working for Chrysler until his death on December 30, 1940.

*1924-
Wills Ste. Claire model A-68 Roadster seen next to Lake Saint Clair
(The car was named after the original French spelling of the lake.)
courtesy AAMA*

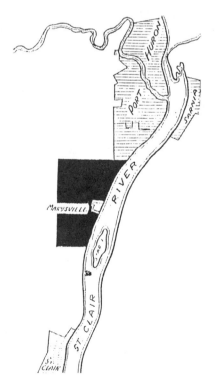

The C. H. Wills and Company property had a frontage of 3.5 miles with a depth of 1.5 miles, totaling 3000 acres. The village of Marysville served as the municipal center.

THE NEW WILLS SAINTE CLAIRE SIX IS THE RECOGNIZED ENGINEERING ACHIEVEMENT OF THE YEAR ◊ ◊ ◊ ◊

ADVANCED engineering and advanced metallurgy have produced no finer car than the new Wills Sainte Claire Six.

It is the recognized engineering achievement of the year—a car designed and built on experience—on *proved engineering*.

It is not a price-built car, but a piece of priceless quality embodying the same advanced principles of engineering; the same supersteels and careful workmanship which have caused the Wills Sainte Claire Eight to be regarded as the most finely engineered and finely built motor car in America.

It has overhead valves and cams—a principle of engineering used in the highest type aviation motors and in several of the finest cars in Europe.

It has the most accessible and the cleanest motor ever designed. The cylinder head and valve and camshaft housing are removable.

Grinding valves, removing carbon and resetting tappets all can be done from the bench.

The cylinders are cast integral with the upper half of the crankcase and adjusting valves is merely a matter of removing the two cover plates on valve housing.

The seven-bearing crankshaft is a remarkable piece of engineering and a superb example of precise workmanship. It is supported by seven main bearings which are 2½ inches in diameter with a total bearing surface of 82½ square inches.

As an example of the infinite care in the manufacture of every working part, all pistons, connecting rods, bearings, cylinder heads, every part of one motor is interchangeable with corresponding parts of every other motor.

Molybdenum steel, the most marvelous alloy steel ever developed, is liberally used throughout the chassis where stresses and strains are encountered, assuring thousands upon thousands of miles of the safest possible and the most luxurious transportation at the minimum maintenance and operation costs.

It has a proved and fully developed system of hydraulic four-wheel brakes; specially designed disc wheels; the most efficient cooling and oiling system on any car; balanced balloon tires and deep yielding upholstery.

The above are but a few of the many advanced principles of engineering and metallurgy embodied in the new Wills Sainte Claire Six. Complete information regarding the construction of the car and also the details of the Wills Sainte Claire Franchise will be sent to any dealer upon request. Address the Sales Department.

WILLS SAINTE CLAIRE, Inc.
Marysville, Michigan

WILLS SAINTE CLAIRE

Motor Cars

The Wills Sainte Claire factory in Marysville, Michigan. picture taken in 1995

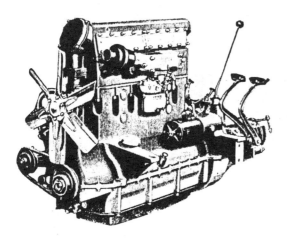
The six-cylinder engine had an overhead camshaft

Childe Harold Wills

DETROIT STEAM MOTOR CORPORATION
DETROIT, MICHIGAN
1922-1923

As a distributor for the Stanley Steamer, O. C. Trask became a producer of steam cars himself. In December 1921 he announced the formation of the Detroit Steam Motor Corporation. The "Trask-Detroit" came as a five passenger phaeton on a 115 inch wb, with genuine black leather upholstering, a one piece windshield, and 22 inch-wide doors. The power plant of the Trask-Detroit had two double-acting cylinders with a 3.25 x 4.25 bore and stroke. It developed 10 hp with a torque equivalent equal to a 50 hp gasoline engine. The engine was mounted directly on the rear axle. The boiler was mounted vertically and was 20 inches in diameter, 14 inches high and contained 74 square feet of heating surface. The burner was an improved bunsen type for kerosene, gasoline or any mixture of them. It carried 17 gallons of water sufficient for 300 to 500 miles. The fuel supply consisted of 15 gallons for the main burner under no pressure and five gallons for pilot. Tanks were at the rear of the car. Gauges were included for steam pressure, combination pilot and main burner fuel, ammeter, Stewart speedometer, and a pilot indicator which showed whether or not the pilot was burning.

Cars were produced at the Schlieder Manufacturing Company, which manufactured valves in Detroit. It was priced at $1585, far higher than an original target of $1000. By mid-1922 Trask had negotiated to manufacture in Windsor, Ontario. In 1923 the Detroit product name was changed to "Detroit Steam Car." Soon after, the entire venture stopped production.

*1922-
Trask-Detroit*

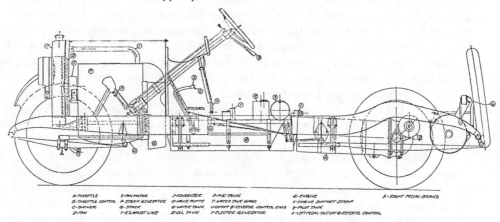

LINCOLN MOTOR COMPANY
DETROIT, MICHIGAN
1917-

As World War I began to take hold, **Henry M. Leland** visited Europe to study aviation as applied to war. He was frequently called to Washington to consult with the council of national defense regarding air problems. **Henry** and **Wilfred Leland**, as managers of the Cadillac Division of GM, proposed to set aside a portion of the plant for the manufacture of airplane motors. GM president **Billy Durant** objected, declaring that he was "not in sympathy with this war; that the war could be stopped tomorrow -- I don't want our company to take part in war work." **Wilfred Leland** later testified to a Senate committee when it visited Detroit that *"we could not pull in the same harness, so we sent in our resignations to the Cadillac company and then began to think what could be done toward lending our experience and ability in this emergency."*

Afterwards, **Durant** did take government contracts as the war effort increased. **Wilfred** was asked if he thought **Durant** would carry them out effectively and he answered, *"I think so. I think he guessed wrong. He was an ambitious man and had plans, the war was upsetting them and therefore he stopped to check himself up. He was not fundamentally disloyal to this country."*

The **Lelands** severed their connection with the Cadillac Company and secured a contract with the US government to build "Liberty" aircraft engines. The **Lincoln Motor Company** was oraganized with cost to the US government of $10,000,000 for a factory and equipment.

The Liberty engine was the idea of **Henry Joy** of Packard Motor Company. At the outbreak of the First World War, he realized the US was poorly prepared to produce motive power for aircraft. There were 37 different kinds of engines in England's aircraft and 46 different kinds in France's. **Joy** reasoned that the US needed a good universal engine with good performance and easy to get spare parts. **J. G. Vincent** of Packard and **E. J. Hall** of the Hall-Scott Motor Company in California collaborated on a design. The project was headed up by **Col. Deeds** of the Dayton Electric Company. **Charles Kettering** designed the ignition and starter system. Contracts to build the engines were issued to Wright Martin Aircraft, Pierce Arrow Motor Company, **Walter P. Chrysler** at Buick, and **Henry Leland**.

After the war, the **Lelands** re-entered the automotive industry with an up-scale car called the "Lincoln." The factory constructed for the Liberty engine was purchased from the US government and production began for the 1920 season. The Lincoln demonstrated Henry Leland's persistence as a quality product manufacturer.

The Lincoln Motor Company went into receivership in late 1921. The specific cause was an erroneous tax bill of $4,500,000. The plant had been built with a government loan and had later been sold to the Lincoln Motor Company for 55 percent cost. The government investigated the transaction and found the **Lelands** guilty of fraud. (They did not take into account that the buildings and equipment were purchased at highly inflated wartime prices. Much of the equipment was so specialized that it was useless later. After contesting in court, the government agreed the company was entitled to a depreciation, reducing its obligation to $500,000. By the time the error was acknowledged, it was too late to save the company.)

William B. Mayo, Ford's chief engineer, was approached by Harrold H. Emmons, attorney for the Lincoln Motor Company. Emmons outlined the desperate condition of his company, and the personal tradegy for the **Lelands**--the loss of their factory, impairment of their reputation, and personal bankruptcy. Mayo relayed this to **Henry Ford**, resulting in a series of meetings with the principals of both companies. On February 4, 1922, **Ford** purchased the Lincoln Motor Company for $8,000,000 from the **Lelands**.

Why **Henry Ford** purchased the Lincoln Motor Company remains obscure. **Ford** was still adamant about the Model T being the universal car. He didn't rescue the **Lelands** out of friendship, because no documentation exists that there was a friendship. Since **Henry Leland** was the replacement for **Ford** in the **Henry Ford Company** 20 years before, **Ford** may have seen an opportunity to become the final victor. Most likely, **Edsel** convinced his father that a luxury car was needed to broaden the range of the Ford Motor Company.

*1920-
Lincoln
courtesy NAHC*

*The Lincoln plant
on Warren Ave. and
Livernois
courtesy MSHC*

*waiting for
shipment
courtesy NAHC*

*1922-
signing over to Ford:
from left to right
Edsel Ford, Henry Ford,
Henry Leland and Wilfred
Leland. Note the picture
of Abraham Lincoln above
Wilfred.*

*The Lincoln plant
along Livernois
picture taken 1994*

Henry Leland chose the name Lincoln because of his lifelong admiration for him. Leland was a collector of Abraham Lincoln memorabilia.

DETROIT AIR COOLED CAR COMPANY
DETROIT & WAYNE, MICHIGAN
1922-1923

A prototype "D.A.C." touring car was first displayed in February 1922 at the Detroit Automobile Show. The public was told "You may have the opportunity to become interested financially in what is going to be one of the largest and most profitable industries. Look this car over and decide for yourself what its advent means in the automotive world. Orders from dealers for thousands of cars are already on hand."

The Detroit Air Cooled Car Company was capitalized at $1,200,000 and was still "perfecting" its organization in May of 1922. W. J. Doughty, the president of the new company, was surrounding himself with several well-known leaders in the automobile industry. Doughty had 18 years automotive experience and was a Franklin distributor in Detroit and previously with the Hupp company. **G. R. Tremolada**, an automotive engineer with experience in Europe, was the designer of the car. John McArthur of Detroit, with 40 years of coach and body building experience, designed the "D.A.C." August Gieseler, Cleveland OH., superintendent and stockholder of the National Tool Company, was in charge of manufacturing operations. Frank Sanders of Chicago, former Franklin distributor and sales representative, was named D.A.C.'s sales manager. L. M. Bradley, a veteran of the industry, was in charge of advertising and publicity. Others with financial interest included: C. H. Bennett, president and general manager of the Daisy Air Rifle Company, Frank Gagnier, manufactures' representative of automobile supplies in Detroit, and Earl B. Newton, secretary treasurer of the Mansfield Steel Corporation and treasurer of the King Trailer Company.

The D.A.C. had a 115 inch wb, weighed 1740 lbs, and was powered by a "twin three" air-cooled engine with 32 hp. The V-6 engine was designed with the cylinder heads staggered, to permit air to be directed to each cylinder in equal volume at the same temperature, to enhance cooling. Each cylinder head had 20 cooling fins. Air conducting chutes brought fan-impelled air onto the radiating surfaces of the engine. The car could get 30 mpg in ordinary driving and was priced at $1250. The low price was partially attributed to the fact that all non-standard parts could be made by die-casting. Also, the valves were located in the heads, and operated by direct pull from the camshaft, located in a separate housing in the "V" between the cylinders. This arrangement obviated the rocker arms, push rods, and tappets, totaling 72 parts.

The organization was not ready until October 1922 when it purchased a small factory at 3745 Cass Avenue in Detroit, with plans to start production in January 1923. The factory was never used for production and was sold to the Detroit Stephens Distributor. In April 1923, the D.A.C. company took over a larger factory located in Wayne, Mich. It was a large two story brick building built in 1910 for the Swift Automobile Company. (The Swift company fell apart before it saw production.) Production began in Wayne, but it was only for a brief time before the line was shut down permanently.

COMMERCE MOTOR CAR COMPANY
DETROIT, MICHIGAN
1922

"THE WONDER OF MOTORDOM"

Walter Edward Parker was born in Ottawa, Ontario in 1873, and educated at Upper Canada College in Toronto. He lived in Canada until he was 21 years old, then spent 12 years in St. Louis in various business enterprises. He moved to Detroit in 1906 and became president of Seamless Rubber Company of New Haven, Conn.

He was president of the Commerce Motor Car Company on Solvay Street in Detroit. (The Commerce Motor Truck Company was a truck producer from 1911 to 1932 in Detroit, then Ypsilanti, and then as Relay Motors Corporation in Lima, Ohio. It was the parent company of the Commerce Motor Car Company.)

Commerce brought out a Char-a-Banc ten-passenger car in 1922. It had a 127 inch wb and was powered by a six-cylinder Continental 9A engine with 73 hp. It stressed particularly luxurious and comfortable body features, low cost of operation and depreciation, and also the fact that the units entering into the construction of the chassis were designed especially to meet the extra requirements placed upon them. It was pointed out that it could be employed for stage routes, rural and urban transportation, sight seeing, depot and hotel transfer, and for club or resort transfer.

Attractive appearance was obtained through the streamline effect. The seats, which were full size and equipped with Marshall springs, were upholstered in long grain leather. The body was equipped with two sets of curtains, one was storm curtains for use in warm, inclement weather. The other set was equipped with crystal glass, constructed to open with doors for use in cold or winter weather. The price was $2350 F. O. B., Detroit.

EDWARD VERNON RICKENBACKER

Eddie *Richenbacher* was born in Columbus, Ohio, the son of a Swiss immigrant construction worker. Eddie had a short childhood, as it ended at 12 years of age when his father died. As one of seven children, he went to work at the Columbus Glass Company to support his family. He worked the night shift and worked 10 hours per day, six days per week. He left the glass factory and held several jobs at a foundry, a shoe factory, and a railroad. He finally found his element at a garage working as a mechanic. He finished his formal education by taking an engineering course from a correspondence school.

At age 16 **Eddie** was such an excellent and reliable mechanic that his employer, **Lee Frayer**, took him to the 1906 Vanderbuilt Cup race in Long Island, N. Y. **Frayer** was a race car driver and manager of the **Frayer-Miller Automobile Company**. He let **Eddie** ride as his mechanic which made a big impression. For the next four years Eddie sold, repaired, and raced automobiles. Eight successive victories at Omaha in 1910 were enough to prompt Frayer to use Eddie as a riding mechanic for the first Indianapolis "500" race in 1911, and as a driver in 1912.

From then on, racing became his full time career. He became a driver for the **Duesenberg team** in 1914 and came in tenth place at Indianapolis, but did much better on other tracks, finishing fifth in the AAA Championship standings that year. The manager of the **Maxwell racing team**, **Paul Bruske**, noticed **Eddie's** driving and signed him up in February 1915. This began the glorious era of the Maxwells, with **Billy Carlson**, **Barney Oldfield** and **Eddie** *Richenbacher* changing the American racing scene with their mounting victories.

Eddie appeared in the 1915 Indianapolis race with a new, lighter, and highly modified 16-valve Maxwell built by **Ray Harroun**. He was up to third place at one time, but blew a connecting rod and was out of the race. His first victory in a Maxwell was scored on a dirt track in Sioux City, Iowa, where he won in four hours with a 74 mph average speed. He repeated the victory in Omaha, and from then on he became one of the biggest money makers in racing. He was known as "lead foot" and "hot chauffeur," always driving fast, but never recklessly.

Carlson was killed about that time, and the Maxwell Company decided to quit racing entirely. The owners of the Indianapolis Motor Speedway bought four of the Maxwell cars and **Eddie** became manager of the "Prest-O-Lite" stable, as the team was known. His organizational skills and leadership qualities were soon made evident. One of his additions was a mobile workshop complete with spare parts and equipment. The 1916 Indianapolis Classic was reduced to 300 miles due to a gasoline shortage during the war. **Eddie** was the favorite, and in the ninth lap, he was in the lead. A steering knuckle broke however, and he took Peter Henderson's Maxwell and managed to place sixth.

Eddie took a trip to England to purchase two Sunbeam racers for millionaire Bill Weightman. With the intention of making a pun, a journalist referred to Eddie as "Baron von *Richenbacher* from Germany." The suspicious British Intelligence caught wind of it and descended upon **Eddie**, subjecting him to search and endless questioning. He was forced to return to the States without the cars. When the United States entered the war in the spring of 1917, **Eddie** was one of the first to enlist in the army. He changed his legal name to **Rickenbacker** to avoid repeating the British incident. Eddie was among those on the U.S.S. Baltic as the first contingent of the American Expeditionary Force (AEF) headed for France.

With the rank of sergeant, he was assigned to drive **General Pershing's** Hudson Super-Six, and by coincidence came across **Billy Mitchell's** stalled car near the front lines. After quickly fixing it, the two became acquainted, and Mitchell helped **Eddie** transfer into the Air Corps. In August Eddie was assigned to the Air Corps as an engineering instructor, and he soon got re-assigned to the flying school. He earned his wings and officer's rank, and was assigned to the famous "Hat-In-The-Ring" 94th squadron. By the time the war ended he shot down 26 German planes, the record for the American army. He was awarded the French Legion of Honor, The Croix de Guerre and three citations, the American distinguished Service Cross and seven citations. The war was over in early 1919, and "**Captain Eddy**," as he was affectionately known, was 28 years old.

*circa 1914-
Eddie Richenbacher
behind the wheel of
one of the three
racers made by Ray
Harroun for Maxwell
courtesy NAHC*

*Captain Eddie
Rickenbacker
courtesy AAMA*

*This photo shows
the "hat-in-a-ring"
emblem used by the
the famous 94th Aero
Squadron. It also
shows the emblem
of the Detroit
Automobile Club.
He was a member of
the club and the
emblem was on his
airplane while he was
in France.*

RICKENBACKER MOTOR COMPANY
DETROIT, MICHIGAN
1921-1928

"A CAR WORTHY OF ITS NAME"

Captain Eddie Rickenbacker returned to the United States with a hero's welcome starting with a "monster banquet" at the Waldorf Astoria in New York, followed by a three day celebration in Los Angeles, and many points in between the two cities. The **Ace of Aces** received publishing and motion picture offers, but was undecided what business to engage in. He retreated to New Mexico to collect his thoughts, and quickly decided not to return to racing. He decided to produce an automobile. By coincidence, someone was thinking the same thing in Detroit.

Barney Everitt had a successful body manufacturing factory on East Jefferson Avenue in Detroit, but was anxious to build a car again. He called his former partner **Billy Metzger,** who was in virtual retirement, but he wasn't interested. When he called **Walt Flanders**, he was not interested either. He had put Maxwell back on its feet after the fall of the United States Motors, and was looking forward to retirement, but **Barney** talked him out of it. **Harry Cunningham** was to become secretary and treasurer. He had a long list of credentials, going back to Winton in Cleveland, and Ford's 999 racer in Detroit. He served as a stand-in for Tom Cooper, and raced Packard's Gray Wolf. On the business side, he worked for Ford as a branch manager and a consulting engineer, then worked for E.M.F. and Maxwell. E. R. Evans of the Canadian Metzger Motor Car Company became chief engineer. Roy M. Hood, former purchasing agent for E.M.F. and Maxwell, became assistant general manager. A. L. Miller, who handled Maxwell finances on the West Coast became comptroller. **Everitt** was made president and general manager, and **Flanders** was VP. **Rickenbacker** declined the presidency and felt his strong point would be in public relations. He assumed the title of VP and director of sales.

With great secrecy, a pilot model was developed with a six cylinder engine and features that were born on America's speedways. It had a double flywheel for reduced engine vibration, four-wheel brakes, and a lower center of gravity for safety. Everybody except **Rickenbacker** wanted to get it in production as quickly as possible. He insisted that the car not be placed on the market until the January automobile show in New York. Everyone else contended there was no need to wait two full years to produce a car, but **Rickenbacker** maintained that he wanted a car "worthy of its name." When the car was ready, **Rickenbacker** took it for several months on cross country roads and drove almost 150,000 miles. As parts broke or wore out, design defects were detected. It was estimated that he used up enough spare parts to build three additional cars. In December 1920 Rickenbacker went to work in California for his friend, **Cliff Durant**, as VP and general manager of a company handling the distribution of a GM car called the Sheridan. While **Rickenbacker** was taking his "refresher course," **Everitt** was gathering the necessary machinery to get the new car going. The Rickenbacker Motor Company was officially incorporated in July 1921 with $5,000,000 capital. The Michigan Avenue plant of Disteel Wheel was taken over, and a new factory designed by **Albert Kahn** was erected later at 4815 Cabot Avenue, in Detroit.

The New York show got underway with Rickenbacker touring, sedan, and coupe models on display. They were medium priced from $1485 to $1985, with plenty of features that justified the two year wait. The six cylinder engine had a 3.125 inch bore and a 4.75 stroke, that displaced 218 cubic inches and developed 58 bhp at 2800 rpm. A speed of 60 mph was assured without vibration, which was unheard of for that price range. Credit was given to having flywheels at both ends of the crankshaft. According to **Rickenbacker**, after shooting down a German aircraft that seemed to be exceptionally maneuverable, he asked if it could be located and brought back to the airdrome for inspection, and that's when he observed the tandem flywheels. The frame channels were wide and eight inches deep to provide rigidity and stength. The front springs were semi-elliptic, 36 inches in front and 57 inches in the rear with the frame cradled between them to provide a lower center of gravity. The ads said that you could get 130 inch wheelbase riding comfort on a 117 inch chassis. In June the Rickenbacker got great publicity when it was chosen by Detroit companies to make the first "transcontinental radio tour" over the Lincoln Highway because it "offers the least resistance to radio because of vibration."

There were a few problems, for example the electric gas gauge was wired to the battery, and although it didn't use much current, it drained the battery if the car sat for several days. After the problem was discovered, the gauge was wired to the dash lamp switch and owners were told to flick on the light when they wanted to check the gas. They didn't think of wiring the gauge to the ignition switch. The tandem flywheel caused a problem with what could be called the fourth main bearing, and there were timing chain problems.

Rickenbacker made sure to inspect all of the modifications for himself, and was always visiting the Everitt Brothers factory where the cars were painted. In the first year and half, he visited the 44 distributorships, 250 dealers, and met over 1000 Rickenbacker car owners. He crossed the continent twice and the Atlantic once when he married Adelaide Frost and took a combined business trip and honeymoon.

The 1923 models were essentially carryover, with a new clutch and an air cleaner to filter "dust, sand, particles of vegetable matter and insects." The company and the automobile industry suffered a great loss when **Walter Flanders** was killed in an automobile accident on June 16th, 1924. The loss was especially hard on **Everitt** considering what good friends they were. On June 27, 1924, the Rickenbacker Motor Company announced it would have four wheel mechanical brakes. It was the first volume-produced, medium-priced car to use them. This was what Eddie wanted on the original 1922 model, but it was vetoed. He became convinced of the concept while driving a Peugeot race car with brakes on all four wheels in 1915. A number of other car companies offered four wheel brakes soon after the announcement, and Packard actually beat them by 16 days. But there were also anti-four wheel brake ads in the papers by competitors, such as Studebaker, who didn't offer them and claimed they were unsafe. Insurance companies offered 10 percent to 15 percent rate reductions for cars with brakes on all four wheels. There were safety problems with two-wheel brakes cars following too close behind. Bumper stickers with "beware-four wheel brakes" were issued, then a warning legend was transferred to the rear spare tire cover. The loss of **Flanders** and the adverse publicty was taking its toll, and by the end of 1924, the Rickenbacker Motor Company had lost $150,000. Something had to be done.

A new six cylinder engine was announced in January 1925, and **Cannon Ball Baker** was hired to race Rickenbackers. He proceeded to run up a series of records that were publicised from coast to coast. A sweeping reduction in price of the eight clylinder was made in July 1925 of two hundred to six hundred dollars, depending on the model. The dealers had stock on hand that was purchased at the previous wholesale prices, and their loss created disenchantment. For 1926 the company turned to styling: "put an evening gown on the lady" was the assignment. The Rickenbacker Super Sport was the outcome that sold for $5000. It had a torpedo rear deck, aerofoil bumpers, cycle fenders made with laminated mahogany, bullet-shaped headlamps, safety glass all the way around, and no running boards. The price of six cylinder and eight cylinder models was reduced in June 1926, resulting in an unusual influx of business. The dealers were furious and it was Eddie's fault. Stockholders and directors began to bicker, and one day in September, **Eddie** scribbled a few lines on a sheet of paper and said: **"Here's where I get off. I can't go along any further because I don't want to be a party to having anyone lose any more money. The Rickenbacker Company is in a ditch and out of the race, and the best way I can let people know that we're out of the running is to walk away from the wreck. You'll find my resignation on the table."**

With assests of $7,000,000 and liabilities of $1,500,000, the Rickenbacker Company was in receivership in November 1926. The plant was sold to **Barney Everitt** in November 1927, and was used to make aircraft. **Barney** died in 1940, and **Captain Eddie** died in Switzerland in 1973.

The Rickenbacker factory on Cabot Street courtesy MSHC

1925- Cannon Ball Baker

1926- The "boat tailed" Supersport courtesy NAHC

DAVIS STEAM MOTORS, INCORPORATED
DETROIT, MICHIGAN
1921

Davis Steam Motors was incorporated in March 1921, capitalized with $100,000. The company officers were Merrill Davis, E. M. Bliss, F. D. Sielberg and A. B. Eggert. Initially, a two-cylinder steam engine for cars and trucks was offered, and by August a touring car was added. The "Davis" had a 120 inch wb and sold for $2300. Production of the car was soon discontinued.

GASOLINE

By the late 1890's most cities used natural gas and electricity for lighting. People in rural communities used candles and kerosene lamps. Kerosene was the main business of the oil industry. Naphtha and gasoline were also produced as unwanted by-products of kerosene manufacturing. Hardware and drug stores usually sold petroleum products. Gasoline was known to be much more volatile than kerosene, and dangerous. Gasoline was extremely expensive until old "spindletop" gushed in Texas on January 10, 1901. Almost immediately speculators went to the Beaumont Texas area to drill for oil. Humble Oil Company got its start there, drilling in Humble field. Gulf Oil also got its start there.

Gasoline was originally purchased by the bucket from the drug stores, grocery stores, or garages. With increased demand, bulk terminals were set up at the edge of town. Some automobile owners set galvanized tanks in their backyard and gasoline was delivered by horse and wagon. Then street vendors with tank carts appeared, but they were messy and dangerous.

In 1905 the nation's first "drive-in" gas station was developed on the northeast corner of First and Fort streets in Detroit. Roy Francis ran out of gasoline at that intersection while driving home from his job at the **C. H. Blomstrom Motor Car Company**. He made a determination at that moment to develop a gas station. He rented land from **Henry B. Joy**, the president of Packard Motor Car Company at the very intersection where he ran out of gasoline. He put up a shed and an iron tank for his gasoline. Automobile tanks were located under the driver's seat at that time, and it was time consuming to transfer the gasoline in pails. He later used a garden hose with a hand pump.

Early petroleum refining processing was derived from whiskey distilling methods perfected on frontier farms, and was not totally suitable for gasoline. Through the early 1920's, 60 percent of gasoline was actually kerosene that fell into the crankcase. Crankcases had to be drained every 500 miles.

With the advent of the self starter in 1912, engine compression could be increased without the constraint of an engine too hard to crank start. However, the higher compression led to incomplete combustion and a resultant pre-ignition and "pinging" because of low octane gasoline. The man who helped create the first successful self starter was tasked by General Motors to eliminate the pre-ignition. **Charles Kettering** developed ethyl gas by adding a small amount of lead, which gave more power and less knock. It was first marketed in Dayton, Ohio, in 1923. In the 1930's a further improvement called "cracking" was developed by subjecting the crude oil to heavy pressure and rearranging the molecules so that more gas could be yielded than from simple distillation.

Portable filling station courtesy AAMA

Portable filling station courtesy AAMA

Early Texaco Gas Station

TIRES

Some of the very first automobile wheels were constructed with iron bands along the outer edge. The self propulsion of the automobile with a pushing effect, compared to the pulling effect of a horse carriage, created the demand for new tire technology. Rubber was needed to derive the desired spring effect and adherence to the road.

The first pneumatic tires were a one piece construction consisting of a "carcass" that held air. These tires were based on **Dunlop's** 1888 patent for an "improvement in tires" using linen covered sheet rubber. By 1905 a two piece construction with a tube and a "casing" was in use. Wheels were typically large in diameter and "clincher" tires required high air pressure to prevent leaks. A 25-inch tire diameter required 50 pounds of air pressure and a 45-inch diameter tire required 90 pounds. These tires resulted in a hard ride, frequent blowouts, and a relatively short typical life of 3,000 miles. In 1908 Seibeling perfected machines to cut grooves in tires and produced "non-skid" tires. Before that, tie ropes, rags, or cable chains were used to reduce skidding.

By 1914 it was known that tires had a "rolling effect" when going around curves. Tires were constructed with fabric and were made oversize with extra fabric and rubber tread to reduce the strain. To strengthen the side walls, "cord" tires started to replace fabric by 1920. In 1923 the **Cole** was the first car to use the "balloon" tire, made by Firestone. It provided better braking, less chance for skidding and less chance for a blowout. It used 25 pounds of air pressure and was 33 x 5 instead of 34 x 7. In March 1924 Ford began using the balloon tires and by November a quarter million cars in the United States were using them. In 1954 the Packard Motor Company introduced the first tubeless tire.

The expected life of a tire was 3000 miles

Non-skid tires became common for rear wheels because front wheels did not have brakes until the 1920's.

CHRYSLER CORPORATION
1925-

Walter Percy Chrysler was born on April 2, 1875, in Wamego, Kansas. His father was a locomotive engineer with the Kansas Pacific Railroad, later called the Union Pacific. Walter inherited a liking for mechanics from his father and quit school to become an apprentice for the Union Pacific, starting out wiping engines in the roundhouse. At age 18 Chrysler designed and built a complete steam locomotive that ran under its own power and was complete with air brakes. He read Scientific American magazine for pleasure, and he saved enough money to take a course in mechanical engineering from the International Correspondence School. He loved music and later took a class where he met his future wife, Della. Chrysler was a two fisted guy that moved up the ranks quickly. At the age of 33, he became superintendent of 10,000 employees at the Chicago Great Western Railway. He then moved and became manager of the American Locomotive Works in Pittsburgh.

Charles Nash worked for the Durant-Dort Carriage Company, starting out polishing lamps. In 1910 Durant was no longer in charge of GM, but he persuaded the directors to place Nash in charge of Buick. Nash looked around for a general manager to help him pull Buick out of its difficulties and hired Walter Chrysler away from the American Locomotive Works. Although Chrysler's salary was halved from $12,000 per year down to $6000 at Buick, a bonus arrangement would eventually make Chrysler a millionaire.

The two men understood each other and worked well together. In 1912 Nash became president of GM, and Chrysler took over the presidency of Buick. In 1916 Durant was in charge of GM again, and Nash decided to leave. In July, Nash bought out the Thomas B. Jeffrey Company and asked Chrysler to join him in Kenosha, Wisconsin.

In July Durant succeeded Nash as president of GM and his first order of business was to persuade Walter Chrysler to remain at GM. Durant received Chrysler's letter of resignation in New York City. Durant left for Flint to talk with Chrysler where he used every argument he could to keep him. He explained there were a great many men who were devoted to Chrysler, mentioning the name K. T. Keller, whom Durant needed in his organization. Durant used every argument that he possessed. Chrysler asked, "What is your proposition? What have you to offer?" Durant replied that he would pay $10,000 a month in cash, and at the end of the year $500,000 in cash, or $500,000 in GM stock. In other words, if Chrysler took the stock he could have the value that GM obtained as the result of Chrysler's managerial influence. Chrysler was to select his own organization without interference. When he finished, Chrysler asked Durant to repeat it. Then Chrylsler said, "I accept. Billy, I didn't think you could win. You have beaten the bankers and have upset their plans." Chrysler made one strong point at the meeting: " I can accept only if I'm to have full authority. I don't want any other boss but you. Just have one channel between Flint and Detroit: From me to you."

Durant was beaming at him then. Chrysler saw Durant touch his fingers lightly to the table top for emphasis. "It's a deal," he said.

Chrysler was made executive VP and was placed in control of all plant production. One of Chrysler's first priorities was to acquire 60 percent of the Fisher Body Company. It soon became apparent that Durant and Chrysler were too headstrong to work together, and Chrysler "retired" as a millionaire in 1919 at 45 years of age. He returned to his home in Great Neck, N. Y., then went to Europe to rest and relax for the rest of his life.

After World War I was over, many car companies were in trouble. GM tried to get Chrysler back, Nash wanted him, Packard made an offer, and Wall Street had several jobs waiting for him. Willys-Overland was once second to Ford in production volume. The Willys plants were located in Toledo, Ohio which was once referred to as the "Second Detroit." John North Willys was in danger of losing his empire because he frequently allowed the people under him to control operations while he provided little guidance.

The company was also experiencing a serious strike that curtailed production. The company was in debt by $46,000,000. Willys' bankers sent a delegation headed by an astute legal expert, named James C. Brady, to offer **Chrysler** the challenge of rescuing the company. That was the sort of trouble **Chrysler** liked. He started by lowering the salary of the company's president, John North Willys. He reduced staff and sold equipment until the debt was lowered to $18,000,000. For that, **Chrysler** made $2,000,000.

While working on the Willys project, **Chrysler** met an engineering team at the organization's Elizabeth, N. J. plant. **Fred Zeder**, **Owen Skelton** and **Carl Breer** were consultants on Willys' new car projects. **Zeder** had designed the EMF cars, and after Studebaker took over he stayed with them. Later, he organized the Zeder, Skelton, Breer Engineering Company. The trio had designed an innovative car with hydraulic brakes, an all-steel body and a high-compression, high speed engine. They hoped to sell the car design to one of the big companies.

Chrysler was still struggling with Willys when he took on the additional load of helping Maxwell which was $35,000,000 in debt. By then, the Willys plant in Elizabeth, N. J. had to be sold and the engineering team moved to Newark, N. J., on contract for **Walter Chrysler**. **Durant** was by then out of GM for the second time and tried to hire them. After **Chrysler** became chairman of Maxwell in 1921, they decided to stay with him. They had designed a car that Willys was not interested in, and **Chrysler** moved them to the Chalmers plant in Detroit in 1922. The coadjutors went to work, and a prototype was completed in 60 days after approval, with engineers testing it on Kercheval Avenue after dark. There was a disheartening point when James C. Brady fell ill and the bankers advised **Chrysler** they were planning on selling out to Studebaker. Fortunately for **Chrysler** the deal fell through. **Walter Chrysler** wanted to display the car in the New York Automobile Show, but was denied because it wasn't in production yet. It was first publicly shown in January 1924 in the lobby of the Hotel Commodore in New York City because many investors and exhibitors were staying there. Finally a Chase Securities banker underwrote a $5,000,000 issue of Maxwell Motor Corporation debenture bonds to finance Chrysler's plans.

The original lineup consisted of six body styles. The six cylinder engine was the first, for a "modern" medium priced motor car, with the desired smoothness via a counterbalanced crankshaft with seven main bearings and a vibration damper. It had 201 cubic inches, and produced 68 hp at 3200 rpms. According to press reports, the high compression engine was built with little regard to cost because the engine would establish the new company's reputation. **The all metal body and Lockheed hydraulic brakes augmented the high speed capability of the "Chrysler." Balloon tires were offered on selected models.**

Walter P. Chrysler was still chairman of Maxwell-Chalmers in 1925 when he decided to purchase the company outright. The Chrysler Corporation was established on June 6, 1925. Chrysler's line expanded and Maxwell continued production until the end of the year. In December 1925 Chrysler had 3800 dealers and posted $17,000,000 profit.

For 1926, Chrysler moved in two directions. A four cylinder Chrysler was offered for $845, and at the opposite extreme the Imperial E-80 was offered at $5495. The name "Imperial" was chosen on the belief that Americans have always had a lingering affection for royalty, its emblems, and its titles. The Chrysler Imperial had an enlarged version of the six-cylinder-head, and featured oil, fuel, and air filters as standard equipment.

Because of his previous tenure at GM as president of Buick, **Walter Chrysler** appreciated the advantages of the multi-car philosophy. In 1928, only GM practiced that philosophy. On July 8, 1928, a new low priced car was announced. It was named Chrysler Plymouth by a Chrysler executive named **Joe Frazer**, and in 1929 was shortened to Plymouth. During the winter of 1928 Plymouth production was moved from the former Maxwell facilities in Highland Park to its own enormous plant on a forty acre site at Mt. Elliot and Lynch Roads in Detroit.

On July 30, 1928, Chrysler formally acquired Dodge Brothers. Dodge was in trouble in 1927 when its dealers could only sell 150,000 of the 178,000 vehicles produced. A full 90 percent of the Dodge shareholders agreed to the sale. In **Walter Chrysler's** own words, "The greatest thing I ever did was to buy Dodge. Without it there would be no Plymouth."

On August 4, 1928, the new DeSoto debuted. It was named for the 16th century Spanish explorer Hernando DeSoto, governor of Cuba and discoverer of the Mississippi River. The name was said to symbolize travel, pioneering, and adventure.

The DeSoto was actually conceived to compete with the Dodge Brothers. Instead of abandoning DeSoto after the Dodge purchase, Chrysler kept both nameplates. It was actually priced below Dodge when it was introduced.

Walter P. Chrysler fell ill in the spring of 1937, ending his management of Chrysler Corporation. He died on August 18, 1940.

*1925-
Walter P. Chrsler
courtesy AAMA*

*1928-
Plymouth
courtesy AAMA*

Fred Zeder(L) and Walter Chrysler(R) in front of the first 6-cylinder Chrysler and Plymouth (R). courtesy AAMA

1929- Plymouth body drop courtesy AAMA

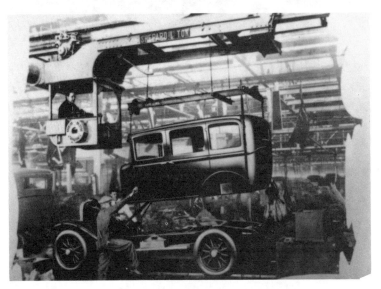

1930- Dodge on a "Belgian roll"- a lifetime of strains and shocks. Remarkable gain in strength, durability. and safety with mono-piece body. (other tests included being rolled down the side of a mountain.) courtesy AAMA

*1929-
A Chrysler Airflow
follows a Plymouth
on the chassis line
courtesy AAMA*

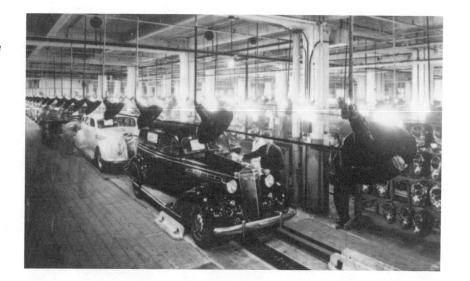

*Interior of the
Chrysler Airflow
shows more visibility.*

*Ann Logan, seen
here at work on
an Airflow DeSoto,
was the first girl to
use a paint stripe
gun in the DeSoto
assembly process.
Also seen here,
Frank Ryan created
all of the stripe
guns for Chrysler
made cars.
courtesy AAMA*

*1936-
The new DeSoto plant at Ford and Wyoming roads.
    courtesy AAMA*

*1958-
Plymouth on the chassis line- "yankee screwdriver" used for applying screws.
    courtesy AAMA*

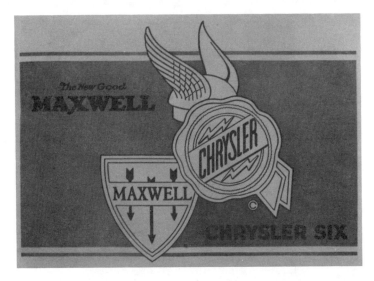

FALCON MOTORS CORPORATION
DETROIT, MICHIGAN
1926-1929

"AMERICA'S FINEST TYPE OF MOTOR"

The Falcon Motors Corporation was formed in late 1926 in Detroit, Mich. to produce the Falcon Knight, which sold for under $1000. John A. Nichols, former VP for the Dodge Brothers, Inc., was president. D.R. Wilson, VP and general manager of the Wilson Foundry in Pontiac, Michigan, was a director. Others who were associated with the new company were **John N. Willys**, John C. Nichols Jr., R. H. Harger, and Ray Allen.

The car was marketed by the Falcon Motor Company's own dealer network with no affiliation to the Willys-Overland Company dealerships (The Willys-Knight built in Toledo used the same type of engine as the Falcon-Knight, but was a more expensive car). Headquarters were established in the Majestic Building in Detroit. The former Knight engine plant of the Willys-Overland Company, in Elyria, Ohio, was selected for manufacturing the Falcon-Knight automobile. The building was of sawtooth construction with a large expanse of windows and over 600,000 square feet of floor space on 25 acres of land. It was located next to the main line of the New York Central Railroad. Chassis and body assembly, along with paint and trim, were done at Elyria. Welding equipment, sheet metal fabricating equipment, and assembly tools were acquired. It was mainly an assembled car, and machine tool requirements were small.

Manufacturing rights were secured from the Knight American Patent Company, for the use of the Knight type double sleeve valve engine. Wilson, who was a principal figure in the project, had long been identified as a volume producer of Knight engines. The Wilson Foundry & Machine Company produced the engines for the Falcon Knight.

By the end of 1927, 10,000 cars were sold, even though production started after the Chicago Automobile Show in January. Production was running at 100 cars per day in 1928, but it was discontinued on March 1, 1929, with the plant reverting back to Willys-Overland, Incorporated.

1928- Falcon Knight roadster courtesy AAMA

GRAHAM-PAIGE MOTOR CORPORATION
DEARBORN, MICHIGAN
1927-1947

Joseph B. Graham was born in 1882, followed by **Robert C. Graham** in 1885, and **Ray A. Graham** in 1887. They were all born on the family farm that James Graham had purchased in 1825, located in Washington, Ind. In 1901, natural gas was discovered near the farm, in Loogootee, and Joseph and his father, Ziba became stockholders in the Lythgoe Bottle Company, which used the gas for fuel. **Joseph** invented a new way for blowing glass bottles, because the shoulders were usually thin and broke easily. He patented a process to blow bottles upside down so the glass would flow toward the shoulder and give it more strength. In 1905, the Grahams, joined by **Robert**, took over the Lythgoe company and renamed it the Southern Indiana Glass Company. Although they were successful from the start, **Joe Graham** would later recall the times he had to contact his banker friend in Indianapolis to meet his payroll. Business flourished until the Grahams were turning out 1000 gross of Coca Cola, Vernor's Ginger Ale and other bottles.

In 1908 **Ray** graduated from the University of Illinois, and had become interested in designing a light truck while he was managing the family's farm properties. In 1916 the Owens Bottle Company of Toledo, Ohio, merged with Graham's Southern Indiana Glass Company. **Joseph** and **Robert** then joined **Ray** in establishing a factory in Evansville, Ind., to build truck bodies that would mount on a passenger car chassis. Eventually, they began to build complete trucks using Continental, Weideley, and Dodge engines.

The president of the Dodge Brothers Motor Car Company, J. Haynes, perceived the Graham trucks as a chance to get into the heavy truck business without disrupting passenger car production. The automobile was still a seller's market with more buyers than production capacity. In April 1921 an agreement was reached for the Grahams to use Dodge engines and drive trains exclusively and to sell their trucks through Dodge dealers. The Grahams added a plant on Meldrum Street in Detroit, and a new company called the Graham Brothers, Incorporated was created, with **Joseph** as president. Additional plants were opened out of state, and the Meldrum plant was replaced by plants on Conant Avenue and Lynch Road in Detroit. By 1926 production reached 37,000, making them the largest exclusive truck manufacturer in the world. In November 1926 Dodge management was re-organized following the sale of the company to Dillon and Read. The Grahams sold 51 percent of their company to the Dodge company for $3,000,000, and in turn reinvested most of it in Dodge stock, making the Grahams one of the largest Dodge stockholders.

In 1927 the Grahams organized a New York-based holding company for their varied interests, including an $11,000,000 share in Libbey-Owens Sheet Glass in Toledo. (In 1930, **Ray Graham** as chairman of Libbey-Owens, brought about a merger with the Edward Ford Plate Glass Company.) The Graham brothers were not content with their role at Dodge, and soon began looking for an opportunity to manufacture automobiles. On June 10, 1927, that opportunity came when they purchased the Paige-Detroit Motor Car Company for $4,000,000 and pledged an equal amount for improvements. The Jewett brothers resigned and **Joseph** became president, **Robert** was VP, and **Ray** was secretary and treasurer. They became Paige directors along with their father, and changed the corporation's name to Graham-Paige Motors Corporation.

The existing line of Paige automobiles were produced by the Graham brothers until they developed their own model. They worked fast and had their first car ready for the 1928 season. The styling was the work of the Lebaron Studios of the Briggs Manufacturing Company. The inspiration for the front-end came from the Hispano-Suiza, with a rounded-off face of the grille and shell. The drawing did not suffice for showing the new look, and a clay model was made for **Joe Graham. This was one of the first instances of a clay model being used to present a styling idea to management.** The radiator shell bore an emblem of three profiles in knights' helmets, representing the Graham brothers. The new cars were launched at the Hotel Roosevelt in New York. The dining hall was decorated in medieval motif, and dealers and salesmen were enlisted in the Graham-Paige "legion." Speakers included Knute Rockne and Gene Tunney. Their first car set a record for its first year with sales of 73,000 in 1928 compared to 21,000 in 1927. The Grahams were even honored by Pope Pius XI for their contributions to church and humanity by introduction into the Roman Catholic Order of St. Gregory the Great. In turn, the brothers presented the Pope with a Lebaron town car, which eventually was displayed at the Vatican Carriage Museum.

By November 1928 employment rose from 2840 in 1927 to 7200, and plant capacity went from 300 to 700 cars per day. In 1929, 77,000 cars were produced, but losses were realized at many dealerships that were in the high rent districts. In 1930 the name of the car was changed to Graham, but the public continued calling it the Graham-Paige; output dropped to 33,500 and the company lost $5,000,000. A notable option for 1930 was non-shattering safety glass, developed in conjunction with Libbey-Owens-Ford, in which the Grahams still retained substantial interest.

Even though output dropped to 20,400 in 1931, the Grahams decided to develop the Blue Streak Eight for 1932. It was designed by Amos Northrup of the Murry Corporation of America. The front end had a sharp rearward slope of the radiator grille and repeated in the slant of the hood louvers, with a one piece windshield. There was no separate radiator shell, the hood ran right up to the grille molding. The radiator cap was concealed beneath the hood to prevent damage from anti-freeze and to improve appearance. **Fenders were deep and fully skirted to conceal mud from view**, a feature that was copied by several other companies the following year. The frame rails had no kick-up at the front end, and the rear axle passed through O-shaped openings in the rails. This was much stronger than conventional frame construction and reduced the tendency for the rear axle to break loose on corduroy roads by reducing flexing and deflection of the frame. The stiff frame, with springs mounted outboard of the side rails, and a two inch wider track made the Blue Streak very stable. Cannon Ball Baker drove one up Mount Washington in New Hampshire in a record time of 13 minutes and 26 seconds. A Graham powered racing car on an O-frame entered the 1932 Indianapolis 500 race and qualified at 109 mph, but was forced to retire after 61 laps with a broken crankshaft. (A similar Graham entered the 1934 race and finished tenth at an average speed of 95.9 mph.) Production declined to a little over 12,000 in 1932, but family tragedy also struck. **Ray** suffered a nervous breakdown, and was being taken home to the East Coast for a complete rest. While en route, he broke away from an accompanying priest and jumped into a creek and died.

For 1934 the Graham became the first moderate priced car with a supercharger, and production increased to 15,000. The 1935 Graham was the worst looking car to bear the name of Graham, but with an upswing in the economy, sales rose to 18,500. In July of 1935, there were merger negotiations between Auburn, Pierce-Arrow, REO, Hupp and Graham, during which REO asked Graham if they wanted to supply motor parts in exchange for body designs and dies. For $7.50 per unit, the 1936 Graham had the body of the REO Flying Cloud. In 1938 a new Graham body was introduced, dubbed the "Spirit of Motion," which became known as the sharknose. It had a forward leaning radiator, flush headlamps in the fenders, concealed door hinges, and door handles that lined up with the belt molding. The price of the sharknose was reduced $100 for 1939, and the running boards were eliminated. Also, the radio offered a choice of five preset stations, and an extension aerial on the left side of the cowl was used. The radio loudspeaker was in the center of the instrument panel. The sharknose models were too radical for the public and were unsuccessful on the market.

Joe Graham spent over $500,000 of his own money to keep the company going, but things were getting desperate with the poor sales of the sharknose. He was approached by Norman De Vaux of Hupp, which had similar problems, and asked if he wanted to build cars for both companies using the discontinued dies from the Cord Beverly sedan. Both companies would convert the chassis to rear wheel drive and use their own engines. The Graham version was called the Hollywood, and the Hupp was the Skylark. The Graham-Paige Corporation was to develop new dies, specially for the trunk floor pan to accommodate a differential for rear wheel drive, and the roof. (The Cord tooling was made for low volume and the roof stamping consisted of seven pieces which had to be welded, leaded, filed, and sanded. These dies were never replaced, and strained the high volume production process.) The 1940 Graham lineup consisted of two sharknose models and the Hollywood. Although it wasn't a new design, it attracted attention. While driving a prototype back from a trip to Indiana, **Joe Graham** was stopped by the Michigan police who wanted to look at the racy new car.

After months of preparation, Graham's Dearborn plant started up in April 1940, building the sharknose at first (renamed Seniors), then turning to Skylarks and Hollywoods in late May at a disappointing rate of 30 per day. In mid-July, the plant changed over to 1941 models and the Seniors were dropped. The price of the Hollywood was dropped to $1045, and minor trim changes were made. A few weeks later, the Hupp Corporation withdrew from the joint operation and went into receivership in October 1940. The Hollywood production continued until September when the plant was temporarily shut down.

Joe Graham announced in November 1940 that the Graham-Paige Corporation suspended automobile production in order to concentrate on the war effort. The parts business was sold to Dallas Winslow who also bought the Hupp parts operation and moved them to Auburn, Ind.

By the spring of 1941, the Graham-Paige Corporation reported its first profit since 1933. In his annual report, **Joe Graham** reported that the capital requirements of an automobile factory were so enormous that the Graham-Paige Corporation should permanently retire from manufacturing complete automobiles. In January 1941 **Joe Graham** retired, and was succeeded by R. J. Hodgson, former manager of the Detroit office of the Reconstruction Finance Corporation.

Joseph(L), Robert(M), and Ray(R) Graham

The body plant in Wayne, Mich. courtesy AAMA

The assembly plant on Warren Ave. and the Terminal R..R.

The assembly plant on Warren Ave. picture taken in 1994

1935- Chassis line courtesy NAHC

1937- Graham series 95 standard coupe. It used a REO body with a supercharged engine, capable of 0-60 mph in 14.5 seconds.
courtesy AAMA

KAISER-FRAZER CORPORATION
WILLOW RUN, MICHIGAN
1945-1953

KAISER-FRAZER CORPORATION
TOLEDO, OHIO
1953-1955

KAISER-JEEP CORPORATION
TOLEDO, OHIO
1955-1970

Joseph Washington Frazer was a descendent of the Virginia Washingtons. Although he was an aristocrat, he shunned high society and enrolled in Hotchkiss and Yale to learn salesmanship. He worked at Packard as an unskilled mechanic in 1912, then transferred into sales. In 1915 he had his own Saxon dealership in Cleveland. He moved to the GM Acceptance Corporation, and in 1923, the New York banks specifically requested GM to lend him to Pierce-Arrow to organize a finance corporation. **Walter Chrysler** brought him to Detroit to head up his sales organization when it started, and he held various jobs including VP. In 1939 he moved to head up Willys-Overland, which was on a steep decline. The Willys "Jeep" helped **Frazer** turn the company around.

After the attack on Pearl Harbor, the automotive industry swung into war work and Detroit became known as the "arsenal of Democracy." The Graham-Paige Company supported the war effort by producing amphibious tractors for transporting troops to the beachheads. It also made precision components for aircraft engines, PT boat engines and torpedoes.

In August 1944 **Joseph Frazer** and his associates assumed control of Graham-Paige by purchasing 530,000 shares of stock from the retired **Joe Graham**. **Frazer** was elected chairman of the board and reversed **Graham's** past decisions by announcing that the **Graham-Paige Company would resume automobile production after the war with a completely new car**.

The new car was called the "Frazer" because the sales people were leery of resurrecting the defunct name of "Graham." A new engineering staff was hired along with Howard "Dutch" Darrin to do the styling. Recruiting began for setting up a 4000 dealer organization.

Henry John Kaiser was born in 1882. From 1914 to 1929 he headed a highway construction company. In 1931 he directed construction of Boulder Dam, followed by the Bonneville Dam in Oregon and the Gran Coulee Dam in Washington. During World War II he managed the Pacific coast shipyards and handled wartime ship building. He increased the speed of ship construction dramatically.

In 1942, **Henry Kaiser** brought together a collection of American and foreign made cars to study and reassemble them in many different combinations. **Kaiser** clearly wanted to market something different and superior compared to what he considered from Detroit as stodgy. He decided to get into the auto business starting with a small economical car with a bubble top that would sell for under $600. **Kaiser** then suggested that the car companies announce their post war plans immediately. The car makers gave the news a supercilious eye and forgot it. However **Joe Frazer**, purporting to speak for the entire auto industry, said he resented a West Coast ship builder asking the car companies if they had the courage to plan post war cars when they were fully engaged in the war effort. He said the public was being misled by pictures of plastic models with glass tops, done by artists who probably wouldn't want to sit under those tops in the summer and sweat. **Joe** said **Henry** was "half baked."

(A potential factory for automobile production was located near Ypsilanti, Mich. Construction of the plant started in March of 1941 on a 985 acre site. The plant was operated by the Ford Motor Company, and became the largest manufacturing site under one roof in the world with almost 2,500,000 square feet. At its peak, close to 46,000 people worked there and built 8685 Consolidated B-24E Liberator bombers in just 21 months. **Eddie Rickenbacker** worked at the plant as a consultant, testing for high altitude operation.) In May 1945 **Henry Ford Sr.** announced that he had no interest in the Willow Run factory for automobile production. He said the $100,000,000 plant, like a battleship, was expendable.

In July 1945, **Joe Frazer** took Dutch Darrin's drawings for the "Frazer" car to the Los Angeles office of Amadeo Peter Giannini, whose bank of America National Trust & Savings was next to the Chase National Bank, the largest bank in the world. **Joe** was looking for financial backing and Giannini's bank was loaded with idle funds. The West Coast feared a depression as shipyards and aircraft plants slacked off. Giannini wanted to lend money to reinforce California prosperity, and suggested "why don't you get together with **Henry Kaiser**?" On July 17, 1945, **Joe Frazer** and **Henry Kaiser** met and shook hands and received a $10,000,000 line of credit.

On July 25 the formation of the Kaiser-Frazer Corporation was announced for the production of automobiles. It was capitalized with 5,000,000 shares at a par value of $1 per share. **Kaiser** planned to produce a low priced automobile called the "Kaiser" on the West Coast and the Frazer was to be produced in the Graham-Paige plant in Dearborn, Mich.

The UAW contacted **Henry Kaiser** about building his cars at the Willow Run bomber plant, and in August 1945 **Henry** and **Joe** began negotiating with the RFC for a five year lease. They ended up with a five year graduated lease starting with $500,000 for the first year. **Joe Frazer** proposed that the government rent his unwanted Graham-Paige factory as storage space. The RFC agreed, and rent for the first two years was $500,000 a year.

More money was needed and **Joe** and **Henry** decided to go public. They planned to issue 2,200,000 shares at $10 per share. Kaiser interests would buy up 250,000 shares and Graham-Paige would buy 250,000. The remaining 1,700,000 would be offered to the public, netting K-F $20,000,000.

The Kaiser-Frazer Corporation was formed in August 1945, with **Joe Frazer** president and **Henry Kaiser** chairman. The Kaiser-Frazer Corporation and the Graham-Paige Motor Corporation were "meshed" but not "merged." The Graham-Paige Motor Corporation still continued with **Joe Frazer** as president.

The Willow Run bomber plant 12 miles from Detroit was used for production. There was a twin assembly line and two Kaisers were built for every one Frazer. The cost of running the plant was split accordingly, with Kaiser funding two thirds and Frazer funding one third. Additional Kaisers were built in the former Douglas Aircraft plant in Long Beach, Calif., for logistical purposes as well as the agreement to help the California economy as war production scaled down. The California plant was headed up by Henry's son, **Edgar**.

In December 1945 the partners went to the well again and sold 1,800,000 more shares of common stock at $15 per share. On January 20, 1946, the first two handmade models of the Kaiser and Frazer, made by striking GM workers, were displayed at the Waldorf-Astoria in Manhattan. The crowd broke a window and lined up two abreast in a blizzardy night to stare at the cars. There were 9000 orders taken without any price set. Business Week magazine said "The good old days are back again."

Production started in June 1946. The Frazer was a smooth looking four door sedan with flow-through fenders, on a 123.5 inch wb. The engine was based on the Continental "Red Seal" six cylinder engine with a 3.31 x 4.38 bore and stroke. It had a 226.5 cubic inch displacement with 100 hp. Most of the engines were made by K-F. The only transmission offered was a three speed manual until 1951.

Although it looked as if Graham-Paige was on the verge of a comeback, the enormous costs associated with designing and manufacturing a brand-new car began to eat away at assets of the Graham-Paige Company. It was also losing its identity because K-F generated most of the publicity and the public soon forgot Graham-Paige.

During 1946 the Graham-Paige Corporation lost nearly $7,000,000 on car production. It was apparent that the Graham-Paige Corporation would not be able to finance its one third share of the Willow Run expenses, and the company decided once again to quit the car business. On February 5, 1947, the stockholders approved the transfer of their automotive assets to K-F in return for 750,000 shares of K-F stock.

Only 6476 Frazers were built by Graham-Paige, plus an additional 2464 with the Graham-Paige labels released after the sale. From Willow Run, the Graham-Paige company moved to York, Penn., and made farm machinery through its subsidiary, the Frazer Farm Equipment Corporation.

The original Kaiser was going to have torsion bar suspension and front wheel drive, but it never saw production. Hard steering, gear whine, and wheel shimmy developed and in May 1946 the decision was made to drop the project. Instead, the Frazer platform was used, with less expensive trim. A unique Kaiser feature was a wide range of colors, such as a vivid pink named Indian Ceramic, Crystal Green, Caribbean Coral, and Arena Yellow. Of 150 different interior fabrics offered by the automobile industry in 1949, K-F owned 62 of them. Of 218 exterior colors, K-F owned 37.

In 1949, **Joe Frazer** could foresee the onrush of new models coming from the "big three." Since K-F had a mere face-lift, he warned **Henry Kaiser** to lower production and retrench. **Kaiser** said "the Kaisers never retrench." **Frazer** yielded the presidency to **Edgar Kaiser**, but remained with the title of vice-chairman on the board of directors. **Edgar** tooled up for 200,000 cars and sold 58,000.

The Frazer was discontinued for 1950, but a new body was introduced for the Kaiser with the famous "widow's peak" windshield. A compact car called the "Henry J" was introduced in 1950 with a companion car called the "Allstate," sold only by Sears and Roebuck. In 1953 a fiberglass two seat roadster was produced called the "Kaiser Darrin." In that same year, **Henry Kaiser** bought the Willys-Overland Corporation. All Kaiser production was relocated to Toledo, Ohio, and re-organized as Kaiser Jeep Corporation. The Willow Run plant was sold to GM for manufacturing transmissions after a major fire destroyed the GM transmission plant in Dearborn, Mich.

All Kaiser production in Toledo that consisted of the Kaiser, Henry J, Allstate, and Aero-Willys ended in 1954. Henry Kaiser moved the tooling to Argentina and produced the Kaiser, renamed "Carabela" from 1955 to 1962. The Jeep continued in production in Toledo and was merged with AMC in 1970.

The first Frazer seen in front of the Willow Run factory with Edgar Kaiser behind the wheel.
courtesy AAMA

Twin body-in-white lines where metal finishers ground the bodies, mounted the doors and trunk lids. The plant had a mile long production process.
courtesy AAMA

Kaiser test car undergoing the splash test. From a test like this, engineers were able to design shields and other protective devices virtually 100% effective against engine "drowning."
courtesy AAMA

1949- A four door convertible was Henry Kaiser's idea. Only 123 were sold.
courtesy MSHC

The Kaiser Marque

The Frazer Marque consisted of a coat of arms with the French motto: "Je Suis Prêt." (I am ready)

EDSEL FORD

Edsel Ford was named in honor of a school boy friend of **Henry Ford**, named Edsel Rudimen, who later married Henry's sister Margaret. **Edsel Ford** attended the University of Detroit and went to work for the Ford Motor Company in 1912. For three years he trained in various departments to learn the business. In 1916 he married Eleanor Lowinthian Clay, daughter of William Clay and the niece of **J. L. Hudson**. In 1915 he was elected as secretary of the company, and in 1917 he was named VP. In December 1918 he became president of the Ford Motor Company.

At the rate **Henry** progressed the Highland Park plant was outmoded before it was ever completed. **Henry** had problems with bankers and suppliers and wanted to control every aspect of his cars, so he built the River Rouge operation, the largest in the world, on 1076 acres inside one fence. He also gained complete ownership of his company in 1919. **Edsel Ford** tried to convince his father to modernize the Model T, and others should have come to **Edsel's** defense during meetings but they never did. Sorensen, brutal with everyone else in the company, was a complete marshmallow in front of **Henry**. In 1924, the Chevrolet had hydraulic brakes and a six cylinder motor. **Edsel** pleaded to his father to modernize the Ford engine. A six, his father retorted, could never be a balanced car. "I've no use for an engine that has more spark plugs than a cow has teats." After all, he built one for the Model K in 1906 and didn't like it.

In 1926 Ford had 110 manufacturing, assembly, mining, and lumbering plants. There were 300,000 people on his payroll, with 120,000 in the Detroit area. But the sales of the Model T was spiraling downward. Finally, Ford shut down for a half year in 1927 to retool for a totally new car. He started back at the beginning of the alphabet and named it the Model A. It was well received with tens of thousands of orders before anyone even knew what it was. In 1932 a cast *en bloc* V-8 was brought out, and styling was improved each year afterwards.

At the Chicago Century of Progress Exposition in 1934, Ford displayed a "car of the future," to see what the public reaction would be. Out of the project came the "Lincoln Zephyr" in 1936. Headed up by **Edsel Ford**, the car did away with the conventional chassis frame and presented streamlined fenders. Running boards were eliminated, and the car had a one piece windshield. It had a 12 cylinder engine, the first in the medium price class.

In 1939 another new automobile conceived by **Edsel Ford** was presented in the medium price class called the "Mercury." It was named for the Greek god associated with nimbleness, fleetness, and flight. It sold for less than the Zephyr and had a V-8 engine with 95 horsepower. Next, **Edsel** championed one of the most significant models offered by any automobile manufacturer called the "Lincoln Continental," first produced in 1940, and produced until 1948.

Edsel died of stomach cancer in 1942. His death came as a blow to his family, friends, peers, and the members of U.A.W. local 600 who knew he tried to treat them with humanity.

On September 4, 1957, the "Edsel" was introduced as a medium priced car. It only lasted until 1961, mainly because of the post Korean War recession. When the subject of a name for the new medium-priced car came up, Edsel's children were united in determination that the name should not be Edsel. As Henry Ford II put it, they did not want their father's name on thousands of spinning hubcaps. The car became known as the "E" car, for experimental. But this precipitated rumors that it was already decided to call it the Edsel. Week after week went by with executive brainstorming trying to name the car.

A poet named Marrianne Moore was solicited to create a name for the new car. Of the many offerings she made over several months, Mongoose Civique, Aeroterre, Turbotorc, Thunder Crester, Magigravure, Pastelogram, Regina-Rex, Varsity Stroke, Angeastro, Cresta Lark, Triskelion, and Pluma Piluma (hairfine, feather-foot), are some examples that were courteously "put on hold."

Suggestions came from everywhere and finally the list was winnowed down to four. In the spring of 1956, the decision went to the executive committee with Ernest Breech, chairman of the board of directors, presiding in the absence of Henry Ford II. Dozens of 40 x 60 inch cards listed the candidate names including the favorites, Citation, Corsair, Pacer, and Ranger. Each candidate's name was presented one by one with commentary. There was fidgeting and the reception to all names was cool. Then after a year of searching, Breech pulled himself out of his chair and said, "I don't like the goddammed things." After looking up and down the table he said, "How about we call it Edsel?"

1915- The Model T was designed and priced to meet the needs of the majority of customers, who happened to be farmers.

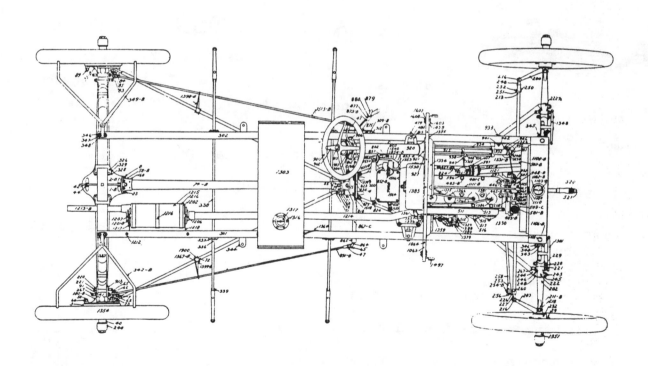

*1933-
Rivard Brothers
showroom in
Centerline, Mich.
courtesy MSHC*

RIVARD BROS.

SALES SERVICE

Cars and Trucks

Phones: Ivanhoe 4242   20955 Van Dyke Avenue
        Center Line 9153

KEY NUMBERS

Door and Ign. FC907

Spare Tire FT235

Motor No. 966182

Base Line, Mich., Jun3 11 1934.

M Ray C. Rivard, Salesman,
    RFD # 1,
    Warren, Michigan.

```
Tudor Sedan (Invoice)                                        $454.23
Federal Taxes                                                  16.79
State Sales Taxes                                              14.13
License & Title                                                 9.75
Minimum UCC Wholesale charge                                    2.00
                                                             $496.90
Down Payment (besides discount)                                96.90
                                                             $400.00
Insurance & Carrying charges                                   37.09
Bal in 12 payments, 11 at $25.00 and 1 at $162.09            $437.09
all payable to Universal Credit Corp, 154 Bagley Ave.,
                         Detroit, Michigan.
                    BILL OF SALE
```

HenryFord (L) Edsel ford (R) courtesy Tinder

Edsel Ford's last personal car, a Lincoln Continental seen parked in his garage.
picture taken in 1994

1947-
The last model seen by Henry: This car is being tested on the cobble stone course at the Dearborn test track by Bob Mallow and Otto Schaefer. Mallow drove at low speed while Schaefer listenened "under the hood."
courtesy AAMA

1959 EDSEL

Above: Corsair 4-door hardtop

The car you hoped for—at the price!

Makes history by making sense

Exciting new kind of car! A full, six-passenger beauty. Roomy without useless length. Solidly built. Powered to save. Priced with the most popular three!

This is a new breed of car. A car with looks, features, power *and price* that make sense. It's styled with beauty and grace you usually find only in expensive cars. It's soundly engineered. Edsel's compact 120-inch wheelbase makes parking a pleasure. Yet there's room for six adults to ride *comfortably*. You get your choice of four new Edsel engines including a thrifty six and a new *economy* V-8 that uses *regular* gas. Plus luxuriously appointed interiors, super-smooth ride, self-adjusting brakes. And the 1959 Edsel is actually priced with the most popular three—Ford, Plymouth and Chevrolet! See it. At your Edsel Dealer now.

EDSEL DIVISION • FORD MOTOR COMPANY

DETROIT AUTOMOTIVE GOLDEN JUBILEE

In 1946, fifty years after Charles Brady King and Henry Ford were recognized as driving the first automobiles on the streets of Detroit, a 10 day long "Golden Jubilee" was held. It celebrated the birth of the Automobile Capital of the world and the resumption of automobile production following the Second World War. The celebration was led by a parade of 300 pioneer automobiles down Woodward Avenue. The pavement was painted gold, which is a story in itself. A special paint mixture was composed that would wear and erode shortly after the festivities were over.

William S. Knudsen was the chairman and George Romney was manager of the Jubilee. A big banquet was staged at Masonic Temple where veterans, such as Apperson, Duryea, Ford, Nash and Olds were on stage. Also there were some of the oldest workers, dealers, suppliers and race drivers of the early era.

There was a revolving display downtown showing an old and a new car encased in an atomic symbol. Trygve Lie, Secretary-General of the United Nations, flew in from a conference and spoke at a "unity" rally at Briggs Stadium, where management and labor representatives joined together.

*Parade down Woodward Avenue.
courtesy AAMA*

*Parade down Woodward Avenue.
courtesy AAMA*

*George Mason of Nash Kelvinator(L), William S. Knudsen, chairman, is 2nd from the right and George Romney is at the far right.
courtesy AAMA*

*Barney Oldfield shows off the 999 racer
courtesy AAMA*

*James Scripps Booth shows off his Bi-Autogo V-8 to George Romney
courtesy AAMA*

*1949-
Detroit
courtesy AAMA*

THE AUTOMOTIVE CAPITAL

As the 19th century came to a close, several cities were heavily involved in the production of automobiles, such as Cleveland, Buffalo, Chicago and especially the New England states. Automobile manufacturing was dominated by the Pope Company which was well ahead of anyone else. It had produced 734 of the total 1575 electrics built in America; 1191 of the 1661 steamers; but, only 171 of the 1207 gasoline engine automobiles. As automobiles were gradually driven farther distances, the market for electrics began to fall due to limited range. Similarly, the market for steamers began to contract because they were complicated to maintain, took a long time to start, and access to water in the countryside had an impact on its range.

By 1903 Albert Pope changed his opinion about the buying public and began producing gasoline automobiles. He was the first to put into practice a varied line of cars. The Pope-Toledo was the luxury model, the Pope-Hartford was the medium priced model and the Pope-Tribune was a low priced model. The Pope-Waiverly was offered as an electric model. At its height the Pope Company was capitalized at $22,000,000 with plants in Hartford, Toledo, Indianapolis, and Hagerstown. There were also bicycle plants in Elyria, Ohio, and Westfield, Mass.

The Pope Company always regarded the automobile as a luxury item and never produced a popular priced car. The majority of the Pope combine fell during the panic of 1907. After a year of receivership, capital was scaled down to $6,500,000 and operations were confined to the Hartford and Westfield factories. The Toledo plant went to Willys; and the Hagerstown, Elyria, and Indianapolis plants were also sold. In 1908 Detroit was being called the Automotive Capital. Colonel Albert A. Pope died in 1909, worn by the struggle to keep his enterprise going. The Pope Company went into receivership again in 1913, selling off the Westfield plant; and the Hartford plant was sold to Pratt and Whitney.

A number of other New England automobile manufacturers inadvertently spent their resources fighting patents and wound up in debt. The key personnel were primarily technical men such as Maxim, Knox, J. F. Duryea and the Stanley brothers, who were not vitally concerned with the business end of automobile manufacturing. In fact, the Stanley brothers never advertised their automobile.

It was difficult to gain financial support from the eastern conservative businessmen who looked askance on the motor car industry. They predicted it would experience the same fictitious growth that the bicycle business had undergone and that it would not establish itself as a permanent fixture. There weren't many capitalists in the East who were willing to invest large sums of money for long periods in the new industry.

Perhaps a more significant factor in New England's failure to obtain the automobile industry was the internal combustion engine. The concentration of the **petroleum** industry in Pennsylvania, Ohio, and Indiana gave the Lake Erie shore and Detroit the **leadership in internal combustion technology**. Due to its ship building expertise, Detroit was the center of this technology. It was the connecting link between the great inland seas of Erie and Huron, and had more than a mile of quay. Detroit was located in close connection with the main line of travel from east to west by rail, giving it unrivaled transportation facilities providing easy distribution that was so necessary to the growth of a manufacturing industry.

Michigan, like Minnesota and the Northwestern states of Montana, Oregon and Washington, was **settled by educated and enterprising Americans** that had built up New England and other eastern states. These were the men who could "make two blades of grass grow where one grew before." A mighty **timber** industry opened in Michigan, the world's richest **ore** deposits were found in the upper ranges and great **salt** deposits in and around Detroit poured forth their riches. The pioneer Detroiters, whose names mark so many streets, wrested from this soil their fortunes and their faith. Detroit was known as the loveliest city in America. There was no river in all the world quite so fine and great shade trees cooled the streets. Detroit led in making stoves, varnishes, paints and electrical furnaces, ranking it 16th among the manufacturing cities of the country.

The demands of the automobile industry were so complex that it was apparent from the beginning that a pooling of talents and resources in a central area was essential. No other place in the world had a **concentration of** such prolific **talent** such as Ford, King, Olds, Barthel, Leland, Marr, Buick, Brush, the Dodge brothers and a host of others. Although the carriage and wagon business was comparatively small, it employed a number of highly skilled workmen whose craftsmanship proved valuable and essential when the making of automobile bodies, frames and wheels began. Ransom E. Olds and the Olds Motor Works initiated Detroit's automobile industry and became a training ground for many of its successful engineers, mechanics and businessmen.

Detroit was one of the first cities with **electricity**, many of the early automobile plants could readily get electric power. The **large tract of level ground** made southeast Michigan excellent for construction of **railroads** and **large factories**. "Beltline" tracks provided access to more of the land which kept land prices affordable for factories. The Cadillac and Packard plants were situated on the Michigan Central Belt Line, which looped around the urban core. In 1904, as the city became congested, a syndicate was organized to build an outer belt line to open up a new area which was completed in 1914. It was known as the Detroit Terminal Railway. It began on the river and extended one mile east of the Belle Isle Bridge. It described a broad arc outside what was then the city limits, crossed Woodward Avenue six miles north of City Hall and joined the main line of the Michigan Central to the west. Not only did the new route facilitate the interchange of traffic among various railways radiating out of Detroit, but it provided a central artery for a vast new industrial complex that sprang up on what had formerly been flat farmlands on the fringe of the city.

Detroit architects like Albert Kahn were innovators who designed and built fireproof buildings using **structural, reinforced concrete**, which reduced the cost of **insurance** and could be constructed cheaper and faster than conventional mill type wooden factories. They also pioneered steel sash windows that allowed more light into the factories, resulting in **improved worker efficiency**. Auto pioneers like Henry Leland and Walter Flanders brought **interchangeable parts** and **progressive assembly** to the automotive industry and Henry Ford created "**mass production**."

The location of the automobile industry was also determined by factors of **personality and business leadership**. Although Detroit was never a lumber town, much of its capital went into the industry and the city acquired its share of sawdust millionaires. Most of them were engaged in other enterprises as well, but it was their capital that helped finance much of the local industry in the latter part of the 19th century, as well as the automobile industry.

As early as May 1891, P. F. Olds & Sons was running ads in Scientific American. They had understood the **value of advertising**. Many magazines wrote articles on companies that were running advertisements. Detroit companies knew how to advertise, and had some of the very best sales people in the business which was key to the success of Detroit as the "Automotive Capital."

Charley King developed the concept of **Freight On Board** (FOB). This was the point where the finished goods were loaded for shipment to the customer. The customer paid freight and took title of the goods. With this arrangement, the manufacturer did not have money tied up in inventory of finished automobiles and more money was available for day-to-day operations, giving Detroit another edge.

As the automobile industry began to grow in Detroit, other industries followed. In 1903 the C. C. Wormer Machine Company warehouse was the largest machine depot in the world. Detroit was the **ship building capital** of the Great Lakes. In 1906 the Great Lakes Engineering Works in Ecorse had 2000 workers and was building six ships at a time.

In 1911 the Selden patent loss represented the last grip that easterners had on the U.S. automobile industry. New York was seeking to take Detroit's place, and held a dinner at the Hotel Brevoort in 1913, at which automobile manufacturers and dealers were present by invitation. The object of the meeting was to discuss fully the possibilities for encouraging the motor car industry to relocate in New York City.

Some of the disadvantages for producing cars in Detroit were discussed. They felt that it was concentrating on a single industry and creating a demand for labor greatly in excess of the supply. Lodging facilities were not able to meet the demand of the increasing population. The street cars were inadequate and many of the workers in the factories were held up for long periods in the evenings before they could get home. There was a shortage of freight cars and the soundness of the Detroit banks was questioned. They felt that too many liberal loans were made to comparatively insecure automobile manufacturers. In short, they felt the automobile industry was in the same situation as the man who put all his eggs in a single basket.

To a degree, they were correct and the answer came in the form of decentralization. As parts became more specialized, all cars became "assembled" and it became impractical to manufacture all of the parts and assemble them into cars under one roof. Gradually, assembly plants were constructed throughout the U.S. as well as Europe.

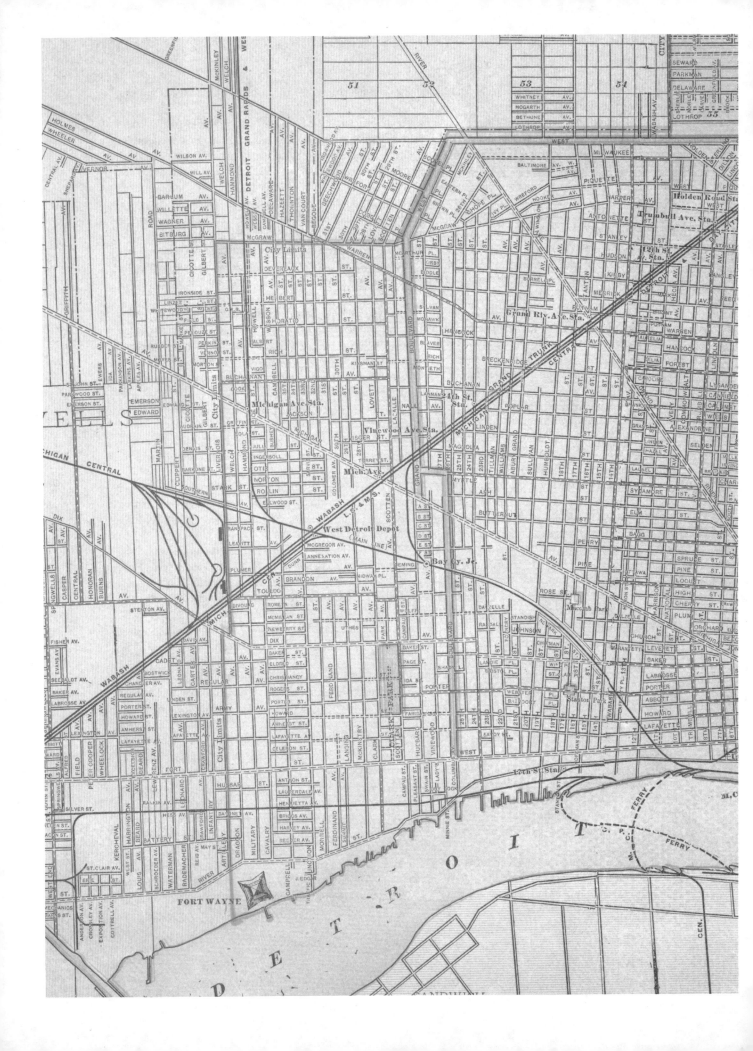

PHOTO AND DOCUMENTRY CREDITS

AAMA
Reprinted with the permission of the American Automobile Manufactures Association.

BENTLEY
Reprinted with permission of the Bentley Historical Library, University of Michigan

BURTON
Reprinted with permission of the Burton Historical Collection of the Detroit Public Library.

FREE PRESS
Oldsmobile factory fire headline (page 28) and the Detroit Driving Course pictures (pages 42 and 44) reprinted with permission of the Detroit Free Press.

KAHN
Reprinted with permission of Albert Kahn and Associates.

MSHC
Reprinted with permission of the State Archives of Michigan, Lansing.

NAHC
Courtesy of the National Automotive History Collection, Detroit Public Library.

NEWS
Reprinted with permission of the Detroit News.

TINDER
Photos courtesy David V. Tinder of Dearborn, Michigan.

ACKNOWLEDGMENTS

I am indebted to the following people for their help because this book would not have been possible without them: My wife Eleanore, daughter Kristina, sons Jeffrey and Michael, and my father Casimer; Marie Bartoszek, Ivan Boivan, Mary Fleming, Bill Brzezinski, Tamy Brady, Belinda Lawson, Andy Matzkin-Bridger, Dave Berrels, Sandy and Mike Skinner, Sallyanne Williams, Jack Rowe, Keven Pogany, Bob Perfetti, Jeff Branch, Zenon Hotra, Andy Franks, Judy Whitacre, and my brother-in-law, Greg Hudson, who is the great-grandson of John Lauer.

Mark Patrick and his staff at the NAHC were very supportive (This is truly the foremost repository of automotive history in the United States). Peggy Dustman at AAMA, Noel VanGorden, David Porenda, John Gibson, Barbara Louie, Dawn Eurich, Janet Nelson, Lillian Stefane, Maryalyce Lubiszewski, Benedict Markowski, Anna Savvides at the Burton Historical Collection, Dick Kolbas, John Dombrowski, John Currie at MSHC, Anne Frantilla at the Bentley Library, and my sister-in-law Dr. Carolyn Fick of Concordia University, Montreal, Canada.

In addition to the institutions listed under Photo and Documentary Credits, a thank you to The British Motor Museum, the Daimler-Benz Museum, the Musee De L' Automobile, the Edison Institute, Antique Automobile Club Of America, Inc., GMI, and the Grosse Pointe Historical Society.

A very special thank you goes to Kim Haddad and all of the personnel at The Typocraft Company in Detroit, Michigan.